AF352808

Advanced Modeling, Control, and Optimization Methods in Power Hybrid Systems - 2021

Advanced Modeling, Control, and Optimization Methods in Power Hybrid Systems - 2021

Editor

Nicu Bizon

MDPI • Basel • Beijing • Wuhan • Barcelona • Belgrade • Manchester • Tokyo • Cluj • Tianjin

Editor
Nicu Bizon
Faculty of Electronics,
Communication and
Computers
University of Pitesti
Pitesti
Romania

Editorial Office
MDPI
St. Alban-Anlage 66
4052 Basel, Switzerland

This is a reprint of articles from the Special Issue published online in the open access journal *Mathematics* (ISSN 2227-7390) (available at: www.mdpi.com/journal/mathematics/special_issues/ Advanced_Modeling_Control_Optimization_Methods_Power_Hybrid_Systems_2021).

For citation purposes, cite each article independently as indicated on the article page online and as indicated below:

LastName, A.A.; LastName, B.B.; LastName, C.C. Article Title. *Journal Name* **Year**, *Volume Number*, Page Range.

ISBN 978-3-0365-4144-0 (Hbk)
ISBN 978-3-0365-4143-3 (PDF)

Contents

About the Editor

Nicu Bizon

Nicu Bizon (senior member, IEEE) was born in Albesti de Muscel, Arges county, Romania, in 1961. He received his 5-year B.S. degree in electronic engineering from the University "Polytechnic"of Bucharest, Romania, in 1986, and his Ph.D. degree in automatic systems and control from the same university in 1996. From 1996 to 1989, he was involved in hardware design with Dacia Renault SA, Romania. Since 2000, he has served as a professor with the University of Pitesti, Romania, and has received two awards from the Romanian Academy, in 2013 and 2016. He is the editor and author of six books published in Springer and the author of 243 scientific papers published in Scopus, which have been cited 2263 times, corresponding to an h-index = 30. His current research interests include power electronic converters, fuel cell and electric vehicles, renewable energy, energy storage system, microgrids, and control and optimization.

Preface to "Advanced Modeling, Control, and Optimization Methods in Power Hybrid Systems - 2021"

This reprint presents the latest solutions in fuel cell (FC) and renewable energy implementation in mobile and stationary applications. The implementation of advanced energy management and optimization strategies are detailed for fuel cell and renewable microgrids, and for the multi-FC stack architecture of FC/electric vehicles to enhance the reliability of these systems and to reduce the costs related to energy production and maintenance. Cyber-security methods based on blockchain technology to increase the resilience of FC renewable hybrid microgrids are also presented. Therefore, this reprint is for all readers interested in these challenging directions of research.

Nicu Bizon
Editor

mathematics

Review

Islanding Detection Methods for Microgrids: A Comprehensive Review

Muhammed Y. Worku [1,*], Mohamed A. Hassan [1], Luqman S. Maraaba [2] and Mohammad A. Abido [1,3,4]

[1] Interdisciplinary Research Center for Renewable Energy and Power Systems (IRC-REPS), Research Institute, King Fahd University of Petroleum and Minerals, Dhahran 31261, Saudi Arabia; mhassan@kfupm.edu.sa (M.A.H.); mabido@kfupm.edu.sa (M.A.A.)

[2] Applied Research Center for Metrology, Standards and Testing, Research Institute, King Fahd University of Petroleum and Minerals, Dhahran 31261, Saudi Arabia; lmaraaba@kfupm.edu.sa

[3] Electrical Engineering Department, King Fahd University of Petroleum and Minerals, Dhahran 31261, Saudi Arabia

[4] K.A.CARE Energy Research & Innovation Center, King Fahd University of Petroleum & Minerals (KFUPM), Dhahran 31261, Saudi Arabia

* Correspondence: muhammedw@kfupm.edu.sa; Tel.: +966-559713973; Fax: +966-138603535

Abstract: Microgrids that are integrated with distributed energy resources (DERs) provide many benefits, including high power quality, energy efficiency and low carbon emissions, to the power grid. Microgrids are operated either in grid-connected or island modes running on different strategies. However, one of the major technical issues in a microgrid is unintentional islanding, where failure to trip the microgrid may lead to serious consequences in terms of protection, security, voltage and frequency stability, and safety. Therefore, fast and efficient islanding detection is necessary for reliable microgrid operations. This paper provides an overview of microgrid islanding detection methods, which are classified as local and remote. Various detection methods in each class are studied, and the advantages and disadvantages of each method are discussed based on performance evaluation indices such as non-detection zone (NDZ), detection time, error detection ratio, power quality and effectiveness in multiple inverter cases. Recent modifications on islanding methods using signal processing techniques and intelligent classifiers are also discussed. Modified passive methods with signal processing and intelligent classifiers are addressing the drawbacks of passive methods and are getting more attention in the recently published works. This comprehensive review of islanding methods will provide power utilities and researchers a reference and guideline to select the best islanding detection method based on their effectiveness and economic feasibility.

Keywords: microgrid; islanding detection; local islanding; remote islanding; signal processing

Citation: Worku, M.Y.; Hassan, M.A.; Maraaba, L.S.; Abido, M.A. Islanding Detection Methods for Microgrids: A Comprehensive Review. *Mathematics* **2021**, *9*, 3174. https://doi.org/10.3390/math9243174

Academic Editor: Nicu Bizon

Received: 3 November 2021
Accepted: 3 December 2021
Published: 9 December 2021

Publisher's Note: MDPI stays neutral with regard to jurisdictional claims in published maps and institutional affiliations.

1. Introduction

Distributed generation (DG) integrated with energy storage, and both renewable and non-renewable energy resources providing power to local loads, forms a microgrid [1,2]. Microgrids increase the reliability and resiliency of the grid by regulating the voltage in medium and low distribution networks. They also offer several advantages and benefits, including a reduction in CO_2 emission, improving energy efficiency, the integration of renewable sources, energy access to remote and developing communities, and a reduction in power transmission losses [3–7].

A microgrid has two modes of operation, namely, grid-connected and island (stand-alone) modes [8,9]. In grid-connected mode, the microgrid operates in parallel with the main utility, and the main grid is responsible for smooth operation by controlling the voltage and frequency. In this mode, the DG units forming the microgrid are controlled and operated in the current control mode, called grid following. In the island mode, the microgrid is operated as an independent power island, controlling its own voltage and frequency. The DG units in this mode are controlled and operated in voltage control mode, commonly called grid forming [10,11].

Microgrid islanding occurs when the main grid power is interrupted but, at the same time, the microgrid keeps on injecting power to the network, which can be intentional or unintentional [12,13]. Intentional islanding is a controllable operation mode required for the maintenance of the main utility, whereas unintentional islanding is an uncontrollable operation caused by regular faults such as line tripping, equipment failure, or other uncertainties in the power system [14–16] and may degrade the power quality, overload the system, damage equipment and cause safety hazards [17–20]. Therefore, detecting the islanding condition and effectively disconnecting the microgrid within a specified time interval from the distribution network is a necessity. Moreover, in the islanding condition, the conventional protection devices might not operate, as the DG units cannot provide the sufficient fault current for its operation [21]. The authors in [22,23] investigated the design and control requirements to safely island a microgrid operating either in grid-connected or island modes.

The IEEE and IEC revise and modify the DG interconnection and islanding codes frequently to accommodate the fast growing renewable integration [24]. The consequences of unintentional islanding can be avoided by safely following the provided standards from the IEEE and IEC. The increase in DG integration makes the need to detect unintentional islanding a hot research topic. Researchers have developed different islanding detection methods (IDMs) to address the challenges associated with unintentional islanding. Many IDMs proposed in the literature claim a high reliability and better accuracy compared to each other.

This paper presents a detailed review of the different IDMs proposed in the literature. The IDMs are studied considering their effectiveness, performances, feasibility and operational capabilities. Their advantages and disadvantages are also critically analyzed. The rest of the paper is organized as follows. Sections 2 and 3 describe the islanding detection standards and the performance evaluation criteria of IDMs, respectively. Section 4 presents the detailed classification of IDMs. Section 5 presents discussion and recommendation, whereas Section 6 concludes the paper.

2. Islanding Detection Standards

Figure 1 [8] shows a distribution network connected with distributed energy resources (DERs) and energy storages. The islanding phenomena shown by the dotted lines occurs when the power supply from the grid is interrupted. Unintentional islanding degrades the power quality, complicates orderly power restoration and endangers the lives of utility personnel.

From Figure 1:

P_{PV} represents PV array generated power;

P_{BAT} is the charging and discharging power of the battery storage system;

P_{GEN} is the power generated from the diesel generator;

P_{LOAD} is the power drawn by the load;

P_{GRID} is the power exchanged between the main grid and the microgrid;

PCC is the point of common coupling;

CB is the circuit breaker.

The IEEE and IEC offer standards on how the DG units are operated and controlled with the main grid. IEEE Std. 1547 [25] defines islanding as a condition in which part of the power system becomes isolated from the rest of the network. Islanding detection is one of the major issues when deciding if a DG unit is being synchronized with a grid. Islanded operation requires fast, precise, and cost-effective IDMs, which does not affect the quality of supply. Thus, detecting the islanding condition accurately and timely are the two most important factors to save a distribution network from collapsing. Operating DERs in island mode are not allowed under existing standards such as IEC 62,116, IEEE 1547, IEEE 929-2000 and AS4777.3-2005 [26]. In fact, the islanding condition should be detected and the microgrid disconnected from the main grid within 2 s, as described in IEEE 1547 [27]. The standards describe in detail the operation of the DG, such as disconnecting the DG unit within 2 s, monitoring the magnitude and direction of power flow, appropriate control of voltage, frequency and power quality.

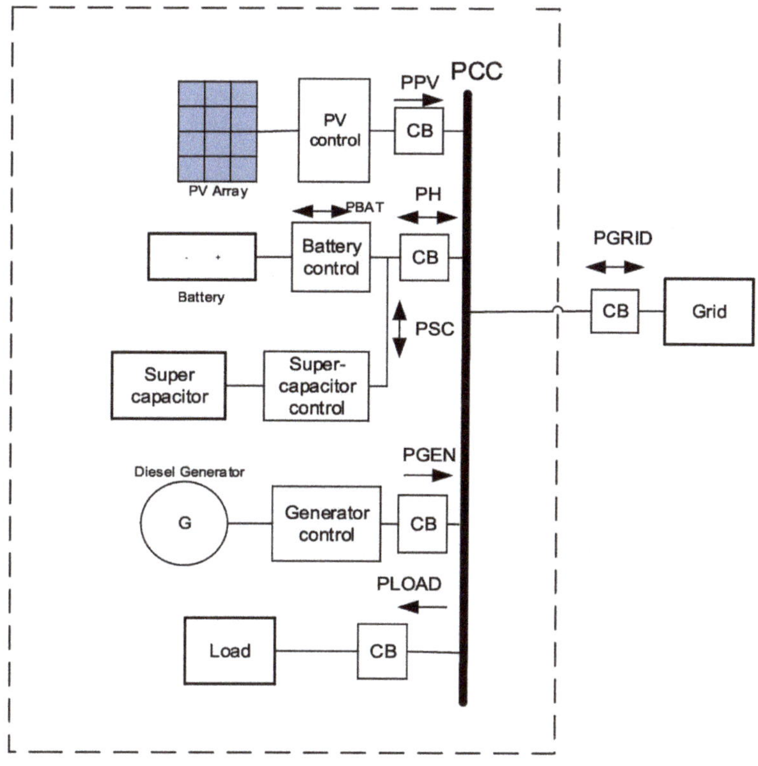

Figure 1. Grid and island operation modes in a DER based microgrid.

Table 1 shows some common standards for islanding detection, voltage and frequency ranges, along with the required detection time.

Table 1. Standards for microgrid islanding.

Standards	Detection Time	Frequency Range	Voltage Range	Quality Factor
IEEE-1547 [28]	$t < 2$ s	$49.3 \text{ Hz} \leq f \leq 50.5 \text{ Hz}$	$0.88 \leq V \leq 1.1$ pu	1
IEEE-929-2000 [29]	$t < 2$ s	$49.3 \text{ Hz} \leq f \leq 50.5 \text{ Hz}$	$0.88 \leq V \leq 1.1$ pu	2.5
IEC-62116 [30]	$t < 2$ s	$48.5 \text{ Hz} \leq f \leq 51.5 \text{ Hz}$	$0.85 \leq V \leq 1.15$ pu	1

3. Performance Evaluation Criteria of IDMs

Power systems with a high penetration of inverter-based resources, such as wind, solar and energy storage, in the distribution network have a reduced inertia, making them prone to an increased risk of frequency instability [31–33]. For a small disturbance at the point of common coupling (PCC), conventional methods fail to detect the islanding condition.

IEEE 1547 will be used to assess the performance of IDMs in this paper. The performances of different IDMs are evaluated on whether they can detect islanding timely, effectively and accurately. Non-detection zone (NDZ), detection time (DT), error detection ratio (EDR) and power quality (PQ) are the most popular performance indices used to evaluate IDMs. These indices are described in detail.

3.1. Non-Detection Zone (NDZ)

The NDZ represents a region of power imbalance between the power generated by the DG units and that dissipated by local loads where the islanding detection method fails [34]. The non-detection zone is the main performance indicator for the implemented IDM and is the main reason IDMs fail to detect islanding. The term "power mismatch space" is used to describe IDMs that are based on monitoring voltage, frequency or phase deviation, whereas IDMs that inject a disturbance are expressed in the "load parameter space". Figure 2 presents an NDZ based on a passive islanding detection method.

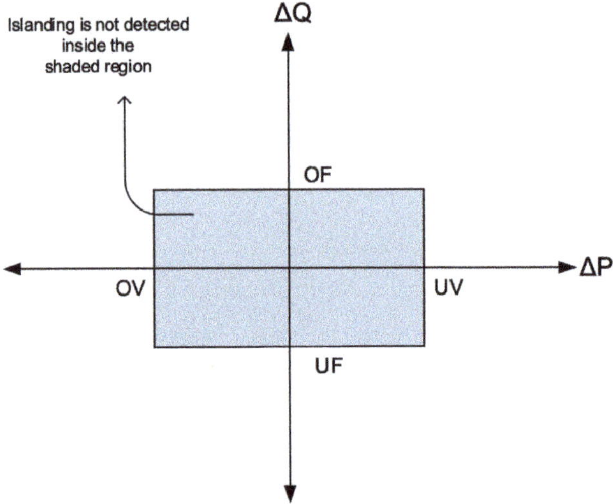

Figure 2. Non-detection zone for over/under voltage (UOV) and over/under frequency (UOF) passive islanding detection method [35].

3.1.1. Power Mismatch Space

For a microgrid operating in island mode, the power imbalance between the power generated from the DG units and that dissipated by the local loads affects the voltage and frequency at the PCC. If the imbalance is nearly equal to zero (ΔP and ΔQ close to zero), the variation of voltage and frequency will not be enough to detect islanding when the microgrid disconnected from the grid [36]. The NDZ in the power mismatch space is defined as the power imbalance ΔP and ΔQ, which cannot cause voltage or frequency to exceed the normal limit to detect islanding and is given as [37]:

$$\left(\frac{V}{V_{max}}\right)^2 - 1 \leq \frac{\Delta P}{P} \leq \left(\frac{V}{V_{min}}\right)^2 - 1 \tag{1}$$

$$Q\left(1 - \left(\frac{f}{f_{min}}\right)^2\right) \leq \frac{\Delta Q}{P} \leq Q\left(1 - \left(\frac{f}{f_{min}}\right)^2\right) \tag{2}$$

where, V and P are the rated voltage and the rated active power, respectively;

V_{min} and V_{max} are the minimum and maximum microgrid voltages, respectively;

Q is the quality factor;

f_{min} and f_{max} are the minimum and maximum frequencies, respectively.

3.1.2. Load Parameter Space

Equation (3) defines the NDZ in parameter space as:

$$F_1(cf, K, Q) < \Delta C_{norm} < F_2(cf, K, Q) \tag{3}$$

where cf is the chopping fraction, K is the accelerating gain, and ΔC_{norm} is the resonate capacitance in the range of NDZ.

3.2. Detection Time (DT)

The detection time is defined as the time taken from the beginning of microgrid disconnection till the end of the IDM detecting islanding.

$$\Delta T = T_{IDM} - T_{trip} \tag{4}$$

where ΔT is the run-on time, T_{IDM} is the moment to detect islanding, and T_{trip} is the moment microgrid disconnects from the grid.

3.3. Error Detection Ratio (EDR)

Due to load switching, or other disturbances that affect measurement parameters to exceed normal limits, IDMs might detect false islanding, called error detection [38]. This is defined as:

$$E = \frac{N_{error}}{N_{error} + N_{correct}} \tag{5}$$

where E is the error detection ratio, N_{error} is the times of error detection, and $N_{correct}$ is the times of correct detection.

3.4. Power Quality (PQ)

Maintaining the power quality of the microgrid is an important index while selecting IDMs. IDMs that inject a disturbance to the system distort the power output and deteriorate the power quality.

4. Classification of Islanding Detection Methods

Islanding detection techniques are mainly classified into local and remote [39–41]. Local islanding techniques are further classified as passive, active and hybrid techniques, based on non-detection zone (NDZ), detection speed, power quality, error detection rate and efficacy in multiple inverter cases. Passive islanding techniques are widely used by utilities because of their low cost and that they do not degrade the power quality. However, these methods have a large NDZ, and setting the threshold setting is a challenge. To overcome the limitations of the passive technique, different signal processing and intelligent classifiers have been used in the literature. Figure 3 [12,42] presents the detail classification of IDMs, and these techniques are discussed in detail in the following sections.

4.1. Local Islanding Detection Techniques

Local islanding detection techniques measure the system parameters at the DG site for islanding detection. The measured parameters include voltage, frequency, active power, reactive power phase angle, impedance and harmonic distortion. Local islanding detection techniques are classified as passive, active and hybrid techniques and are described as follows.

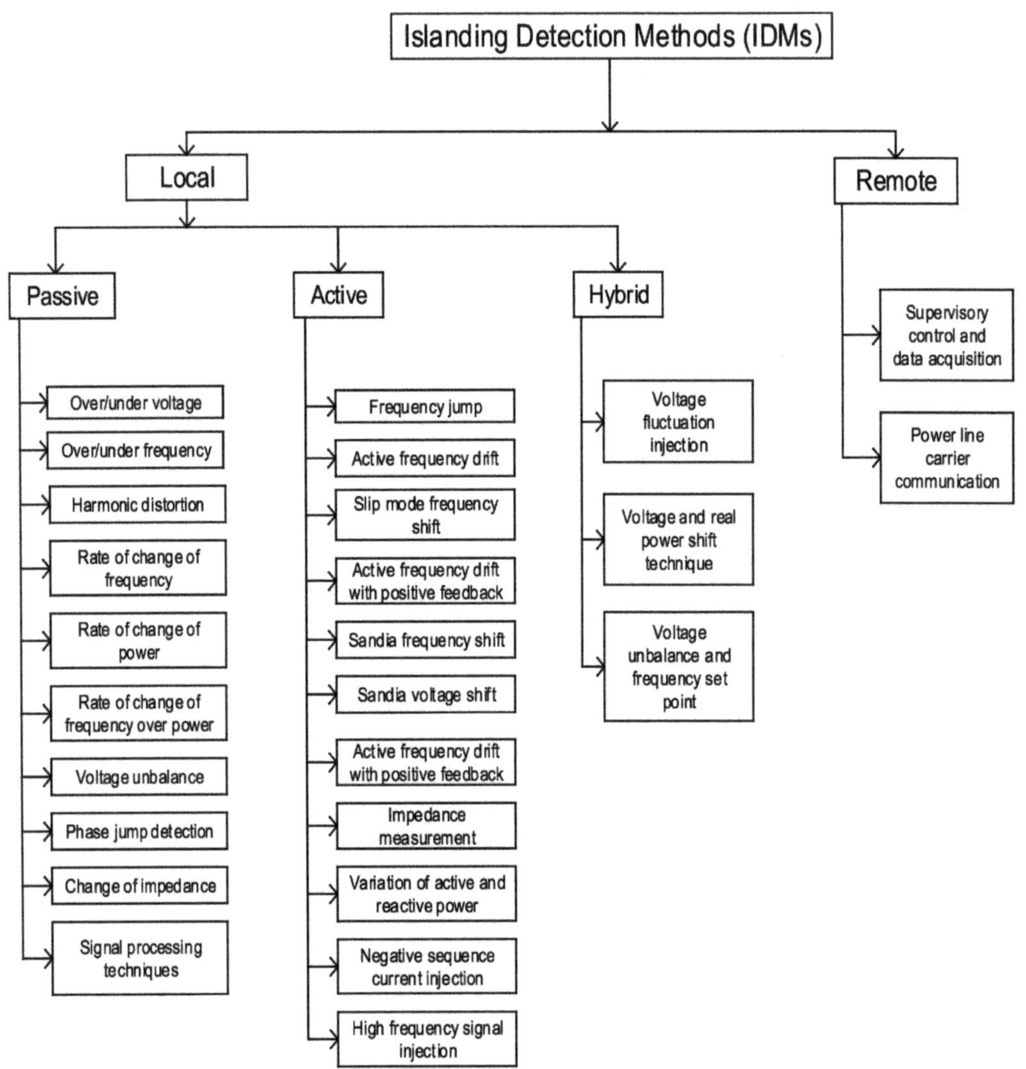

Figure 3. Classification of islanding detection methods.

4.1.1. Passive Islanding Detection Techniques

This method measures the system parameters and compares them with a predetermined threshold value for islanding detection. The measured system parameters at the DG terminal or PCC include voltage, frequency, phase angle and harmonics. The passive islanding detection techniques working principle is depicted in Figure 4. Passive islanding detection techniques are mostly used by power utilities as they are simple, low cost, do not degrade the power quality and have a fast detection speed within 2 s, as recommended by IEEE 1547. However, these methods have a large NDZ, the error detection rate is high and setting the threshold requires special consideration. Some of the popular passive IDMs are described below.

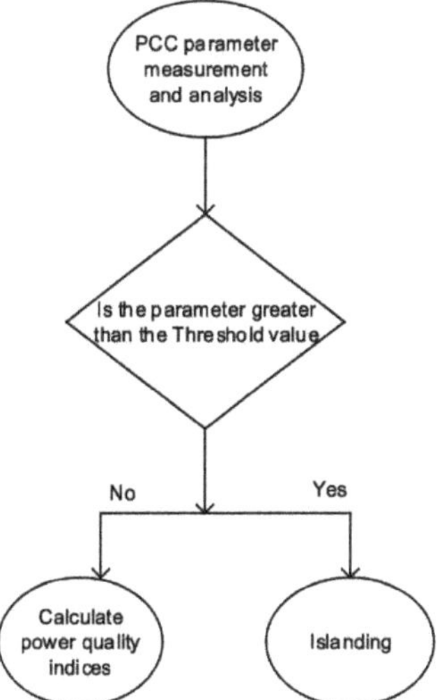

Figure 4. Passive islanding detection technique.

Harmonic Detection (HD)

The HD method is based on comparing the total harmonic distortion (THD) measured at the PCC and a predefined THD to detect islanding. When the microgrid is operated in grid-connected mode, the PCC voltage is a normal sine wave, and the harmonics generated by the load and the inverter are negligible. However, during islanding mode of operation, the harmonics produced by the inverter will distort the PCC voltage and, hence, islanding will be detected [42–45].

This method is easy to implement and is also effective for multiple DGs connected to the same PCC with a detection time of 45 ms. However, selecting the threshold is difficult since grid disturbance can cause error detection, and it might fail to detect islanding for loads with a large quality factor Q and a large NDZ. Q is defined in Equation (6) as [45,46]:

$$Q = R\sqrt{\frac{C}{L}} \tag{6}$$

Over/under Voltage and over/under Frequency (OUV/OUF)

This method works by comparing the PCC voltage and frequency with a predefined threshold voltage and frequency to detect islanding. The microgrid will be disconnected from the main grid if the measured voltage and frequency at the PCC exceed the thresholds. The microgrid disconnection from the grid deviates the frequency and voltage at the PCC due to an active and reactive power mismatch between the power generated in the DG units and dissipated in the loads.

$$\Delta P = P_{load} - P_{DG} \tag{7}$$

$$\Delta Q = Q_{load} - Q_{DG} \tag{8}$$

In grid-connected operation, the main grid injects ΔP and ΔQ, and the balance of active and reactive power will be kept. When islanding occurs, to keep the active and reactive power balance, the voltage and frequency will drift until $\Delta P = 0$ and $\Delta Q = 0$. By detecting the deviation in voltage and frequency, OUV/OUF can detect islanding [47–49]. The advantage of this method is that it does not affect the power quality and the cost is low. The disadvantages are that it is difficult to predict the detection time and it has a relatively large NDZ, with a detection time between 4 ms to 2 s [50]. This method is suitable for microgrids with some power imbalance between DG and loads.

Rate of Change of Frequency (ROCOF)

When the microgrid is disconnected from the main grid with a power mismatch, the frequency will change. The ROCOF method works by measuring df/dt for a few cycles and comparing it with a setting threshold. Islanding will be detected if the measured df/dt exceeds the predefined threshold [51–54]. Compared to OUV/OUF, this method has a fast detection time of 24 ms, is more sensitive and highly reliable. However, it has a high error detection rate for systems with high load switching and fluctuation. Hence, ROCOF is best suited for loads with less fluctuation as it cannot discriminate whether the frequency change comes from load changes or by islanding [55]. An extension of ROCOF that considers the dynamic behavior of the load is proposed in [56,57]. The authors incorporated the exponential static load model to incorporate the dynamic behavior of the load and determine the threshold to detect islanding.

Rate of Change of Frequency over Power (ROCOFOP)

This method works by measuring $\partial f/\partial PL$, where P_L is the load power, to detect whether or not islanding occurs. It has a lower error detection ratio, smaller NDZ and higher reliability than ROCOF. This method has a detection time of 100 ms and works efficiently for microgrids that have a small power imbalance between the DG units and the load [58,59].

Rate of Change of Power Output (ROCOP)

This method measures the changes in the DG power output (dP/dt) over a few cycles and compares it with a setting threshold to detect islanding. Generally, a loss of the main grid produces load changes, and dP/dt measured after the microgrid is islanded is greater than dP/dt measured before the microgrid is islanded. This method has a fast detection, with a detection time of between 24 and 26 ms, and the power imbalance between the DG units and the load does not affect the detection speed.

Phase Jump Detection (PJD)

The working principle of PJD is to monitor the phase jump between the inverter's terminal voltage and the current for islanding detection. During grid-connected mode, the inverter's current will be synchronized with the voltage at the PCC using a phase-locked loop (PLL) to detect the zero crossing of the voltage. In islanding operation, since PLL works only at the zero crossing of the voltage, the inverter output current remains unchanged. However, the voltage will have a sudden jump due to the load phase angle. Comparing the measured phase difference with a predefined threshold can detect islanding. The advantages of this method are that it has a fast detection speed with a detection time of between 10 to 20 ms, does not affect the power quality, works for multiple inverters and is easy to implement [60,61]. However, it is difficult to choose thresholds for microgrids with frequent load switching.

Voltage Unbalance (VU)

A microgrid disconnected from the main grid changes the topology of the network that, in turn, causes a voltage unbalance at the DG output. This voltage unbalance can be used for islanding detection if it exceeds the setting threshold. The voltage unbalance at the time t is defined as:

$$VU_t = \frac{NS_t}{VS_t} \tag{9}$$

where NS_t and PS_t are the negative and positive sequence voltage amplitudes, respectively.

The authors in [62] proposed a variational mode decomposition method to obtain the principal modes from the measured three phase voltage signal and employed the mode singular entropy to determine the index for islanding detection. This method is not sensitive to system disturbances caused by variation in normal loads and has a detection time of about 53 ms [63]. However, the system harmonics affects the extraction of the negative sequence voltage component, making the threshold calculation difficult. This method has better applications for systems with frequent load fluctuations, such as motor starting and capacitor bank switching.

Table 2 compares the different passive islanding detection techniques described, with respect to their performance evaluation, such as NDZ, DT, EDR and power quality.

Table 2. Summary of different passive techniques.

Method	NDZ	Detection Time	Impact on Power Quality	Error Detection Rate
Harmonic distortion	Large for high Q	45 ms	No	High
OUV/OUF	Large	4 ms to 2 s	No	Low
ROCOF	Small	24 ms	No	High
ROCOFOP	Smaller than ROCOF	100 ms	No	Low
ROCOP	Small	24–26 ms	No	High
Phase jump	Large	10–20 ms	No	Low
Voltage unbalance	Large	53 ms	No	Low

4.1.2. Active Islanding Detection Techniques

The performance of active detection methods is based on the perturbation and observation concept. These methods perturb system parameters such as frequency, voltage, currents and harmonics. In the presence of a stiff grid, the amplitude of the variation at the PCC is negligible since the grid parameters are dominant. However, during the islanding phenomenon, injecting a disturbance at the PCC results in a significant variation in the DG parameters. Figure 5 shows the basic working principle of active islanding detection techniques.

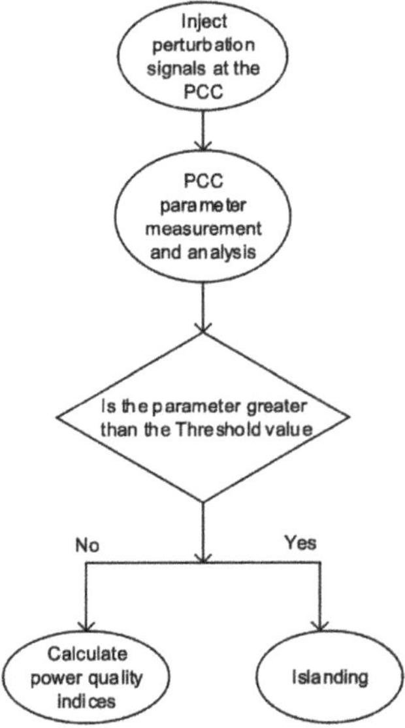

Figure 5. Active islanding detection technique.

Compared to passive islanding techniques, active techniques have a reduced NDZ and low error detection rate. However, active techniques deteriorate the power quality, and additional power electronic circuits are required to inject the perturbations. To observe the effect of perturbation, additional detection time is required, which can affect the stability of the system. Moreover, most of the active detection methods are developed for inverter-based DG units and are not applicable for synchronous generators. Some of the popular active IDMs are described below.

Active Frequency Drift (AFD)

An AFD works by slightly distorting the inverter current waveform injected into the PCC. In grid-connected mode, the voltage and frequency are controlled by the grid and are stable. When islanding occurs, the voltage zero crossing occurs earlier than expected because of the distortion of the injected current waveform. This results in a phase error between the inverter's output current and the voltage, which makes the frequency of the inverter output current drift to eliminate the phase error. This drift in frequency again causes an earlier zero crossing than expected. The frequency of the inverter output current will continue to drift until the voltage frequency measured at the PCC exceeds the threshold of OUF to detect islanding.

An islanding detection method based on a low-frequency current injection disturbance, injected in the conventional dq controller of the distribution generator, is proposed in [64]. The frequency deviation is processed using the estimation of a signal parameter via the rotational invariance technique to extract the dominant mode of the oscillations present in the PCC frequency signal to detect islanding. The method has a detection time of 0.12 sec, eliminates NDZ and does not affect the power quality.

The chopping fraction given in Equation (10) describes the distortion of the inverter's injected current [65,66].

$$cf = \frac{2t_z}{Tvutil} \tag{10}$$

where t_z is the dead time and T_{Vutil} is the voltage period.

The advantages of AFD are that it has a small NDZ and is easy to implement, with a detection time within 2 s. The disadvantages are that it degrades the power quality if the injected current is heavily distorted, and the method might also fail to detect islanding for multiple inverter cases.

Frequency Jump (FJ)

Frequency jump is a modified version of AFD, as it also inserts dead zones into the current waveform. However, unlike AFD, where dead zones are inserted into every cycle, in FJ, it is inserted in every three cycles. In grid-connected mode, the waveform of the voltage at the PCC is not distorted, despite the inverter's distorted current. During islanding, there will be a variation in voltage frequency that will be used to detect islanding [67]. Similar to AFD, this method might fail to detect islanding for multiple inverters working in parallel.

Active Frequency Drift with Positive Feedback (AFDPF)

This method is an extension of AFD and works by applying a positive feedback to increase the chopping fraction, which in turn accelerates the frequency deviation to detect islanding more effectively.

$$cf_k = cf_{k-1} + F(\Delta\omega_k) \tag{11}$$

where cf_k and cf_{k-1} are the kth and $k-1$th cycle chopping fractions, respectively, and can be positive or negative;

F is usually a linear function;

ω_k is the frequency of the kth cycle;

$\Delta\omega_k = \omega_{k-1} - \omega_0$.

As compared to AFD, this method has a small NDZ however, it still affects the power quality [68].

Sandia Frequency Shift (SFS)

SFS is also an extension of AFD and works by applying a perturbation to the frequency of the inverter's voltage with a positive feedback. The modified chopping fraction used in SFS is given in [69] as:

$$cf = cf_0 + K(f_{pcc} - f_{grid})\qquad(12)$$

where cf_0 is the no deviation in frequency chopping factor;

K is the accelerating gain;

f_{pcc} is the frequency of the PCC voltage;

f_{grid} is the grid frequency.

In grid-connected mode, the voltage frequency of the PCC is maintained by the grid, even if the method attempts to change it. However, during islanding the chopping fraction increases with the increase of f at the PCC, which also increases the frequency of the inverter. The process continues until islanding is detected. This method has a detection time of 0.5 s and has the smallest NDZ compared to other active methods [70,71].

Sandia Voltage Shift (SVS)

The working principle of SVS and SFS is similar, in that it perturbs the voltage amplitude of the PCC with a positive feedback to change the inverter's output current and power. In grid-connected mode, the power change does not affect the voltage amplitude of the PCC, whereas in island mode, the power change affects the voltage amplitude, which can be used to detect islanding [72]. SVS is easy to implement; however, its disadvantage are that it lightly degrades the power quality, and the inverter's operation efficiency might be reduced because of the change in the output power.

Sliding Mode Frequency Shift (SMS)

SMS perturbs the voltage phase of the PCC with a positive feedback and monitors the frequency deviation to detect islanding. In SMS, the current–voltage phase angle of the inverter is set as [73]:

$$\theta = \theta_m \sin\left(\frac{\pi}{2}\frac{f^{k-1} - f_n}{f_m - f_n}\right)\qquad(13)$$

where θ_m is the maximum phase angle at the frequency f_m, f_n is the rated frequency, and f^{k-1} is the previous cycle frequency.

In grid-connected mode, the microgrid injects active power to the main grid, and its power factor is close to unity, with the phase angle between the inverter current and the PCC voltage close to zero. During islanding operation, the phase angle of the load and the frequency will vary, and if the frequency variation exceeds the threshold, islanding can be detected. The advantages of the SMS method are that it is easy to implement, is highly effective for multiple inverter systems and has a smaller NDZ with a detection time of about 0.4 s [74]. However, this method reduces the power quality and has an impact on the transient stability of the system.

Variation of Active and Reactive Power

This works by varying the injected inverter power and monitors the voltage amplitude and frequency variation for islanding detection. During islanding, the active power generated in the DG units will be dissipated in the loads, and the voltage variation must satisfy Equation (14) to balance the active power between DG and the loads.

$$P_{DG} = P_{load} = \frac{V^2}{R}\qquad(14)$$

When the voltage variation exceeds the threshold of OUV, islanding can be detected. Similarly, the frequency will be affected by the reactive power disturbance, and islanding can be detected when the frequency variation exceeds the threshold. This method is easy to implement and has a small NDZ with a detection time between 0.3 s and 0.75 s.

However, the method greatly affects the power quality and transient stability since it varies the inverter output power continuously. The method also might not work effectively for multiple inverters working in parallel.

Negative-Sequence Current Injection

The basic working principle of this method is to perturb the three-phase voltage-sourced converter with a negative-sequence current and monitor the negative-sequence voltage at the PCC for islanding detection. During normal grid-connected operation, the injected negative-sequence current will not affect the PCC voltage and will flow into the grid since the grid has low impedance. However, during islanding operation, the injected negative-sequence current will flow into the load, creating an unbalance in the PCC voltage, and islanding can be detected if the voltage unbalance exceeds the threshold. The advantages of this method are that it has a very short detection time of 60 ms (3.5 cycles), it is insensitive to load change, has no NDZ and has a higher accuracy than positive-sequence voltage detection [75,76].

Impedance Measurement (IM)

This method works by changing the amplitude of the output inverter current. During islanding operation, the current perturbation varies the voltage, and this variation will be calculated as *dv/di*, an equivalent impedance seen from the inverter. Islanding can be detected if the impedance calculated exceeds the threshold [77]. This method has a detection time of between 0.77 s and 0.95 s and has a small NDZ for a single DG system. However, the detection efficiency will decrease in multiple-inverter cases, and the setting the impedance threshold is difficult since it requires the exact grid impedance.

Detection of Impedance at Specific Frequency

This method works by injecting specific frequency harmonics and is a special case of harmonic detection method. During grid-connected mode, the injected harmonic current will not affect the PCC voltage and will flow into the grid. During islanding operation, the injected harmonic current will flow into the local load and produce a harmonic voltage at the PCC. Islanding can be detected if the produced harmonic voltage is large enough. The disadvantage of this method is that it affects the operation of equipment such as transformers.

Table 3 describes the different active islanding techniques found in the literature.

Table 3. Summary of different active techniques.

Method	NDZ	Detection Time	Impact on Power Quality	Error Detection Rate
AFD	Large if value of Q is high	With 2 s	Degrades	High
FJ	Small	75 ms	Degrades	Low
AFDPF	Smaller than AFD	1 s	Slightly degrades	Lower than AFD
SFS	Smallest	0.5 s	Slightly degrades	Low
SVS	Smallest	0.5 s	Slightly degrades	Low
SMS	Small	0.4 s	Degrades	Low
Variation of active and reactive power	Small	0.3–0.75 s	Degrades	High
Negative sequence current injection	None	60 ms	Degrades	Low
Impedance measurement	Small	0.77–0.95 s	Degrades	Low
High frequency signal injection	Smallest	Few ms	Slightly degrades	Low
Virtual capacitor	Smallest	20–51 ms	Slightly degrades	Low
Virtual inductor	Smallest	13–59 ms	Slightly degrades	Low
Phase PLL perturbation	Smallest	120 ms	Negligible	Low

4.1.3. Hybrid Islanding Detection Techniques

Hybrid islanding detection techniques are developed from the combination of passive and active detection techniques, and are implemented with two steps. The first step utilizes a passive technique, primarily to detect islanding. If islanding is suspected in the first step, an active technique is employed to accurately detect the islanding [78–81]. Figure 6 depicts the flow chart of the hybrid islanding detection technique.

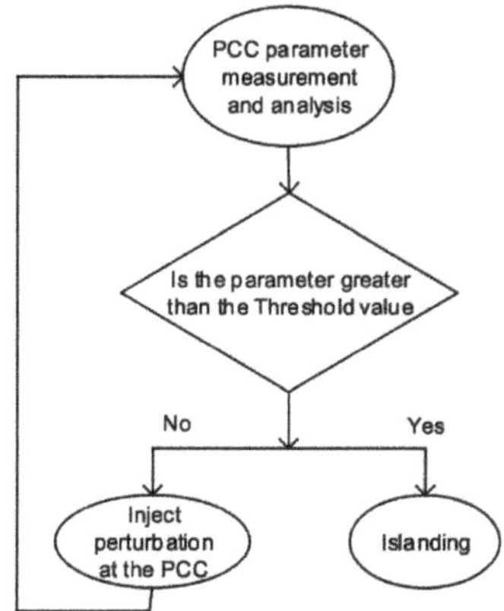

Figure 6. Hybrid islanding detection technique.

The performance indices will improve from the combination of these methods; they generally have a small NDZ, and the power quality degradation is low. However, the cost of the system is high, and the method is time consuming, which makes their real implementation infeasible. Authors in [82,83] described the recent literature on hybrid islanding detection techniques. Some of the hybrid detection methods discussed in the literature are described below.

Voltage Unbalance and Frequency Set-Point

This works by combining the voltage unbalance and the positive feedback-based methods. Computing the average voltage unbalance caused by changes in the system and load is the first step of this method. To differentiate whether the voltage unbalance is caused by islanding or system variation, the second step employs a positive feedback-based method to lower the frequency set point gradually, from 60 to 59 Hz in one second, if the measured voltage unbalance is greater than 35% of the average voltage unbalance [84]. If the nominal frequency is maintained at the PCC, then islanding was not the cause of the voltage variation. However, islanding will be detected if the frequency falls below 59.2 Hz in the following 1.5 s. This method has a detection time of 0.15–0.21 s and works best for microgrids with a low penetration of non-synchronous generation units.

Voltage Change and Power Shift

This works by combining the rate of change of voltage and the variation of active power methods. Firstly, to suspect islanding, the average rate of change of voltage is calculated for five cycles. If this calculated voltage exceeds the predefined threshold value, islanding is suspected, and the second method injects extra active power into the system to confirm whether or not islanding has occurred. For normal grid-connected operation, the grid regulates the PCC voltage to be within a predefined interval and compensates the extra injected active power. However, if the microgrid is in islanding operation, the extra active power will increase the voltage amplitude at the PCC. Therefore, islanding can be detected by monitoring the PCC voltage with a detection time of less than 0.5 s. Using reactive power instead of active power for the second step was proposed in [85–87].

The injected reactive power causes an increase in the PCC frequency and islanding can be detected if the frequency variation is more than a predefined threshold.

Voltage Fluctuation Injection

This works by combining the rate of change of frequency and voltage methods to detect islanding [88]. The rate of change of voltage and the rate of change of frequency at the PCC are monitored as a first step to detect if one of them exceeds the predefined threshold to indicate islanding might have occurred. Then, the second step employs a voltage perturbation by applying a periodically switching high-impedance load for confirmation. During normal grid-connected operation, the grid stabilizes the PCC voltage perturbation caused by the switching of the high impedance load. However, during the islanding operation, the effect of the periodic perturbation is observed at the PCC voltage to detect islanding. This method does not depend on quality factor and has a detection time of less than 0.216 s, but it might be less efficient for large scale DG units [89,90].

Hybrid Sandia Frequency Shift and Q_g-f Method

This technique modifies the Sandia frequency shift method to reduce the NDZ by adding a Q_g-f droop curve to maintain the optimal gain K_f at a stable value [91]. The optimal gain K_f is directly proportional to the quality factor of the load; however, when the quality factor is more than five, K_f will be too large to create a false detection and can even cause system instability. The authors in [88] proposed a Q_g-f droop curve method to keep K_f to a safe value and monitor the change of frequency for islanding detection. During normal operation, the reactive power is controlled by the grid; however, during islanding operation, since the DG units operate at unity power factor and produce no reactive power, this creates a frequency difference between the actual and the rated system frequency. This method monitors this change in frequency for islanding detection and has a detection time of 1.4 s.

Rate of Change of Reactive Power and Load-Connecting Strategy

This works by combining the change of reactive power and load connection to detect islanding. If the change in reactive power monitored in the first step is more than the predefined threshold, the second step varies the reactive power by connecting an appropriate load to the microgrid [92,93]. During normal grid-connected operation, the grid regulates the reactive power at the PCC, and the rate of change of reactive power is small. However, during islanding operation, any load change affects the generated reactive power from the DG units. The extra connected load affects the rate of change of reactive power and is used to detect islanding. This method can effectively detect islanding, even in the presence of a small load change, and has a detection time of 40 ms. The disadvantage of this method is that selecting the appropriate extra load is not straightforward.

Table 4 presents the different hybrid islanding detection methods.

Table 4. Summary of different hybrid techniques.

Method	NDZ	Detection Time	Impact on Power Quality	Error Detection Rate
ROCOV and power variation	-	Low	Small	Small
ROCOF and IM	0.216 s	Low	-	Small
VU and SFS, SVS	-	None	Reduce negative impact	Very small

4.2. Remote Methods

The remote methods utilize advanced signal processing and communication infrastructure for islanding detection. Remote methods do not have a non-detection zone (NDZ), error detection can be eliminated, and they do not affect the power quality; therefore, they are very sound approaches for islanding detection. Whereas remote methods tend to be expensive to implement for small microgrids, they are very beneficial for large microgrid applications. Some of the remote methods described in the literature are discussed below.

4.2.1. Power Line Carrier Communication (PLCC)

In the PLCC method, transmitters and receivers are set at the grid and DG side, respectively. Transmitters produce a communication signal along with the power line, and if this signal is interrupted, it indicates that islanding has occurred [94]. The PLCC method has a signal period of four consecutive cycles, and the method can detect islanding if the signal is lost in three consecutive periods. This method has a detection time of 200 ms, has no NDZ, has no impact on power quality and is proven to work on multiple inverter system [95]. However, the transmitter set at the grid side is costly, and this method is not economically feasible for low density DG systems.

4.2.2. Signal Produced by Disconnect (SPD)

Similar to the PLCC method, this method also uses signal transmission between the grid and the DG units to detect islanding. However, this method utilizes microwave, telephone and other forms of signal transmission, rather than the power line. This method has no NDZ; however, it needs more investment to set up the communication line. The SPS method can be extended to add additional control of the DG by the main grid, such as coordinating the power generated between the DG units and the main grid, which is beneficial to black start.

4.2.3. Supervisory Control and Data Acquisition (SCADA)

The SCADA system is based on monitoring main grid parameters such as voltages, currents and frequency, which can also be used for monitoring the status of the circuit breakers and sending them to the microgrid. With proper installation, the NDZ can be eliminated with better efficiency. However, the detection speed of this method is slow and requires a high investment to install a separate instrumentation link. Similar to SPD, this method allows additional control of DG by the main grid.

Table 5 describes the different remote islanding detection techniques.

Table 5. Summary of different remote techniques.

Method	NDZ	Detection Time	Impact on Power Quality	Error Detection Rate
PLCC	Without NDZ in range of normal loads	200 ms	None	None
SPD	None	100–300 ms	None	None
SCADA	None	Detection speed is slow for busy systems	None	None

4.3. Passive and Signal Processing

The local, remote and hybrid IDMs discussed above have their own advantages and disadvantages. However, accuracy, high detection speed and detecting islanding for multiple inverter system are the most unresolved issues that need more research. Compared to local IDMs, remote-based IDMs are highly reliable and have a negligible NDZ. However, they have a high cost, are complex for implementation and are not preferred [96]. Similarly, passive techniques do not perform well in inverter-based DGs with different system configurations. Passive IDMs are better in terms of them not degrading power quality, their fast detection time and being compatible for all DG types. However, they have a large NDZ, and selecting the threshold is not straightforward in most cases. Modifying passive IDMs using advanced signal processing in time-domain, frequency-domain, and time-frequency-domain can address these two limitations.

These modified passive islanding detection methods improve detection time, reduce NDZ and improve detection performance [97–99]. The signal processing tools help extract and analyze the measured signal in order to perform the required operation. However, these methods require a time consuming data training process. The following section describes the signal processing techniques used for islanding detection.

4.3.1. Fourier Transform (FT)-Based Method

This method extracts the features of a signal at specific frequencies using a frequency domain. Short time Fourier transform (STFT), discrete Fourier transform (DFT) and fast Fourier transform (FFT) have evolved from FT to develop efficient and fast IDMs [100–104]. This method has some limitations, such as low-frequency resolution and reduced spectral estimation [105].

4.3.2. Wavelet Transform (WT)-Based Method

To extract the features of a distorted voltage, current or frequency signal, a wavelet transform is the best signal processing method [106]. This method compares the measured signal wavelet coefficients with a predefined threshold value, and islanding will be detected if these values are larger than the predefined values. The disadvantages of this method are that it can only analyze low frequency band, selecting the threshold is not straightforward, and that the different sampling frequencies and the mother wavelet selection have an impact on the wavelet transform. To analyze the high frequency components, the wavelet packet transform (WPT) is applied using the d-q axis of three-phase apparent power [107].

4.3.3. S-Transform (ST)-Based Method

This method is an extension of the WT method and converts a time-domain function into a two-dimensional frequency-domain function. The ST method generates the S-matrix and the equivalent time-frequency contours from the measured PCC voltage or current signals. To detect islanding, the spectral energy content of the time-frequency contours is calculated containing the frequency and magnitude deviations [108]. However, the ST method requires more computational memory, and the processing time is large compared to other similar techniques.

4.3.4. Time-Time Transform (TTT)-Based Method

This method works by transforming a one-dimensional time-domain signal into a two-dimensional time-domain signal by giving a time–time distribution in a particular window. This method distributes the low and high frequency components in different positions. The TTT method performs well for noisy signals, as the method allows a time-local view of the signal through the scaled window [109].

4.3.5. Auto Correlation Function (ACF)-Based Method

This method measures the power or energy signals and extracts information using finite summation limits. The authors in [110] proposed a modified ACF of the modal current envelope to extract the transient features of sample variance that will be used to detect islanding.

4.3.6. Kalman Filter (KF)-Based Methods

This method uses measured voltage or current signals to extract harmonic features using a time–frequency domain. The authors in [111] proposed a voltage harmonics and selected harmonic distortion (SHD) technique, based on KF, for islanding detection. The residual signal and SHD are used as a condition for islanding detection, where the residual signal is used for the islanding detection and the SHD is used for timely detection.

4.4. Intelligent IDMs

Passive islanding techniques combined with artificial intelligence provide the most effective and economical method to detect islanding. They have a high accuracy, high reliability, are less complex and have a higher computational efficiency than other methods.

Intelligent IDMs do not require threshold selection. Different intelligent IDMs associated with signal processing, such as artificial neural network (ANN), fuzzy logic (FL), support vector machine (SVM), decision tree (DT) and probabilistic neural network (PNN), are commonly used for islanding detection. The only disadvantage of intelligent IDMs is that they suffer from a large computational burden. Figure 7 [12] shows the basic operation of an intelligent IDM. The method starts offline using a training algorithm to train the system from the PCC measured voltage or current signal. Then, to make the final decision, the online process is activated using an intelligent classifier. Some of the intelligent IDMs are discussed below.

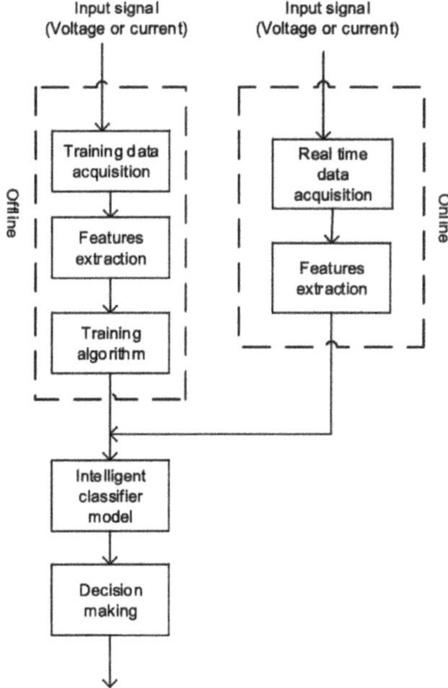

Figure 7. Intelligent IDMs.

4.4.1. Artificial Neural Network (ANN)-Based Method

The ANN-based approaches extract important features from the measuring data, which are used for identifying variations in power system parameters [112,113]. ANN-based IDMs perform well for multi-inverters and have a high accuracy and efficiency, but the data processing time is large [114]. The authors in [115] reported an islanding detection technique based on an adaptive neuro-fuzzy inference system (ANFIS).

4.4.2. Decision Tree (DT)-Based Method

This method, with a combination of WPT or discrete wavelet transform (DWT), are used for islanding detection in different intelligent IDMs [116]. WPT or DWT are used to extract information from the measured voltage or current signals, and then the DT classifier further processes these features to detect islanding as tested on the CIGRE distribution system [117].

4.4.3. Probabilistic Neural Network (PNN)-Based Method

This method uses artificial neural hardware in traditional pattern recognition schemes. It can compute non-linear decision boundaries based on a Bayesian classification technique and has four layers as the input, pattern, summation and output [118]. These layers do not need learning and can perform their functions. PNN-based IDMs are reliable for islanding detection.

4.4.4. Support Vector Machine (SVM)-Based Method

The PCC-measured voltage or current signals are used to extract signature features to indicate islanding occurrence using the SVM classifier and autoregressive modelling [119,120]. IDMs based on SVM have a fast detection speed and high accuracy, but the data training and the algorithm make it too complex for practical application.

4.4.5. Fuzzy Logic (FL) Based Method

FL using DT transformation shows a promising and efficient result in islanding detection [121]. However, the disadvantages of FL is that, because of maximum and minimum combinations, the fuzzy classifiers are highly abstract and sensitive to noisy data [122].

5. Discussion and Recommendation

The IDMs based on different techniques discussed above have many critical technical issues. These issues have to be fixed to improve their performance and make the IDMs more reliable and efficient. IDMs based on a threshold setting have an NDZ that is hard to eliminate, whereas methods based on a disturbance injection degrade the power quality. On the other hand, signal processing-based techniques have a higher precision and are more robust, versatile, reliable and efficient than the existing IDMs. However, they have a high computational burden. There is always a trade off when selecting the performance indices of the IDMs, and it is important to fully consider the practical operation of the microgrid, with all key performance indices taken into account while selecting the appropriate IDM. Researchers should focus on improving the performance of signal processing and intelligent classifier techniques to come up with the best IDM with a high detection speed, smaller NDZ and lower error detection ratio that does not degrade the power quality.

6. Conclusions

A comprehensive review of various islanding detection methods (IDMs) is presented in this paper. IDMs are broadly classified into two types: remote and local. Remote-type IDMs use communication signals between the microgrid and the main grid and are fast, reliable and effective with zero non-detection zones. These techniques do not degrade the power quality and can be applied to multi-inverter microgrids; however, they are complex and expensive. On the other hand, local methods are classified as passive, active and hybrid. Passive-based IDMs measure microgrid parameters such as voltage, current, frequency and phase angle and monitor their changes to detect islanding. Passive methods are preferred as they are easy and cheap for practical implementation. However, passive techniques have a large non-detection zone. Active-based IDMs inject a perturbation into the system that affects the power quality, whereas hybrid techniques are a combination of passive and active techniques. Active and hybrid techniques need additional devices to introduce the perturbation, which increases the complexity and implementation cost. Compared to the passive, active, and hybrid techniques, IDMs based on signal processing have been gaining more attention recently for islanding detection to detect the islanding condition accurately and precisely within the shortest period without affecting power quality. Moreover, they have the potential of working for multiple inverters and can also overcome the non-detection zone and threshold setting requirements of conventional techniques. Several methods have been studied and a comparison has been provided based on important performance indices including NDZ, detection time, error detection ration and power quality.

Author Contributions: M.Y.W. contributed in identifying and classifying the IDMs, manuscript writing and concluding; M.A.H. contributed in identifying the IDMs, manuscript writing and revising. L.S.M.; contributed in classifying IDMS, manuscript writing and editing; and M.A.A. participated in revising the manuscript, editing and concluding. All authors have read and agreed to the published version of the manuscript.

Funding: The authors acknowledge the support provided by the Interdisciplinary Research Center for Renewable Energy and Power Systems (IRC-REPS), Research Institute, King Fahd University of Petroleum and Minerals, through Project #INRE2111.

Conflicts of Interest: The authors declare no conflict of interest.

References

1. Parhizi, S.; Lotfi, H.; Khodaei, A.; Bahramirad, S. State of the Art in Research on Microgrids: A Review. *IEEE Access* **2015**, *3*, 890–925. [CrossRef]
2. Worku, M.Y.; Hassan, M.A.; Abido, M.A. Power Management, Voltage Control and Grid Synchronization of Microgrids in Real Time. *Arab. J. Sci. Eng.* **2021**, *46*, 1411–1429. [CrossRef]
3. Li, J.; Liu, Y.; Wu, L. Optimal Operation for Community-Based Multi-Party Microgrid in Grid-Connected and Islanded Modes. *IEEE Trans. Smart Grid* **2016**, *9*, 756–765. [CrossRef]
4. Jia, Y.; Wen, P.; Yan, Y.; Huo, L. Joint Operation and Transaction Mode of Rural Multi Microgrid and Distribution Network. *IEEE Access* **2021**, *9*, 14409–14421. [CrossRef]
5. Nasser, N.; Fazeli, M. Buffered-Microgrid Structure for Future Power Networks; a Seamless Microgrid Control. *IEEE Trans. Smart Grid* **2021**, *12*, 131–140. [CrossRef]
6. Yan, B.; Luh, P.B.; Warner, G.; Zhang, P. Operation and Design Optimization of Microgrids with Renewables. *IEEE Trans. Autom. Sci. Eng.* **2017**, *14*, 573–585. [CrossRef]
7. Hussain, A.; Bui, V.-H.; Kim, H.-M. Resilience-Oriented Optimal Operation of Networked Hybrid Microgrids. *IEEE Trans. Smart Grid* **2017**, *10*, 204–215. [CrossRef]
8. Worku, M.Y.; Hassan, M.A.; Abido, M.A. Real Time Energy Management and Control of Renewable Energy based Microgrid in Grid Connected and Island Modes. *Energies* **2019**, *12*, 276. [CrossRef]
9. Che, L.; Shahidehpour, M.; AlAbdulwahab, A.; Al-Turki, Y. Hierarchical Coordination of a Community Microgrid with AC and DC Microgrids. *IEEE Trans. Smart Grid* **2015**, *6*, 3042–3051. [CrossRef]
10. Han, H.; Hou, X.; Yang, J.; Wu, J.; Su, M.; Guerrero, J. Review of Power Sharing Control Strategies for Islanding Operation of AC Microgrids. *IEEE Trans. Smart Grid* **2016**, *7*, 200–215. [CrossRef]
11. Gush, T.; Bukhari, S.B.A.; Haider, R.; Admasie, S.; Oh, Y.-S.; Cho, G.-J.; Kim, C.-H. Fault detection and location in a microgrid using mathematical morphology and recursive least square methods. *Int. J. Electr. Power Energy Syst.* **2018**, *102*, 324–331. [CrossRef]
12. Kim, M.-S.; Haider, R.; Cho, G.-J.; Kim, C.-H.; Won, C.-Y.; Chai, J.-S. Comprehensive Review of Islanding Detection Methods for Distributed Generation Systems. *Energies* **2019**, *12*, 837. [CrossRef]
13. Sivadas, D.; Vasudevan, K. An Active Islanding Detection Strategy with Zero Non detection Zone for Operation in Single and Multiple Inverter Mode Using GPS Synchronized Pattern. *IEEE Trans. Ind. Electron.* **2020**, *67*, 5554–5564. [CrossRef]
14. Chiang, W.J.; Jou, H.L.; Wu, J.C. Active islanding detection method for inverter- based distribution generation power system. *Int. J. Electr. Power* **2012**, *42*, 158–166. [CrossRef]
15. Worku, M.Y.; Hassan, M.; Abido, M.A. Real Time-Based under Frequency Control and Energy Management of Microgrids. *Electronics* **2020**, *9*, 1487. [CrossRef]
16. Khamis, A.; Xu, Y.; Dong, Z.Y.; Zhang, R. Faster Detection of Microgrid Islanding Events using an Adaptive Ensemble Classifier. *IEEE Trans. Smart Grid* **2016**, *9*, 1889–1899. [CrossRef]
17. Ghalavand, F.; Al-Omari, I.; Karegar, H.K.; Karimipour, H. Hybrid Islanding Detection for AC/DC Network Using DC-link Voltage. In Proceedings of the IEEE International Conference on Smart Energy Grid Engineering (SEGE), Oshawa, ON, Canada, 12–15 August 2018.
18. Issa, W.R.; Abusara, M.A.; Sharkh, S.M. Control of Transient Power during Unintentional Islanding of Microgrids. *IEEE Trans. Power Electron.* **2014**, *30*, 4573–4584. [CrossRef]
19. Salles, D.; Freitas, W.; Vieira, J.C.M.; Venkatesh, B. A practical method for non-detection zone estimation of passive anti-islanding schemes applied to synchronous distributed generators. *IEEE Trans. Power Deliv.* **2015**, *30*, 2066–2076. [CrossRef]
20. Huang, J.; Jiang, C.; Xu, R. A review on distributed energy resources and MicroGrid. *Renew. Sustain. Energy Rev.* **2008**, *12*, 2472–2483.
21. Bukhari, S.B.A.; Zaman, M.S.U.; Haider, R.; Oh, Y.-S.; Kim, C.-H. A protection scheme for microgrid with multiple distributed generations using superimposed reactive energy. *Int. J. Electr. Power Energy Syst.* **2017**, *92*, 156–166. [CrossRef]
22. Cagnano, A.; De Tuglie, E.; Cicognani, L. Prince—Electrical energy systems lab: A pilot project for smart microgrids. *Electr. Power Syst. Res.* **2017**, *148*, 10–17. [CrossRef]

23. Cagnano, A.; De Tuglie, E.; Trovato, M.; Cicognani, L.; Vona, V. A simple circuit model for the islanding transition of microgrids. In Proceedings of the IEEE 2nd International Forum on Research and Technologies for Society and Industry Leveraging a Better Tomorrow (RTSI), Bologna, Italy, 7–9 September 2016.

24. Ghanbari, T.; Farjah, E.; Naseri, F. Power quality improvement of radial feeders using an efficient method. *Electr. Power Syst. Res.* **2018**, *163*, 140–153. [CrossRef]

25. IEEE Std 929-2000 (Revision of IEEE Std 929–1988). IEEE Recommended Practice for Utility Interface of Photovoltaic (PV) Systems, USA, IEEE Std 929-2000. Available online: https://standards.ieee.org/standard/929-2000.html (accessed on 25 June 2021).

26. Zeineldin, H.; Abdel-Galil, T.; El-Saadany, E.; Salama, M. Islanding detection of grid connected distributed generators using TLS-ESPRIT. *Electr. Power Syst. Res.* **2007**, *77*, 155–162. [CrossRef]

27. Ku Ahmad, K.N.E.; Selvaraj, J.; Rahim, N.A. A review of the islanding detection methods in grid-connected PV inverters. *Renew. Sustain. Energy Rev.* **2013**, *21*, 756–766. [CrossRef]

28. IEEE Standard. *1547–2018—IEEE Standard for Interconnection and Interoperability of Distributed Energy Resources with Associated Electric Power Systems Interfaces (Revision of IEEE Std 1547–2003)*; IEEE Standards Association: Piscataway, NJ, USA, 2018.

29. Reis, M.V.; Barros, T.A.; Moreira, A.B.; Ruppert, E.; Villalva, M.G. Analysis of the Sandia Frequency Shift (SFS) islanding detection method with a single-phase photovoltaic distributed generation system. In Proceedings of the 2015 IEEE PES Innovative Smart Grid Technologies Latin America (ISGT LATAM), Montevideo, Uruguay, 5–7 October 2015.

30. Rawat, M.S.; Vadhera, S. Evolution of Islanding Detection Methods for Microgrid Systems. In *Optimizing and Measuring Smart Grid Operation and Control*; IGI Global: Hershey, PA, USA, 2021; pp. 221–257. [CrossRef]

31. Yap, K.Y.; Sarimuthu, C.R.; Lim, J.M.-Y. Virtual Inertia-Based Inverters for Mitigating Frequency Instability in Grid-Connected Renewable Energy System: A Review. *Appl. Sci.* **2019**, *9*, 5300. [CrossRef]

32. Shin, H.; Jung, J.; Lee, B. Determining the Capacity Limit of Inverter-Based Distributed Generators in High-Generation Areas Considering Transient and Frequency Stability. *IEEE Access* **2020**, *8*, 34071–34079. [CrossRef]

33. Song, Y.; Hill, D.J.; Liu, T. Impact of DG Connection Topology on the Stability of Inverter-Based Microgrids. *IEEE Trans. Power Syst.* **2019**, *34*, 3970–3972. [CrossRef]

34. Cui, Q.; El-Arroudi, K.; Joos, G. Islanding Detection of Hybrid Distributed Generation Under Reduced Non-Detection Zone. *IEEE Trans. Smart Grid* **2017**, *9*, 5027–5037. [CrossRef]

35. Karimi, M.; Farshad, M.; Hong, Q.; Laaksonen, H.; Kauhaniemi, K. An Islanding Detection Technique for Inverter-Based Distributed Generation in Microgrids. *Energies* **2020**, *14*, 130. [CrossRef]

36. Ropp, M.E.; Begovic, M.; Rohatgi Ajeet Kern, G.A.; Bonn, R.H.; Gonzalez, S. Determining the relative effectiveness of islanding detection methods using phase criteria and non-detection zones. *IEEE Trans. Energy Convers.* **2000**, *15*, 290–296. [CrossRef]

37. Ye, Z.; Kolwalkar, A.; Zhang, Y.; Du, P.; Walling, R. Evaluation of anti-islanding schemes based on non-detection zone concept. *IEEE Trans. Power Electr.* **2004**, *19*, 1171–1176. [CrossRef]

38. Vieira, J.C.; Salles, D.; Freitas, W. Power imbalance application region method for distributed synchronous generator anti-islanding protection design and evaluation. *Electr. Power Syst. Res.* **2011**, *81*, 1952–1960. [CrossRef]

39. Hosseinzadeh, M.; Salmasi, F.R.R. Islanding Fault Detection in Microgrids—A Survey. *Energies* **2020**, *13*, 3479. [CrossRef]

40. Choudhury, B.K.; Jena, P. Superimposed Impedance Based Passive Islanding Detection Scheme for DC Microgrids. *IEEE J. Emerg. Sel. Top. Power Electron.* **2021**. [CrossRef]

41. Khamis, A.; Shareef, H.; Bizkevelci, E.; Khatib, T. A review of islanding detection techniques for renewable distributed generation systems. *Renew. Sustain. Energy Rev.* **2013**, *28*, 483–493. [CrossRef]

42. Velasco, D.; Trujillo, C.; Garcerá, G.; Figueres, E. Review of anti-islanding techniques in distributed generators. *Renew. Sustain. Energy Rev.* **2010**, *14*, 1608–1614. [CrossRef]

43. PVPS IEA. Evaluation of Islanding Detection Methods for Photovoltaic Utility-Interactive Power Systems. Report IEA PVPST5-09.2002. Available online: https://iea-pvps.org/wp-content/uploads/2020/01/rep5_09.pdf (accessed on 12 April 2021).

44. Mango, F.; Liserre, M.; Dell'Aquila, A.; Pigazo, A. Overview of Anti-Islanding Algorithms for PV Systems. Part I: Passive Methods. In Proceedings of the 2006 12th International Power Electronics and Motion Control Conference, Portorož, Slovenia, 30 August–1 September 2006; pp. 1878–1883.

45. Hu, S.-H.; Tsai, H.-T.; Lee, T.-L. Islanding detection method based on second-order harmonic injection for voltage-controlled inverter. In Proceedings of the 2015 IEEE 2nd International Future Energy Electronics Conference (IFEEC), Taipei, Taiwan, 1–4 November 2015; pp. 1–5.

46. Iii, J.W.S.; Bonn, R.H.; Ginn, J.W.; Gonzalez, S.; Kern, G. *Development and Testing of an Approach to Anti-Islanding in Utility-Interconnected Photovoltaic Systems*; Office of Scientific and Technical Information (OSTI): Albuquerque, NM, USA, 2000.

47. Kulkarni, N.K.; Khedkar, M. Performance Study of OUV, OUF and THD Passive Islanding Detection Methods for Single and Multiple Inverter based DGs and PCCs. In Proceedings of the 2019 Innovations in Power and Advanced Computing Technologies (i-PACT), Vellore, India, 22–23 March 2019; pp. 1–7. [CrossRef]

48. Yingram, M.; Premrudeepreechacharn, S. Over/Undervoltage and Undervoltage Shift of Hybrid Islanding Detection Method of Distributed Generation. *Sci. World J.* **2015**, *2015*, 654942. [CrossRef]

49. Manikonda, S.K.; Gaonkar, D.N. Comprehensive review of IDMs in DG systems. *IET Smart Grid* **2019**, *2*, 11–24. [CrossRef]

50. Almas, M.S.; Vanfretti, L. RT-HIL Implementation of the Hybrid Synchrophasor and GOOSE-Based Passive Islanding Schemes. *IEEE Trans. Power Deliv.* **2016**, *31*, 1299–1309. [CrossRef]

51. Jones, R.; Sims, T.; Imece, A. Investigation of potential islanding of a self-commutated static power converter in photovoltaic systems. *IEEE Trans. Energy Convers.* **1990**, *5*, 624–631. [CrossRef]

52. Mahat, P.; Chen, Z.; Bak-Jensen, B. Review of islanding detection methods for distributed generation. In Proceedings of the 2008 Third International Conference on Electric Utility Deregulation and Restructuring and Power Technologies, Nanjing, China, 6–9 April 2008; pp. 2743–2748.

53. Freitas, W.; Wilsun, X.; Affonso, C.M.; Zhenyu, H. Comparative analysis between ROCOF and vector surge relays for distributed generation applications. *IEEE Trans. Power Deliv.* **2005**, *20*, 1315–1324. [CrossRef]

54. Jia, K.; Bi, T.; Liu, B.; Thomas, D.; Goodman, A. Advanced islanding detection utilized in distribution systems with DFIG. *Int. J. Electr. Power Energy Syst.* **2014**, *63*, 113–123. [CrossRef]

55. Dysko, A.; Booth, C.; Anaya-Lara, O.; Burt, G.M. Reducing unnecessary disconnection of renewable generation from the power system. *IET Renew. Power Gener.* **2007**, *1*, 41–48. [CrossRef]

56. Xie, X.; Huang, C.; Li, D. A new passive islanding detection approach considering the dynamic behavior of load in microgrid. *Int. J. Electr. Power Energy Syst.* **2020**, *117*, 105619. [CrossRef]

57. Xie, X.; Xu, W.; Huang, C.; Fan, X. New islanding detection method with adaptively threshold for microgrid. *Electr. Power Syst. Res.* **2021**, *195*, 107167. [CrossRef]

58. Abyaz, A.; Panahi, H.; Zamani, R.; Alhelou, H.H.; Siano, P.; Shafie-Khah, M.; Parente, M. An Effective Passive Islanding Detection Algorithm for Distributed Generations. *Energies* **2019**, *12*, 3160. [CrossRef]

59. Pai, F.-S.; Huang, S.-J. A detection algorithm for islanding-prevention of dispersed consumer-owned storage and generating units. *IEEE Trans. Energy Convers.* **2001**, *16*, 346–351. [CrossRef]

60. Singam, B.; HuiL, Y. Assessing SMS and PJD schemes of anti-islanding with varying quality factor. In Proceedings of the 2006 IEEE International Power and Energy Conference, Putra Jaya, Malaysia, 28–29 November 2006; pp. 196–201.

61. O'Kane, P.; Fox, B. Loss of mains detection for embedded generation by system impedance monitoring. In Proceedings of the Sixth International Conference on Developments in Power System Protection (Conf. Publ. No. 434), Nottingham, UK, 25–27 March 1997; pp. 95–98.

62. Admasie, S.; Bukhari, S.B.A.; Haider, R.; Gush, T.; Kim, C.-H. A passive islanding detection scheme using variational mode decomposition-based mode singular entropy for integrated microgrids. *Electr. Power Syst. Res.* **2019**, *177*, 105983. [CrossRef]

63. Jang, S.-I.; Kim, K.-H. An Islanding Detection Method for Distributed Generations Using Voltage Unbalance and Total Harmonic Distortion of Current. *IEEE Trans. Power Deliv.* **2004**, *19*, 745–752. [CrossRef]

64. Kumar, A.; Panda, R.; Mohapatra, A.; Singh, S.; Srivastava, S. Mode of oscillation based islanding detection of inverter interfaced DG using ESPRIT. *Electr. Power Syst. Res.* **2021**, *200*, 107479. [CrossRef]

65. Mango, F.; Liserre, M.; Dell'Aquila, A. Overview of Anti-Islanding Algorithms for PV Systems. Part II: Active Methods. In Proceedings of the 2006 12th International Power Electronics and Motion Control Conference, Portorož, Slovenia, 30 August–1 September 2006; pp. 1884–1889.

66. Ropp, M.; Begovic, M.; Rohatgi, A. Analysis and performance assessment of the active frequency drift method of islanding prevention. *IEEE Trans. Energy Convers.* **1999**, *14*, 810–816. [CrossRef]

67. Li, C.; Cao, C.; Cao, Y.; Kuang, Y.; Zeng, L.; Fang, B. A review of islanding detection methods for microgrid. *Renew. Sustain. Energy Rev.* **2014**, *35*, 211–220. [CrossRef]

68. Reis, M.V.; Villalva, M.G.; Barros, T.A.; Moreira, A.B.; Ruppert, E. Active frequency drift with positive feedback anti-islanding method for a single phase two-stage grid-tied photovoltaic system. In Proceedings of the 2015 IEEE 13th Brazilian Power Electronics Conference and 1st Southern Power Electronics Conference (COBEP/SPEC), Fortaleza, Brazil, 29 November–2 December 2015; pp. 1–6.

69. Trujillo, C.L.; Velasco, D.; Figueres, E.; Garcerá, G. Analysis of active islanding detection methods for grid-connected microinverters for renewable energy processing. *Appl. Energy* **2010**, *87*, 3591–3605. [CrossRef]

70. Zeineldin, H.H.; Kennedy, S. Sandia Frequency-Shift Parameter Selection to Eliminate Nondetection Zones. *IEEE Trans. Power Deliv.* **2009**, *24*, 486–487. [CrossRef]

71. Zeineldin, H.; Conti, S. Sandia frequency shift parameter selection for multi-inverter systems to eliminate non-detection zone. *IET Renew. Power Gener.* **2011**, *5*, 175–183. [CrossRef]

72. Hatata, A.Y.; Abd-Raboh, E.l.-H.; Sedhom, B.E. Proposed Sandia frequency shift for anti-islanding detection method based on artificial immune system. *Alex. Eng. J.* **2018**, *57*, 235–245. [CrossRef]

73. Liu, F.; Kang, Y.; Zhang, Y.; Duan, S.; Lin, X. Improved SMS islanding detection method forgrid-connected converters. *IET Renew. Power Gener.* **2010**, *4*, 36–42. [CrossRef]

74. Lopes, L.A.C.; Huili, S. Performance assessment of active frequency drifting islanding detection methods. *IEEE Trans. Energy Convers.* **2006**, *21*, 171–180. [CrossRef]

75. Karimi, H.; Yazdani, A.; Iravani, R. Negative-Sequence Current Injection for Fast Islanding Detection of a Distributed Resource Unit. *IEEE Trans. Power Electron.* **2008**, *23*, 298–307. [CrossRef]

76. Tuyen, N.D.; Fujita, G. Negative-sequence current injection of dispersed generation for islanding detection and unbalanced fault ride-through. In Proceedings of the 2011 46th International Universities' Power Engineering Conference, Soest, Germany, 5–8 September 2011; pp. 1–6.

77. Papadimitriou, C.; Kleftakis, V.; Hatziargyriou, N. A novel islanding detection method for microgrids based on variable impedance insertion. *Electr. Power Syst. Res.* **2015**, *121*, 58–66. [CrossRef]

78. Khodaparastan, M.; Vahedi, H.; Khazaeli, F.; Oraee, H. A Novel Hybrid Islanding Detection Method for Inverter-Based DGs Using SFS and ROCOF. *IEEE Trans. Power Deliv.* **2017**, *32*, 2162–2170. [CrossRef]

79. Dhar, S.; Dash, P.K. Harmonic Profile Injection-Based Hybrid Active Islanding Detection Technique for PV-VSC-Based Microgrid System. *IEEE Trans. Sustain. Energy* **2016**, *7*, 1473–1481. [CrossRef]

80. Azim, R.; Li, F.; Xue, Y.; Starke, M.; Wang, H. An islanding detection methodology combining decision trees and Sandia frequency shift for inverter-based distributed generations. *IET Gener. Transm. Distrib.* **2017**, *11*, 4104–4113. [CrossRef]

81. Murugesan, S.; Murali, V.; Daniel, S.A. Hybrid Analyzing Technique for Active Islanding Detection Based on d-Axis Current Injection. *IEEE Syst. J.* **2018**, *12*, 3608–3617. [CrossRef]

82. Chang, W.-Y. A hybrid islanding detection method for distributed synchronous generators. In Proceedings of the 2010 International Power Electronics Conference-ECCE ASIA, Sapporo, Japan, 21–24 June 2010; pp. 1326–1330.

83. Laghari, J.; Mokhlis, H.; Bakar, A.; Mohamad, H. Application of computational intelligence techniques for load shedding in power systems: A review. *Energy Convers. Manag.* **2013**, *75*, 130–140. [CrossRef]

84. Menon, V.; Nehrir, M.H. A Hybrid Islanding Detection Technique Using Voltage Unbalance and Frequency Set Point. *IEEE Trans. Power Syst.* **2007**, *22*, 442–448. [CrossRef]

85. Yin, J.; Chang, L.; Diduch, C. A New Hybrid Anti-Islanding Algorithm in Grid Connected Three-Phase Inverter System. In Proceedings of the 37th IEEE Power Electronics Specialists Conference, Jeju, Korea, 18–22 June 2006.

86. Bakhshi-Jafarabadi, R.; Sadeh, J.; Chavez, J.D.J.; Popov, M. Two-Level Islanding Detection Method for Grid-Connected Photovoltaic System-Based Microgrid With Small Non-Detection Zone. *IEEE Trans. Smart Grid* **2021**, *12*, 1063–1072. [CrossRef]

87. Bakhshi-Jafarabadi, R.; Sadeh, J.; Popov, M. Maximum Power Point Tracking Injection Method for Islanding Detection of Grid-Connected Photovoltaic Systems in Microgrid. *IEEE Trans. Power Deliv.* **2021**, *36*, 168–179. [CrossRef]

88. Arif, A.; Imran, K.; Cui, Q.; Weng, Y. Islanding Detection for Inverter-Based Distributed Generation Using Unsupervised Anomaly Detection. *IEEE Access* **2021**, *9*, 90947–90963. [CrossRef]

89. Mahat, P.; Chen, Z.; Bak-Jensen, B. Review on islanding operation of distribution system with distributed generation. In Proceedings of the IEEE Power and Energy Society General Meeting, Detroit, MI, USA, 24–28 July 2011.

90. Takeshita, K.; Sakamoto, O.; Maruyama, M.; Harimoto, T. Consideration on Voltage Fluctuation caused by Active Method of Islanding Detection of Photovoltaic Generation. In Proceedings of the 2018 7th International Conference on Renewable Energy Research and Applications (ICRERA), Paris, France, 14–17 October 2018; pp. 1–6.

91. Vahedi, H.; Noroozian, R.; Jalilvand, A.; Gharehpetian, G.B. Hybrid SFS and Q-f Islanding Detection Method for inverter-based DG. In Proceedings of the IEEE International Conference on Power and Energy, Kuala Lumpur, Malaysia, 29 November–1 December 2010; pp. 672–676.

92. Laghari, J.; Mokhlis, H.; Bakar, A.; Karimi, M. A new islanding detection technique for multiple mini hydro based on rate of change of reactive power and load connecting strategy. *Energy Convers. Manag.* **2013**, *76*, 215–224. [CrossRef]

93. Chen, X.; Li, Y.; Crossley, P. A Novel Hybrid Islanding Detection Method for Grid-Connected Microgrids With Multiple Inverter-Based Distributed Generators Based on Adaptive Reactive Power Disturbance and Passive Criteria. *IEEE Trans. Power Electron.* **2018**, *34*, 9342–9356. [CrossRef]

94. Xu, W.; Zhang, G.; Li, C.; Wang, W.; Wang, G.; Kliber, J. A power line signaling based technique for anti-islanding protection of distributed generators—Part I: Scheme and analysis. *IEEE Trans. Power Deliv.* **2007**, *22*, 1758–1766. [CrossRef]

95. Wang, W.; Kliber, J.; Zhang, G.; Xu, W.; Howell, B.; Palladino, T. A power line signaling based scheme for anti-islanding protection of distributed generators—Part II: Field test results. *IEEE Trans. Power Deliv.* **2007**, *22*, 1767–1772. [CrossRef]

96. Do, H.T.; Zhang, X.; Nguyen, N.V.; Li, S.S.; Chu, T.T.-T. Passive islanding detection method using Wavelet Packet Transform in Grid Connected Photovoltaic Systems. *IEEE Trans. Power Electron.* **2015**, *31*, 6955–6967. [CrossRef]

97. Samui, A.; Samantaray, S.R. Wavelet Singular Entropy-Based Islanding Detection in Distributed Generation. *IEEE Trans. Power Deliv.* **2013**, *28*, 411–418. [CrossRef]

98. Hartmann, N.B.; Dos Santos, R.C.; Grilo, A.P.; Vieira, J.C.M. Hardware Implementation and Real-Time Evaluation of an ANN-Based Algorithm for Anti-Islanding Protection of Distributed Generators. *IEEE Trans. Ind. Electron.* **2017**, *65*, 5051–5059. [CrossRef]

99. Hussain, N.; Nasir, M.; Vasquez, J.C.; Guerrero, J.M. Recent Developments and Challenges on AC Microgrids Fault Detection and Protection Systems—A Review. *Energies* **2020**, *13*, 2149. [CrossRef]

100. Raza, S.; Mokhlis, H.; Arof, H.; Laghari, J.; Wang, L. Application of signal processing techniques for islanding detection of distributed generation in distribution network: A review. *Energy Convers. Manag.* **2015**, *96*, 613–624. [CrossRef]

101. Abdelsalam, A.A.; Salem, A.A.; Oda, E.S.; Eldesouky, A.A. Islanding Detection of Microgrid Incorporating Inverter Based DGs Using Long Short-Term Memory Network. *IEEE Access* **2020**, *8*, 106471–106486. [CrossRef]

102. Rostami, A.; Jalilian, A.; Zabihi, S.; Olamaei, J.; Pouresmaeil, E. Islanding Detection of Distributed Generation Based on Parallel Inductive Impedance Switching. *IEEE Syst. J.* **2020**, *14*, 813–823. [CrossRef]

103. Ezzat, A.; Elnaghi, B.E.; Abdelsalam, A.A. Microgrids islanding detection using Fourier transform and machine learning algorithm. *Electr. Power Syst. Res.* **2021**, *196*, 107224. [CrossRef]

104. Dutta, S.; Olla, S.; Sadhu, P.K. A secured, reliable and accurate unplanned island detection method in a renewable energy based microgrid. *Eng. Sci. Technol. Int. J.* **2021**, *24*, 1102–1115. [CrossRef]
105. Hashemi, F.; Mohammadi, M.; Kargarian, A. Islanding detection method for microgrid based on extracted features from differential transient rate of change of frequency. *IET Gener. Transm. Distrib.* **2017**, *11*, 891–904. [CrossRef]
106. Paiva, S.C.; Ribeiro, R.L.D.A.; Alves, D.K.; Costa, F.B.; Rocha, T.D.O.A. A wavelet-based hybrid islanding detection system applied for distributed generators interconnected to AC microgrids. *Int. J. Electr. Power Energy Syst.* **2020**, *121*, 106032. [CrossRef]
107. Barros, J.; Diego, R. Application of the Wavelet-Packet Transform to the Estimation of Harmonic Groups in Current and Voltage Waveforms. *IEEE Trans. Power Deliv.* **2005**, *21*, 533–535. [CrossRef]
108. Samantaray, S.; Samui, A.; Babu, B.C. Time-frequency transform-based islanding detection in distributed generation. *IET Renew. Power Gener.* **2011**, *5*, 431–438. [CrossRef]
109. Mohanty, S.R.; Kishor, N.; Ray, P.K.; Catalão, J.P.S. Comparative Study of Advanced Signal Processing Techniques for Islanding Detection in a Hybrid Distributed Generation System. *IEEE Trans. Sustain. Energy* **2015**, *6*, 122–131. [CrossRef]
110. Haider, R.; Kim, C.H.; Ghanbari, T.; Bukhari, S.B.A.; Zaman, M.S.U.; Baloch, S.; Oh, Y.S. Passive islanding detection scheme based on autocorrelation function of modal current envelope for photovoltaic units. *IET Gener. Transm. Distrib.* **2018**, *12*, 726–736. [CrossRef]
111. Haider, R.; Kim, C.H.; Ghanbari, T.; Bukhari, S.B.A. Harmonic-signature-based islanding detection in grid-connected distributed generation systems using Kalman filter. *IET Renew. Power Gener.* **2018**, *12*, 1813–1822. [CrossRef]
112. Merlin, V.L.; Santos, R.C.; Pavani, A.P.G.; Coury, D.V.; Oleskovicz, M.; Vieira, J.C.M. A methodology for training artificial neural networks for islanding detection of distributed generators. In Proceedings of the 2013 IEEE PES Conference on Innovative Smart Grid Technologies (ISGT Latin America), Sao Paulo, Brazil, 15–17 April 2013; pp. 1–6. [CrossRef]
113. Admasie, S.; Bukhari, S.B.A.; Gush, T.; Haider, R.; Kim, C.H. Intelligent Islanding Detection of Multi-distributed Generation Using Artificial Neural Network Based on Intrinsic Mode Function Feature. *J. Mod. Power Syst. Clean Energy* **2020**, *8*, 511–520. [CrossRef]
114. Merlin, V.; Santos, R.; Grilo, A.; Vieira, J.; Coury, D.; Oleskovicz, M. A new artificial neural network based method for islanding detection of distributed generators. *Int. J. Electr. Power Energy Syst.* **2016**, *75*, 139–151. [CrossRef]
115. Mlakić, D.; Baghaee, H.R.; Nikolovski, S. A Novel ANFIS-Based Islanding Detection for Inverter-Interfaced Microgrids. *IEEE Trans. Smart Grid* **2019**, *10*, 4411–4424. [CrossRef]
116. El-Arroudi, K.; Joos, G.; Kamwa, I.; McGillis, D.T. Intelligent-Based Approach to Islanding Detection in Distributed Generation. *IEEE Trans. Power Deliv.* **2007**, *22*, 828–835. [CrossRef]
117. Heidari, M.; Seifossadat, G.; Razaz, M. Application of decision tree and discrete wavelet transform for an optimized intelligent-based islanding detection method in distributed systems with distributed generations. *Renew. Sustain. Energy Rev.* **2013**, *27*, 525–532. [CrossRef]
118. Khamis, A.; Shareef, H.; Mohamed, A.; Bizkevelci, E. Islanding detection in a distributed generation integrated power system using phase space technique and probabilistic neural network. *Neurocomputing* **2015**, *148*, 587–599. [CrossRef]
119. Matic-Cuka, B.; Kezunovic, M. Islanding Detection for Inverter-Based Distributed Generation Using Support Vector Machine Method. *IEEE Trans. Smart Grid* **2014**, *5*, 2676–2686. [CrossRef]
120. Baghaee, H.R.; Mlakić, D.; Nikolovski, S.; Dragicčvić, T. Anti-Islanding Protection of PV-Based Microgrids Consisting of PHEVs Using SVMs. *IEEE Trans. Smart Grid* **2021**, *11*, 483–500. [CrossRef]
121. Dash, P.; Padhee, M.; Panigrahi, T. A hybrid time–frequency approach based fuzzy logic system for power island detection in grid connected distributed generation. *Int. J. Electr. Power Energy Syst.* **2012**, *42*, 453–464. [CrossRef]
122. Hashemi, F.; Ghadimi, N.; Sobhani, B. Islanding detection for inverter-based DG coupled with using an adaptive neuro-fuzzy inference system. *Int. J. Electr. Power Energy Syst.* **2013**, *45*, 443–455. [CrossRef]

mathematics

Article

Optimising Energy Management in Hybrid Microgrids

Javier Bilbao *, Eugenio Bravo, Olatz García, Carolina Rebollar and Concepción Varela

Applied Mathematics Department, Engineering School of Bilbao, University of the Basque Country (UPV/EHU), Pl. Ing. Torres Quevedo, 1, 48013 Bilbao, Spain; eugenio.bravo@ehu.eus (E.B.); olatz.garcia@ehu.eus (O.G.); carolina.rebollar@ehu.eus (C.R.); concepcion.varela@ehu.eus (C.V.)
* Correspondence: javier.bilbao@ehu.eus

Abstract: This article deals with the optimization of the operation of hybrid microgrids. Both the problem of controlling the management of load sharing between the different generators and energy storage and possible solutions for the integration of the microgrid into the electricity market will be discussed. Solar and wind energy as well as hybrid storage with hydrogen, as renewable sources, will be considered, which allows management of the energy balance on different time scales. The Machine Learning method of Decision Trees, combined with ensemble methods, will also be introduced to study the optimization of microgrids. The conclusions obtained indicate that the development of suitable controllers can facilitate a competitive participation of renewable energies and the integration of microgrids in the electricity system.

Keywords: hybrid microgrids; renewable energies; energy management; electricity system

Citation: Bilbao, J.; Bravo, E.; García, O.; Rebollar, C.; Varela, C. Optimising Energy Management in Hybrid Microgrids. *Mathematics* **2022**, *10*, 214. https://doi.org/10.3390/math10020214

Academic Editor: Nicu Bizon

Received: 12 December 2021
Accepted: 5 January 2022
Published: 11 January 2022

Publisher's Note: MDPI stays neutral with regard to jurisdictional claims in published maps and institutional affiliations.

1. Introduction

In recent years, the microgrid paradigm has emerged, introduced in the early 21st century by Lasseter [1] as an approach that considers generation and associated loads as a subsystem or microgrid. A microgrid can be considered as a set of loads, generators and storage that can be managed in isolation or connected to the rest of the grid in a coordinated manner to supply electricity reliably [2–7]. In emergency situations (faults, disturbances, etc.), the generators and the corresponding loads can be separated from the distribution grid, maintaining service without damaging the integrity of the system. Although originally associated with electricity grids, the concept has been broadened to any set of equipment, such as loads, storage systems and generators, which operates as a unique manageable system that can provide both electrical and thermal power or fuel to a given area [8]. Today, the operation of Distributed Energy Resources (DER) together with manageable loads (domestic consumption or electric vehicles) and various forms of storage such as batteries, supercapacitors or flywheels, is at the core of the hybrid microgrid concept [9]. A microgrid can operate interconnected to the utility through the main distribution grid, using the so-called Point of Common Coupling (PCC), or in island mode, and can also be interconnected with other microgrid systems, which can lead to more complex structures.

The management of hybrid microgrids presents many challenges [10,11], as they can operate either in island mode or connected to the main grid through the PCC. Proper management of the microgrid is, therefore, necessary for stable and economically efficient operation in both situations. The management system must control and adjust both frequency and voltage in either operating mode, share all the loads between the different Distributed Generators (DG) and storage, control the flow with the main grid and optimize operating costs. In grid-connected mode, voltage and frequency will be set by the main grid, which has synchronous generators and large rolling storage systems.

A necessary step in the difficult process of managing a hybrid microgrid is the mathematical modeling of the aspects (power flow, generation, storage, etc.) of that grid, for a next analysis with method. Several methods are, today, applying for this aim: optimizing energy management. Different methods and algorithms have been developed, and are developing [12,13], such as heuristic methods, optimization methods and Machine Learning methods, which will be predominant in the near future since they can compete in accuracy with the others, and because they can be adapted to different topologies, having enough data.

We present in this article a detailed and deep mathematical modeling development, which is not normally found in the bibliography, and is the basis of two heuristic methods (Hysteresis Band Control and control by means of Fuzzy Logic) and the Decision Trees Machine Learning method, combined with ensemble methods, concluding with a comparison of these methods in a microgrid.

This article is organized as follows: the next section presents the main challenges and functions of hybrid microgrids, along with some benefits that can be derived from their correct operation. Section 3 introduces energy management systems and their need to be modeled mathematically. Section 4 then mathematically models the main aspects of a hybrid microgrid to be taken into account when managing them, such as power flow equations, the generation of electrical energy, its storage, power converters, energy consumption and CO_2 emissions and other factors. In Section 5, the metaheuristic techniques are developed: they will be defined and their types will be classified according to the process they follow. The main ones (the Hysteresis Band Control and control by means of Fuzzy Logic) will be developed, although at the end of the chapter, mention will be made of some techniques that are either evolutions of other techniques or that are less well known but are widely used due to their good results. Section 6 introduces the Machine Learning method of Decision Trees, their basis, classification and some applications, to continue with the development of Decision Trees. In that section, bagging and boosting ensemble methods are also introduced. Section 7 details the design and experimental results obtained from the comparison operation of a laboratory microgrid. Section 8 discusses some of the open lines of research, and finally ends with the conclusions, where the two main ones are the mathematical modeling compiled in a single article, including hybrid components (renewable energy and storage), and that the Decision Tree method can be applied to the energy management of a hybrid microgrid, but without a great advantage over more classical methods such as Hysteresis Band Control or the application of Fuzzy Logic.

2. Management of Hybrid Microgrids

The objective of the management and control of a microgrid is to provide the energy demanded by the loads, using generation and storage systems efficiently and reliably, both under regular conditions and when contingencies occur, whether or not there is a connection to the external network.

Hybrid microgrids introduce a number of operational challenges that must be taken into account in the design of their management and protection systems, due to certain particularities that distinguish them from other systems. According to Olivares et al. [14], the most relevant are:

- Power flows. In contrast to conventional grids, the integration of DGs in low voltage can result in bi-directional power flows and lead to difficulties in protection systems or undesirable flow patterns.
- Stability. Local oscillations may occur as a result of the interaction of DG management systems and problematic transitions between stand-alone and grid-connected mode.
- Network model. The generally accepted assumptions of three balanced phases, inductive transmission lines and constant loads become meaningless in this type of network, leading to the need to adapt the models to the new situation. A hybrid microgrid is intrinsically subject to load unbalance by the DGs themselves.
- Low inertia. The dynamic characteristics of DG equipment, fundamentally those that are electronically coupled, are different from those based on large generation turbines. If appropriate monitoring and management measures are not implemented, the low

inertia of the system can lead to considerable frequency deviations in the isolated mode of operation.

- Uncertainty. In hybrid microgrids there is greater uncertainty regarding demand and, above all, generation, as the use of renewable energies means that generation is linked to environmental conditions. Therefore, reliable and economic operation must take into account weather forecasting.

Under these circumstances, the management system must ensure reliable operation of the microgrid. The main functions that can be requested from the management system in the microgrid are [14–16]:

- Control of voltages and currents in the various DGs, according to the standards and adequately reducing oscillations.
- Frequency and voltage regulation in both stand-alone and grid-connected modes.
- Power balancing, when changes are produced in both generation and load, while maintaining voltage and frequency within acceptable limits.
- Demand Side Management (DSM) mechanisms that allow some fluctuation in the demand of a part of the loads to adapt to the requirements of the hybrid microgrid.
- Smooth transition between operating modes, using the most appropriate strategy for each of them and promptly identifying the situations that produce the switching. Resynchronization with the main network.
- Economic dispatch, distributing the load between the different DGs and storage systems in such a way as to reduce the cost of operation, while maintaining reliability. Optimization of the cost of operation will include maximizing the economic benefit in the case of grid connection.
- Management of power flows between the microgrid and the main network and, where appropriate, with other microgrids.

3. Energy Management System (EMS)

The architecture of a system is defined as the fundamental organization of a system, including its components, the relationships between them and the environment and the principles that govern its design and evolution [17,18]. Among the different control architectures, centralized, decentralized and distributed control architectures have been widely used in industry. On the one hand, the centralized implementation stands out for having greater precision, since the control of the process in question is carried out by a single controller, which receives all the signals provided by the network sensors and, after the control process, issues the values to be taken by the different actuators to achieve correct operation [19]. It is therefore a master–slave configuration in which the controller tries to optimize the operation of a set formed by all the subsystems of the network or process, leaving aside the interest that the subsystems themselves may have in optimizing their own operation at the expense of the common good. On the other hand, the decentralized implementation delegates the control problem to several controllers, reducing the computational expense, but also reducing the accuracy of the controller, as input/output data may overlap. Undoubtedly, the great benefit of this type of architecture lies in the ease of implementation with respect to centralized implementation, due to the reduction of a problem into multiple problems of less difficulty. Halfway between centralized and decentralized implementation is the distributed control architecture. In this architecture, there are problems that are related to each other, allowing the coordination of subsystems. This topology is characteristic of microgrids in which each subsystem has a control objective that is different from the others. The higher the number of components in the microgrid, the higher the data traffic between the different controllers and subsystems, thus requiring more bandwidth in the communication system. However, a distributed implementation can reduce data traffic compared to a centralized one, due to the reduced difficulty of the 'local subproblems' that make up the optimization problem.

A fundamental part of a microgrid is the control system, and more specifically, the control strategy or method that will manage the operation of the microgrid in terms of energy generation and demand, so that the energy storage and the external distribution network can satisfy, at all times, the energy balance in the system as a whole. The Energy Management System (EMS) is the system that performs this task, trying to achieve an efficient use of the different components of the microgrid [20,21]. In order to achieve this goal of efficient use, mathematical modeling of the parts of the system is essential.

4. Mathematical Modeling

4.1. Power Flow Equations

Each interconnecting component of an electrical network is called a branch or line and links a node n to another node m in the network. A line can be modeled by its single-phase equivalent circuit. This equivalent circuit accounts for the electrical properties of the conductors (conductivity and insulation) and the physical properties (diameter and distance between conductors). The most commonly used equivalent circuit of a line is the equivalent Π, although there are other models such as T. The complex admittance (inverse of impedance) values are represented by the letter Y together with an arbitrary number and each bus can have a generator connected to it at a voltage represented by the letter V + bus number.

Expressing the magnitudes in complex form, assuming a permanent and balanced sinusoidal regime, the system can be represented, in compact matrix notation, as follows:

$$\mathbf{I} = \mathbf{Y}_{BUS}\mathbf{V} \tag{1}$$

where $\mathbf{I}$ is the column vector of currents at each node, $\mathbf{V}$ is the column vector of voltages at each node and $\mathbf{Y}_{BUS}$ is the admittance matrix. The admittance matrix is composed of complex numbers and has well-known properties: it is symmetrical, each element Y_{nn} of the diagonal is the sum of the admittances of the equivalent circuits Π that are connected to node n, and the off-diagonal elements Y_{nm} are the negative of the admittance of the equivalent Π connected between nodes n and m. Therefore, the admittance matrix is a square matrix of the same dimension as the number of buses. For each current n of the column vector, the power of bus n can be calculated as a factor of one, as follows:

$$s_n = \frac{V_n I_n^*}{S_{base}} = \frac{V_n}{S_{base}}\left(\sum_{m=1}^{N} Y_{nm}V_m\right)^* = \frac{V_n}{S_{base}}\sum_{m=1}^{N} Y_{nm}^* V_m^* = \sum_{m=1}^{N} v_n v_m e^{j\theta_{nm}}(G_{nm} - jB_{nm}), n = 1,\ldots,N \tag{2}$$

where v_n is the modulus of V_n in per unit, θ_{nm} is the angle difference $\theta_n - \theta_m$, and Y_{nm} is the element nm of the admittance matrix $G_{nm} + jB_{nm}$ also in per unit (pu). With Euler's formula, one can write the above equation in rectangular coordinates in the complex plane as shown below:

$$s_n = \sum_{m=1}^{N} v_n v_m(\cos\theta_{nm} + j\sin\theta_{nm})(G_{nm} - jB_{nm}) \; n = 1,\ldots,N \tag{3}$$

Remembering that the complex part of the apparent power is the reactive power and the real part is the active power, the two can be separated as follows:

$$p_n = \sum_{m=1}^{N} v_n v_m(G_{nm}\cos\theta_{nm} + B_{nm}\sin\theta_{nm}) \; n = 1,\ldots,N$$
$$q_n = \sum_{m=1}^{N} v_n v_m(G_{nm}\sin\theta_{nm} - B_{nm}\cos\theta_{nm}) \; n = 1,\ldots,N \tag{4}$$

The above representation is compact and allows observation of the asymmetric and non-linear character of the power flow equations, but to apply the relevant approximations and obtain a linearization, the special structure of the admittance matrix is considered: the elements of the diagonal Y_{nn} are the negative of the sum of the off-diagonal elements (negative of the admittance of the equivalent Π connected between nodes n and m) of the corresponding rows and of the shunt admittances of the bus (superscript *sh*). This can be seen in the three bus example below:

$$Y_{BUS} = \begin{pmatrix} Y_1^{sh} + Y_2 + Y_8 + Y_9^{sh} & -Y_2 & -Y_8 \\ -Y_2 & Y_2 + Y_3^{sh} + Y_5 + Y_4^{sh} & -Y_5 \\ -Y_8 & -Y_5 & Y_8 + Y_7^{sh} + Y_5 + Y_6^{sh} \end{pmatrix} = \begin{pmatrix} Y_{11} & Y_{12} & Y_{13} \\ Y_{21} & Y_{22} & Y_{23} \\ Y_{31} & Y_{32} & Y_{33} \end{pmatrix} \tag{5}$$

In the notation currently considered, with $y_{nm} = y_{mn}$, the admittance matrix is as follows:

$$Y_{BUS} = \begin{pmatrix} y_{12}^{sh} + y_{12} + y_{13} + y_{13}^{sh} & -y_{12} & -y_{13} \\ -y_{21} & y_{21}^{sh} + y_{21} + y_{23} + y_{23}^{sh} & -y_{23} \\ -y_{31} & -y_{32} & y_{31}^{sh} + y_{31} + y_{32} + y_{32}^{sh} \end{pmatrix} \tag{6}$$

$$Y_{nm} = \begin{cases} -y_{nm} & if\ m \neq n \\ \sum\limits_{m=1,m\neq n}^{N} y_{nm} + y_{nm}^{sh} & if\ m = n \end{cases} \tag{7}$$

$$\sum_{m=1,m\neq n}^{N} y_{nm} + y_{nm}^{sh} = y_n^{sh} + \sum_{m=1,m\neq n}^{N} g_{nm} + jb_{nm} = j\left(b_n^{sh} + \sum_{m=1,m\neq n}^{N} b_{nm} \right) + g_n^{sh} \sum_{m=1,m\neq n}^{N} g_{nm} \tag{8}$$

Then, the active and reactive power equations for each node can be rewritten, based on the admittance of each line between bus n and bus m, $Y_{nm} = -y_{nm} = -g_{nm} - jb_{nm}$, and of the shunt of bus n as:

$$p_n = \left(g_n^{sh} + \sum_{m=1,m\neq n}^{N} g_{nm} \right) v_n^2 - \sum_{m=1,m\neq n}^{N} v_n v_m (g_{nm} \cos\theta_{nm} + b_{nm} \sin\theta_{nm})\ n = 1,\dots,N \tag{9}$$

$$q_n = -\left(b_n^{sh} + \sum_{m=1,m\neq n}^{N} b_{nm} \right) v_n^2 - \sum_{m=1,m\neq n}^{N} v_n v_m (g_{nm} \sin\theta_{nm} - b_{nm} \cos\theta_{nm})\ n = 1,\dots,N \tag{10}$$

You can group the summation terms, since the sum is over the same set, and take out common factor v_n in both equations by grouping the conductance and susceptance coefficients:

$$p_n = g_n^{sh} v_n^2 + \sum_{m=1,m\neq n}^{N} g_{nm} v_n (v_n - v_m \cos\theta_{nm}) - b_{nm} v_n v_m \sin\theta_{nm}\ n = 1,\dots,N \tag{11}$$

$$q_n = -b_n^{sh} v_n^2 + \sum_{m=1,m\neq n}^{N} -b_{nm} v_n (v_n - v_m \cos\theta_{nm}) - g_{nm} v_n v_m \sin\theta_{nm}\ n = 1,\dots,N \tag{12}$$

Different assumptions can now be made to linearize the power flow. Each approach assumes a different approach to the problem; however, the following assumptions are common according to the normal operation of an electrical power system [22,23]:

1. The voltage values, expressed in per unit (pu), are very close to 1.
2. The difference between the angles of two interconnected buses is a small number close to 0.

Based on the above, to eliminate the non-linearity of the trigonometric functions, they are approximated by their Taylor series centered at zero and neglecting terms of order equal to or greater than three:

$$\cos\theta_{nm} \sim 1 + \frac{\theta_{nm}^2}{2}, \ \sin\theta_{nm} \sim \theta_{nm} \tag{13}$$

However, the quadratic term is non-linear and, in addition, there are products of several variables, which is also a non-linear function. The power equations at each bus n take the following form:

$$p_n = g_n^{sh} v_n^2 + \sum_{m=1, m \neq n}^{N} g_{nm} v_n \left(v_n - v_m - v_m \frac{\theta_{nm}^2}{2} \right) - b_{nm} v_n v_m \theta_{nm} \quad n = 1, \ldots, N \tag{14}$$

$$q_n = -b_n^{sh} v_n^2 + \sum_{m=1, m \neq n}^{N} -b_{nm} v_n \left(v_n - v_m - v_m \frac{\theta_{nm}^2}{2} \right) - g_{nm} v_n v_m \theta_{nm} \quad n = 1, \ldots, N \tag{15}$$

At this point, there is currently no consensus on the best approximation of the non-linear terms. Several approaches consider a second-order approximation based on the Taylor series of the products of two variables, which results in a linear form of the equations, except for the losses. In fact, power losses are non-convex quadratic functions which forces relaxations, such as piecewise linear function approximation or binary expansion, which give rise to a mixed integer linear programming problem (known by its acronym as MILP or MIP), or also convex relaxation, which generates a second-order conic programming problem; both are linear cases considering v_n^2 as a variable [24–27].

Since MIP-type problems are more costly to solve in terms of resources and time than continuous linear problems, other authors propose not considering the second order terms of the Taylor series expansion, both of the trigonometric functions and of the products of the variables, in such a way that the resulting flow is symmetric and allows calculating the voltage and angle value at each bus, as well as the powers, however in this approximation, losses are not represented [28].

Taking the first term of the trigonometric functions expansion, and according to Yang et al. [29], three linear approximations of the term $v_n(v_n - v_m)$ are compared, being the one with the least error, in terms of voltage and active power flow, the one that considers a decomposition of the bus voltages as $v_n = 1 + \Delta v_n$, where Δv_n is an order of magnitude smaller than v_n, therefore:

$$v_n(v_n - v_m) = (1 + \Delta v_n)(\Delta v_n - \Delta v_m) = \Delta v_n - \Delta v_m + \Delta v_n \Delta v_n - \Delta v_m \Delta v_n \sim \Delta v_n - \Delta v_m \tag{16}$$

In the resulting product of the expansion, the multiplication of the difference of the voltages, $\Delta v_n \Delta v_n$, is neglected with respect to its nominal value of 1 pu, since the result is at most two orders of magnitude smaller than v_n; thus $v_n(v_n - v_m) \simeq v_n - v_m$. Furthermore, the voltage squared multiplying the shunt terms is simply approximated by the value of the voltage at that bus, and in the case of reactive power the shunt conductance is assumed to be negligible compared to the shunt susceptance. This gives rise to a linear problem in the voltage and angle variables [30,31].

Expanding the product terms of variables by their first order Taylor series gives the following:

$$\begin{aligned} v_n v_m \theta_{nm} &\simeq v_{n,0} v_{m,0} \theta_{nm} + (v_n v_m - v_{n,0} v_{m,0}) \theta_{nm,0} = \theta_{nm} \\ v_n v_m \theta_{nm}^2 &\simeq v_{n,0} v_{m,0} \theta_{nm}^2 + (v_n v_m - v_{n,0} v_{m,0}) \theta_{nm,0}^2 = \theta_{nm}^2 \end{aligned} \tag{17}$$

where the subscript 0 denotes the point around which the expansion is performed, which in the case of the voltages is $v_{n,0} = v_{m,0} = 1$, and in the case of the angle difference is $\theta_{nm,0} = \theta_{nm,0}^2 = 0$, justified by the usual operating conditions in most power systems [23]. The power equations at each bus are then as follows

$$p_n = g_n^{sh} v_n^2 + \sum_{m=1, m \neq n}^{N} g_{nm} \left(v_n^2 - v_n v_m - \frac{\theta_{nm}^2}{2} \right) - b_{nm} \theta_{nm} \tag{18}$$

$$q_n = -b_n^{sh} v_n^2 + \sum_{m=1, m \neq n}^{N} -b_{nm} \left(v_n^2 - v_n v_m - \frac{\theta_{nm}^2}{2} \right) - g_{nm} \theta_{nm} \tag{19}$$

Neglecting $v_n v_m$ and θ_{nm}^2, together with the approximation $v_n^2 \simeq v_n$, the model of (29) is obtained, while taking the voltage squared as independent variable, the above equations can be rewritten, considering the following voltage product transformation:

$$v_n v_m = \frac{v_n^2 + v_m^2}{2} - \frac{(v_n - v_m)^2}{2} \tag{20}$$

As

$$p_n = g_n^{sh} v_n^2 + \sum_{m=1, m \neq n}^{N} g_{nm} \left(\frac{v_n^2 - v_m^2}{2} + \frac{(v_n - v_m)^2}{2} - \frac{\theta_{nm}^2}{2} \right) - b_{nm} \theta_{nm} \tag{21}$$

$$q_n = -b_n^{sh} v_n^2 + \sum_{m=1, m \neq n}^{N} -b_{nm} \left(\frac{v_n^2 - v_m^2}{2} + \frac{(v_n - v_m)^2}{2} - \frac{\theta_{nm}^2}{2} \right) - g_{nm} \theta_{nm} \tag{22}$$

Approximating the terms with θ_{nm}^2 and $(v_n - v_m)^2$, which account for the losses, by linear functions of θ_{nm} and $v_n^2 - v_m^2$, we would now yield the method proposed by Yang et al. [32], which is linear with respect to the voltage squared and the angle value. However, rearranging the above expressions,

$$p_n = g_n^{sh} v_n^2 + \sum_{m=1, m \neq n}^{N} g_{nm} \frac{v_n^2 - v_m^2}{2} - b_{nm} \theta_{nm} + g_{nm} \left(\frac{(v_n - v_m)^2}{2} - \frac{\theta_{nm}^2}{2} \right) \tag{23}$$

$$q_n = -b_n^{sh} v_n^2 + \sum_{m=1, m \neq n}^{N} -b_{nm} \frac{v_n^2 - v_m^2}{2} - g_{nm} \theta_{nm} - b_{nm} \left(\frac{(v_n - v_m)^2}{2} - \frac{\theta_{nm}^2}{2} \right) \tag{24}$$

and considering the following approximation of $(v_n - v_m)^2 / 2$ around the point $v_n = v_m = 1$

$$\frac{(v_n - v_m)^2}{2} \simeq \frac{1}{2} \left[(v_n - v_m) \frac{v_n + v_m}{2} \right]^2 = \frac{(v_n^2 - v_m^2)^2}{8} \tag{25}$$

the formulation of (26) is obtained:

$$p_n = g_n^{sh} v_n^2 + \sum_{m=1, m \neq n}^{N} g_{nm} \frac{v_n^2 - v_m^2}{2} - b_{nm} \theta_{nm} + g_{nm} \left(\frac{(v_n^2 - v_m^2)^2}{8} - \frac{\theta_{nm}^2}{2} \right) \tag{26}$$

$$q_n = -b_n^{sh} v_n^2 + \sum_{m=1, m \neq n}^{N} -b_{nm} \frac{v_n^2 - v_m^2}{2} - g_{nm} \theta_{nm} - b_{nm} \left(\frac{(v_n^2 - v_m^2)^2}{8} - \frac{\theta_{nm}^2}{2} \right) \tag{27}$$

In this case, given that the equations show the value of the angle difference θ_{nm} and its square θ_{nm}^2, as well as the square of the squared voltage difference $(v_n^2 - v_m^2)^2$, it is necessary to linearize these two terms by piecewise linear functions to properly estimate the losses.

At this point it is necessary to make a decision on the approximation of the power flow equations to be implemented in solving the problem. However, it requires two additional linearizations by piecewise functions, which makes it slower to solve than the simplified method of (30), which does not include terms that account for losses but is computationally faster since it only involves one linearization corresponding to the power limitation of the lines.

Since the objective of implementing these equations is to ensure the correct operation of the power system under study under different scenarios, the procedure involves the resolution of these flows in all cases, as the number of scenarios is very high, a more elaborate description results in possibly unaffordable computation times, which depends on the number of scenarios considered and the time intervals contained in each one.

In order not to limit the scenarios to be considered and the time steps, we implement the model evaluated by Yang et al. [29] and Morvaj et al. [33], originating from (30). As the case considered is a distribution network whose lines have a length of less than 5 km, without committing significant error, the shunt admittance of the lines can be considered null [34,35]. Naming $v_{nm} = v_n - v_m$, the linearized power flow equations are as shown below:

$$p_n = p_n^G - p_n^D = \sum_{m=1, m \neq n}^{N} g_{nm} v_{nm} - b_{nm} \theta_{nm} = \sum_{m=1, m \neq n}^{N} p_{nm} \quad n = 1, \ldots, N \qquad (28)$$

$$q_n = q_n^G - q_n^D = \sum_{m=1, m \neq n}^{N} -b_{nm} v_{nm} - g_{nm} \theta_{nm} = \sum_{m=1, m \neq n}^{N} q_{nm} \quad n = 1, \ldots, N \qquad (29)$$

where the superscripts denote power generation, G, and power demand, D. Each summand nm of the active and reactive power expressions is the power flow per each line linking bus n to bus m.

On the other hand, the maximum power constraint through the lines is defined by the inequality:

$$p_{nm}^2 + q_{nm}^2 \leq s_{max,nm}^2 = i_{max,nm}^2 v_x^2 \quad \forall x = n, m \qquad (30)$$

where S_{max}, is the maximum apparent power in per unit (pu) that can circulate through each line, i_{max}, is the maximum current that can pass through the conductor and v_x is the voltage at bus n or m. The manufacturers provide the maximum current limit because the limiting factor is the temperature of the conductor due to the heat caused by the passage of the current. The model of the conductors is given by the standards and technical requirements that can be found in the technical literature [36], and the value of the maximum current in [37,38], which is 285 A for underground conductors (20 °C ground temperature and 70 °C conductor temperature), and 262 A for overhead conductors (at 75 °C conductor temperature and 35 °C ambient temperature). This is what is known as the thermal limit of the conductors and is a factor to be considered in lines shorter than 80 km, as is the case; while between 80 and 320 km the limiting factor is the voltage drop, in lines longer than 320 km it is the stability of the angle [34]. The current limit depends on the conductor temperature, since the electrical parameters of the conductor vary with temperature, however, the change in these parameters would modify the power flow and therefore an iterative calculation would be necessary, in fact, this maximum current decreases with ambient temperature and the dynamics of heat transfer would have to be considered, which gives rise to a non-linear problem: the current limit of the conductors increases non-linearly with conductor temperature and decreases with ambient temperature [38]. This variation is not considered to be significant and the maximum current is assumed to be fixed and equal to that given above, at the given temperature.

4.2. Generation of Electrical Energy

The parameter representing the photovoltaic active power leaving the solar field and entering the inverter at each bus, in each scenario and time instant, is given by the following formula [39]:

$$P_{pv}(t) = P_{nom} \frac{G(t)}{G_n} \left[1 - \alpha \left(T(t) + \frac{G(t)}{800} [NOCT - 20] - 25 \right) \right] \qquad (31)$$

where T is the ambient temperature, P_{nom} is the power at nominal conditions, G_n is the nominal irradiance in W/m^2 (this value is sometimes generalized to 1000 [40]), G is the incident irradiance in W/m^2, α is the parameter of the power-temperature characteristic in %/°C, $NOCT$ takes the value of 45 °C and is the nominal operating temperature of the cell at 800 W/m^2 with 20 °C ambient temperature and 1 m/s wind. The alpha factor of the efficiency is negative, which implies an increase in efficiency with decreasing temperature. The inverter efficiency and other characteristics are introduced in more detail in later sections.

In the case of wind turbines, the parameter denoting the maximum energy extractable from the device is calculated using its power curve, but applying the inverter efficiency to the proportional part of the power flowing through the inverter:

$$P_{wt}^{max}(t) = P_{wt}(t)(0.7 + 0.3\eta_{b2b}(t)) \simeq P_{wt}(t)(0.7 + 0.3 \cdot 0.965) = 0.9895 P_{wt}(t) \tag{32}$$

$$P_{wt}(t) = \begin{cases} 2100\left[1 - e^{-\left(\frac{c(t)}{5.692}\right)^{3.398}}\right] & 2\,\text{m/s} \le c(t) < 20.5\,\text{m/s} \\ 4378 - 111c(t) & 20.5\,\text{m/s} \le c(t) \le 25\,\text{m/s} \\ 0 & \text{any other case} \end{cases} \tag{33}$$

This gives the upper limits of active power injection to the grid, those of reactive power are related by means of the equations of the converters presented in the section corresponding to power converters.

4.3. Electrical Energy Storage

This section presents the equations that give the state of charge of each storage system on each bus, in each scenario and at each moment in time. The active and reactive power variables that appear represent the energy output/input of the system through the converter that connects the *ESS* (Energy Storage System) to the grid.

The following stochastic variables are defined for the load power, the discharge power and the state of charge of each *ESS* at each bus n for time t:

$$P_{in}^{ESS,n}(t), P_{out}^{ESS,n}(t), SOC(t) \ge 0 \tag{34}$$

These powers are the real effective powers received or delivered by the *ESSs*, that is, they include the load and converter efficiency in their definition, which is explained in the following development.

The weight of the problem lies in the binary part, not in the linear part, therefore it is convenient to create a variable for loading and another for unloading and to use only a binary variable and its complementary to avoid simultaneous loading and unloading. This restriction is as follows:

$$P_{in}^{ESS,n}(t) \le b^{ESS,n}(t)P_{in,max}^{ESS,n}, \; P_{out}^{ESS,n}(t) \le (1 - b^{ESS,n}(t))P_{out,max}^{ESS,n} \; \forall ESS, n \tag{35}$$

where the subscript max denotes the nominal power of each storage system (*ESS*) on each bus n and the binary variable is $b^{ESS,n}$ which takes the value 1 or 0 at each time t for each *ESS* on each bus n and each scenario. This constraint forces the load to zero if there is unloading and vice versa.

On the other hand, the load state model is estimated to be linear and without capacity reduction due to unloading depth or gradation. The correctness of this assumption is ensured by adding a number of constraints to limit the state of charge to a safe range of each *ESS*. The state of charge of each *ESS* on each bus n at each time t and scenario is given by the following expression:

$$SOC_{ESS,n}(t) = (1 - \alpha\Delta t)SOC_{ESS,n}(t-1) + P_{in}^{ESS,n}(t)\Delta t - P_{out}^{ESS,n}(t)\Delta t \; \forall ESS, n \tag{36}$$

where *SOC* represents the state of charge, the parameter α is the relative self-discharge per unit time and Δt is the time step. The upper limit of the state of charge of each *ESS* is defined by the following constraints:

$$SOC_{ESS,n}^{max}DOD_{ESS,n} \le SOC_{ESS,n}(t) \le SOC_{ESS,n}^{max}(1 - DOD_{ESS,n}) \tag{37}$$

where *DOD* is the relative depth of discharge parameter of each *ESS* on each bus, set to zero for flow batteries and 0.1 for lithium-ion batteries and hydrogen cells, as already introduced in the section on system components.

All these quantities are expressed in International System units, but to link these subsystems to the grid they must be expressed in per unit, and for this purpose they are simply redefined through the quotient between the base power. In addition, the definition of the charging and discharging power variables relate them to the converter that connects these systems to the grid and is where the charging and discharging efficiencies are applied, as well as those of each converter, as noted at the beginning:

$$p_{out}^{ESS,n} = \frac{P_{out}^{ESS,n}}{S_{base}} = \frac{p_{ESS2net,n}(t)}{\eta_{out}^{ESS,n}\eta_{conv}^{ESS,n}}, \; p_{in}^{ESS,n}(t) = \frac{P_{in}^{ESS,n}(t)}{S_{base}} = p_{net2ESS,n}(t)\eta_{in}^{ESS,n}\eta_{conv}^{ESS,n} \quad (38)$$

where *conv* refers to the power converters and the subscripts *ESS2net* indicate power transfer from the storage system to the grid and *net2ESS* from the grid to the storage system.

In the application case, the charging and discharging efficiencies are those justified in the system components section and the efficiency of the converters is approximated in a constant way based on the commercial model also chosen in the aforementioned section, with a value of 0.98.

Finally, constraints are necessary to force the initial SOC equal to the final SOC, during the time horizon, in order to increase the lifetime of the *ESS*. This variable, $SOC_{ESS,n}^0$, is not stochastic:

$$SOC_{ESS,n}(t_0) = SOC_{ESS,n}^0, \; SOC_{ESS,n}(T) = SOC_{ESS,n}^0 \quad (39)$$

4.4. Power Converters

One aspect to consider for power converters is their performance curve: PV plants and storage systems need to be able to deliver more than a certain percentage of the inverter's rated power for the power output to be effective and work above the bend of the performance curve, whenever it is considered appropriate for them to inject power.

These curves describe a potential behavior and the three parameters that characterize them can be adjusted from a number of real samples or from the curve itself to obtain the continuous version. In other words, they are power functions that introduce non-linearities to the problem and must therefore be linearized. This approximation error is not remarkable since the performance reaches high values at relatively low powers. In this case, a discretization of the curve into four intervals is chosen. In order to model the active power output, parameters are created that represent the maximum power limit that can be generated at any given time by each system on each bus and in each scenario. These parameters are characterized by the super index max. Then, for example, the maximum PV active power produced at each time *t* is given by the following formula:

$$P_{pv}^{max}(t) = \begin{cases} \eta_1 P_{pv}(t) & P_{pv}(t) \in [P_0, P_1] \\ \eta_2 P_{pv}(t) & P_{pv}(t) \in (P_1, P_2] \\ \eta_3 P_{pv}(t) & P_{pv}(t) \in (P_2, P_3] \\ \quad\vdots \\ \eta_u P_{pv}(t) & P_{pv}(t) \in (P_{u-1}, P_u] \end{cases} = \begin{cases} 0.98 P_{pv}(t) & 0.4 \leq P_{pv}(t)/P_{nom} \\ 0.972 P_{pv}(t) & 0.2 \leq P_{pv}(t)/P_{nom} < 0.4 \\ 0.955 P_{pv}(t) & 0.08 \leq P_{pv}(t)/P_{nom} < 0.2 \\ 0 & \text{any other case} \end{cases} \quad (40)$$

where $P_{pv}(t)$ is the active power coming from the solar field at time *t*, η_n are the yields corresponding to the average of the discretized interval, P_{nom} is the nominal power of the inverter of the solar field and P_u, P_{u-1} are the powers of each interval. This expression is applied to each PV system in each bus and scenario. In the specific case of the wind turbine with doubly fed induction machine (DFIG) this efficiency only applies to the percentage of the power that circulates through the converter, that is, about 30% of the generated power.

Another important component of these machines is the ability to absorb/inject reactive power. Based on the specific characteristics of each manufacturer's model and for each application, the operating region of the converters is limited differently. In the case of PV inverters, the active power is limited to non-negative values so that the portion of the circle with the positive semi-axis of abscissa (right half of the circle) is obtained, in other words, the following constraints apply:

$$\left[\sin\left(\tfrac{2\pi l}{k}\right) - \sin\left(\tfrac{2\pi(l-1)}{k}\right)\right] p_{pv}(t) - \left[\cos\left(\tfrac{2\pi l}{k}\right) - \cos\left(\tfrac{2\pi(l-1)}{k}\right)\right] q_{pv}(t) \leq s_{inv}\sin\left(\tfrac{2\pi}{k}\right)$$

$$0 \leq p_{pv}(t) \leq P_{pv}^{max}(t)/S_{base}, \quad -s_{inv} \leq q_{pv}(t) \leq s_{inv} \tag{41}$$

where p_{pv} is the decision variable of active power per unit that the inverter or group of PV inverters deliver to the grid, q_{pv} is the variable of reactive power per unit that the PV inverter delivers to the grid and s_{inv} is the parameter corresponding to the nominal or maximum apparent power of the inverter also expressed in per unit (pu). Here s_{inv} is a deterministic parameter while p_{pv} and q_{pv} are stochastic variables. Again, the above expression is applied to each bus where there is a PV installation.

For wind turbines, the constraint is similar except that the power factor fp_{wt}, both inductive and capacitive, is limited to maximum values of 0.95, as indicated by the manufacturer [41]:

$$\left[\sin\left(\tfrac{2\pi l}{k}\right) - \sin\left(\tfrac{2\pi(l-1)}{k}\right)\right] p_{wt}(t) - \left[\cos\left(\tfrac{2\pi l}{k}\right) - \cos\left(\tfrac{2\pi(l-1)}{k}\right)\right] q_{wt}(t) \leq s_{wt}\sin\left(\tfrac{2\pi}{k}\right)$$

$$0 \leq p_{wt}(t) \leq P_{wt}(t)/S_{base}, \quad -\tan(\cos^{-1}(fp_{wt}))s_{wt} \leq q_{wt}(t) \leq \tan(\cos^{-1}(fp_{wt}))s_{wt} \tag{42}$$

The following consideration should be noted: inverters only provide/absorb reactive power from the grid if there is active power available in the solar field. Likewise, the DFIG only manages reactive power when it is possible to generate active power. In principle, this limitation is not due to the design of the inverters, but is normal practice in this type of installation. In algebraic terms, this is achieved by modifying the restriction of the reactive power limits. The photovoltaic case is shown as an example, but for all other converters it is analogous:

$$\left.\begin{array}{ll} -s_{inv} & P_{pv}^{max}(t) > 0 \\ 0 & P_{pv}^{max}(t) = 0 \end{array}\right\} \leq q_{pv}(t) \leq \left\{\begin{array}{ll} s_{inv} & P_{pv}^{max}(t) > 0 \\ 0 & P_{pv}^{max}(t) = 0 \end{array}\right. \tag{43}$$

The converters used in storage systems are a special case since the active power flow can be both positive and negative (discharge and load), however, the approach applied is to separate these two processes (to save binary variables and thus speed up the resolution) and apply a similar constraint to each one together with the additional reactive limitations presented by these converters:

$$\left[\sin\left(\tfrac{2\pi l}{k}\right) - \sin\left(\tfrac{2\pi(l-1)}{k}\right)\right] p_{ESS2net}(t) - \left[\cos\left(\tfrac{2\pi l}{k}\right) - \cos\left(\tfrac{2\pi(l-1)}{k}\right)\right] q_{ESS}(t) \leq s_{ESS}\sin\left(\tfrac{2\pi}{k}\right)$$

$$\left[\sin\left(\tfrac{2\pi l}{k}\right) - \sin\left(\tfrac{2\pi(l-1)}{k}\right)\right] p_{net2ESS}(t) - \left[\cos\left(\tfrac{2\pi l}{k}\right) - \cos\left(\tfrac{2\pi(l-1)}{k}\right)\right] q_{ESS}(t) \leq s_{ESS}\sin\left(\tfrac{2\pi}{k}\right) \tag{44}$$

where the subscript *ESS2net* indicates power transfer from the storage system to the grid and *net2ESS* from the grid to the storage system. These and the following constraints are for each *ESS*, at each bus, at each time and in each scenario. The capacitive (*cap*) and inductive (*ind*) reactive limits are defined by giving a range to the variable as follows:

$$-\sin(\cos^{-1}(fp_{ESS,ind}))s_{ESS} \leq q_{ESS}(t) \leq \sin(\cos^{-1}(fp_{ESS,cap}))s_{ESS} \tag{45}$$

where s_{ESS} is the nominal apparent power of the converter of each storage system.

In this way, the P-Q operating curve of these converters is approximated in such a way that there is a circle cut by a horizontal straight line at the top (capacitive power factor) and by another horizontal straight line at the bottom (inductive power factor). This approximation is conservative in the sense that it does not consider power peaks slightly above nominal, as these converters allow according to the manufacturer, nor does it represent the performance variation in a non-linear way in two quadrants: first and fourth. Note that these considerations could be added by combining circles that cut in such a way that the operating area is only that enclosed by all of them at the same time, however, this increases the computational time and it is not considered meaningful to implement this detail in this case.

The complete energy balance corresponding to the pu power generation can now be expressed as shown below:

$$p_n^G(t) = p_{wt}^n(t) + p_{pv}^n(t) + p_{ESS2net}^n(t) - p_{net2ESS}^n(t) \tag{46}$$

$$q_n^G(t) = q_{wt}^n(t) + q_{pv}^n(t) + q_{ESS2net}^n(t) - q_{net2ESS}^n(t) \tag{47}$$

Here all variables are expressed for each time point t, each bus n and stochastic scenario.

4.5. Energy Consumption

The consumption data are stochastic parameters as they vary over time, for each bus and each scenario. They are defined on the basis of the demand curve proposed in the literature [42,43], but modified to make it similar in shape to the Spanish typical curve. In turn, in the aforementioned literature, it is proposed that different loads depend on whether they correspond to the residential or industrial sector, and we have sought to maintain this distinction. The power factor of each bus is also known from the benchmark and the nominal power is given in its apparent form. The network demand is given as a percentage of the nominal power of each bus. Then, the active power to be supplied at each bus n is given by the following expression:

$$p_n^D(t) = \frac{1}{S_{base}}(S_{indus}(n)fp_{indus}(n)rate_{indus}(t) + S_{res}(n)fp_{res}(n)rate_{res}(t)) \tag{48}$$

where *res* denotes residential and *indus* denotes industrial, furthermore here S corresponds to the nominal apparent power of each bus. Equivalently, the reactive demand is constructed as follows:

$$q_n^D(t) = \frac{1}{S_{base}}\left(S_{indus}(n)\sqrt{1-fp_{indus}(n)^2}rate_{indus}(t) + S_{res}(n)\sqrt{1-fp_{res}(n)^2}rate_{res}(t)\right) \tag{49}$$

4.6. CO_2 Emissions and Other Factors

Finally, to define the objective function, it is necessary to estimate the CO_2 emissions corresponding to the import of energy from the transmission grid. This is achieved by applying a time-varying emissions factor that represents the number of metric tons of CO_2 that it costs to produce one MWh unit of energy. This factor is considered to be similar to the Spanish factor and depending on the case study will be constant over time or may vary. This factor is deterministic because there is no correlation with other variables and there are no studies of its prediction.

The CO_2 emissions corresponding to the import of energy from the slack bus over a period T are calculated using the following expression:

$$Emissions = \sum_{t}^{T} f_{CO_2}(t)p_{slack}(t)S_{base}\Delta t \tag{50}$$

The effect of this factor is studied in different sections. Note that the slack bus powers are free variables and if, for example, energy export is not allowed or the power factor is to be limited, constraints such as the following must be added:

$$p_{slack}(t) \geq 0, \; q_{slack}(t) \geq 0 \tag{51}$$

5. Heuristic Methods

Heuristic methods have been commonly used in the field of optimization, being able to obtain solutions quickly to complex problems that, even with the use of clusters of computers, may not have an optimal solution. To do so, they sacrifice solution accuracy at the cost of reduced computational cost and time. Nevertheless, they are capable of finding an exact solution to problems of relative simplicity.

These methods are often used in off-grids to find a suitable location for the development of this type of microgrids [44,45]. In recent years, research is being carried out on certain metaheuristic methods, such as Particle Swarm Optimization and Salp Swarm Algorithm, among others, which are population-based methods with promising results [46].

Heuristic methods are also applied, with success, to improve the effectiveness of the distribution network by means of reconfiguration. Reconfiguration of the distribution network aims to find the optimal combination of all switches in the distribution, mainly determining proper sizing and siting of DG together with network reconfiguration. In this way, some researchers, like Muhammad et al. [47], use the discrete network reconfiguration of the data set method, employing the Water Cycle Algorithm (WCA) together with dataset approach to reduce the complexity of search space. This type of method has good convergence performance, and can obtain a global optimal solution for single-objective optimization problems. Others, such as Helmi et al. [48], propose the Harris Hawks Optimization (HHO) to minimize the power losses of the network.

However, all heuristic methods that are inspired by natural processes have parameters that are highly dependent on their own algorithm; therefore, the algorithm may behave differently affecting its performance [49]. Furthermore, according to Yang, and in relation to computational cost, no consensus has been reached on what are the best values or configurations of an algorithm, nor on possible ways to adjust these parameters to achieve the best performance [50]. On the other hand, the selection of a method as the most appropriate for solving a problem such as energy management in hybrid microgrids is an open problem [51]. This is largely based on the "no free lunch" theorem [52] of mathematical optimization, which shows that at the same time that a heuristic is very efficient for one collection of problems, it is very inefficient for another collection.

So, there are a multitude of methods that can be applied in this field, without any of them being clearly better than any other for any topology and type of grid, as we have discussed in previous sections. Even though the optimal power flow is a non-convex problem [53], convex approximations for the power flow equations have been studied [54,55], but generally assuming strong approximations, such as all generators are constant current injections, which is far from real microgrids and further from hybrid microgrids [56].

In this paper, we try to optimize energy management without reconfiguring the network, assuming that the location and schedule of DG and storage banks does not depend on the utility, but on the consumers, and therefore, it is not in their hands to reconfigure the network. Thus, we will not take into account, for example, the temporary shifting of loads, which is currently possible with electric vehicles. The study of controllable loads could be attacked by stochastic optimization algorithms, according to Hosseini et al. [57] or Barbato et al. [58]. In the following, two heuristic methods known in the study of microgrids will be described, namely: Hysteresis Band Control and Fuzzy Logic Control.

5.1. Hysteresis Band Control

The heuristic method of Hysteresis Band Control has been a certain success in the field of microgrids. According to the example proposed by Ipsakis et al. [59], from which this control strategy will be explained, a hysteresis band consists of the operating range existing between two limit values of a problem variable, in this case, the State of Charge (SOC) or State of Charge of the accumulator. In this way, the storage units of the microgrid in the example, electrolyzer and fuel cell, absorb energy or give it up according to the band defined by the limit values mentioned:

- in case of lack of power, it will be the fuel cell that will give up power; and
- in case of reaching the maximum state of SOC, it will be the turn of the electrolyzer, which will start to operate at that moment of excess power.

This way of operating has benefits such as reducing the number of start-ups and shutdowns of the electrolyzer, which can reduce its life expectancy. Also, the example method achieves a protection of the accumulator by excluding it from excessively long operation.

Appropriate SOC limit values must be calculated to achieve effective control and in accordance with the operating requirements of the problem.

5.2. Control by Means of Fuzzy Logic

The Fuzzy Logic Control (FLC) method is characterized by reaching solutions to problems in which the data, variables, or in short, the available information, is ambiguous or imprecise, hence the term fuzzy. Therefore, fuzzy control is characterized by being described by what could be called discard logic, that is: If a certain event occurs, then the control signal will take the value 'X'.

This is why Fuzzy Logic Control (FLC) lacks accuracy when it comes to providing robust solutions. However, Fuzzy Logic Control has some advantages that make it attractive for tackling certain types of problems, these are [60,61]:

- It provides an orderly and efficient working structure from information given orally and fuzzily by human experts.
- Due to its simplicity, it is easy to understand and simplifies the design of the problem, which gives it a quick implementation and a lower cost compared to other methods.
- It is capable of generating numerous output signals from any reasonable number of inputs.
- It does not require a model to find approximate solutions to the control problem and provides non-linear controllers.

We can brief the concept by explaining its methodology with the following steps: firstly, the input data provided are processed and a smearing or merging is performed on them. In this first step, certain qualitative characteristics are given a numerical value. Once this is done, decisions are made in accordance with logical relationships called Fuzzy Rules. Finally, the defuzzification process takes place, in which concrete data are obtained that will be used to generate the appropriate control signals required by the problem.

Despite the ease of implementation of the heuristic methods described above, the large number of restrictions and variables that appear in the microgrid under study make them a bad strategy to follow for its control, as it is difficult to find optimal solutions to the problem [62,63].

6. Machine Learning Methods

Machine Learning is a scientific discipline that tries to make systems learn automatically. Learning, in this context, means identifying complex patterns in millions of pieces of data [64,65]. The machine that actually learns is an algorithm that reviews the data and is able to predict future behavior in some fields of knowledge [66–68]. Machine Learning is therefore a process of knowledge induction, that is, a method of deriving a general statement by generalizing from statements describing particular cases.

Machine Learning is learning from data, it is discovering the structure and patterns underlying the data. The main objective of Machine Learning is to extract the information contained in a dataset to acquire knowledge to make decisions about new datasets [69].

Formally, and according to Mitchell [70], we can define the algorithms used by Machine Learning as:

"A computer program is said to learn from experience E with respect to some class of tasks T and performance measure P if its performance at tasks in T, as measured by P, improves with experience E".

These learning algorithms are based on a set of data on which to learn and then apply the experience gained on other sets. It is necessary to evaluate its performance on a set other than the one on which the system has been trained in order to obtain a valid estimate of its generalizability to new examples. Thus, the available data set is divided into two subsets: on the one hand, we have the training set and, on the other one, the validation set or test set. In this way, the model is generated from the training data and evaluated in the test set, in which the accuracy of the model can be measured. The obtained result on this set is a good approximation to the expected one for the new data [71].

Therefore, generalization is one of the key aspects in the design of Machine Learning algorithms [72]. At the same time, the models must fit the training set and capture all its information. In this way, the problem of balancing bias and variance arises: bias measures the average error of the model using different training sets, while variance measures the sensitivity of the model to small changes in the training data [73]. In other words, very complex models have a low bias and a high variance, which is known as overfitting. On the other hand, simple models have a high bias but a very low variance. Overfitting occurs when, by adding levels to the Decision Tree, the hypotheses are so refined that they describe the examples used in the learning process very well; however, when evaluating the examples, an error occurs. That is, it classifies the training data very well, but then it fails to generalize the test set. This is because it learns down to the noise of the training set, adapting to the regularities of the training set [74]. Therefore, overfitting will be an important evaluation indicator to take into account in the study.

Machine Learning algorithms are usually divided into three categories, the first two being the most common:

- Supervised learning: this type of algorithm is based on prior learning, usually related to a system of labels associated with the data. This allows them to make decisions based on the previous data or even make predictions from these data. An example could be a spam detector, that is, a system that thinks it detects spam and labels an email as spam based on the patterns it has learned from the email history (keywords in the subject line, sender, text/image ratio, etc.).
- Unsupervised learning: unlike the previous type, these algorithms do not use prior knowledge. What they use is all the available data with the aim of finding patterns among them. If they find such patterns, they try to organize them in some way. For example, unsupervised learning is applied when you want to extract patterns from massive social media data, to recommend products or create advertising campaigns
- Reinforcement learning: in this less common case, an algorithm learns from its own experience. A trial-and-error process is normally used in which correct decisions are rewarded in some way (with reinforcement factors). In this way, the aim is to make the best decision in different situations. Examples include: DNA classifications, facial recognition, etc.

6.1. Operation with Machine Learning Models

The process to be followed for the construction of a Machine Learning system can be divided into:

1. Data collection. This is usually a tedious process that takes up a large part of the development of the system, since it is generally necessary to collect large amounts of data in order to ensure that the used sample is representative of the set under study.

2. Feature selection. This is a critical step since it is necessary to extract those variables that are useful to distinguish the patterns of each category.

3. Choice of model. In this step, we will choose the model that best fits our problem and that achieves the expected performance on the test set. This model, among other tasks, must maintain the bias-variance balance explained above.

4. Model training. In this phase, the classifier is built, whose parameters are adjusted from the training data set. Finding the parameters that fit our model is an optimization problem since the objective is always to minimize a certain objective function.

5. Evaluation of the model. Using the test set, an error measure is set and the performance of the model is obtained. If the result is not expected, it is necessary to test by going back to each of the previous points and go through the process again.

In mathematical terms, the principle of Machine Learning, in a supervised learning context, consists of starting from a sample of learning:

$$\mathcal{L} = \left\{ (x_n, y_n) | n = 1, 2, \ldots, N, x_n \in \Re^d, y_n \in \{1, 2, \ldots, C\} \right\} \tag{52}$$

constituted by n realizations of a pair of random variables (X, Y), to construct a $f : \Re^d \to \{1, 2, \ldots, C\}$ function which, given a new X input vector, can predict with some degree of certainty the variable $Y = f(X)$. For each observation (x_i, y_i) of $\mathcal{L}$, the variable $x_i \in X$ is called the input variable or explanatory variable and $y_i \in Y$ the dependent or output variable [75]. When the dependent variable is discrete or categorical, we speak about a classification problem; and when it is continuous, about a regression problem. That is to say, depending on the type of objects that we are trying to predict, there are two types of problems:

- Classification problems: They try to predict the classification of objects on a set of pre-fixed classes. For example, classifying whether a news is about sports, entertainment, politics, etc.
- Regression problems: They try to predict a real value. For example, predict the value of the stock market tomorrow from the stock market behavior that is stored (past).

One of the most widely used Machine Learning methods is Decision Tree Learning. This is a method for approximation of discrete-valued functions, robust to noisy data and able to learn disjoint expressions. There is a family of Decision Tree Learning algorithms: ID3, C4.5, … In turn, based on these Decision Trees, hybrid methods have been created that build more than one Decision Tree: Bagging, Boosting and Random Forest.

6.2. Decision Trees

Learning through Decision Trees is based on the principle of divide and conquer. Let $\mathcal{L}$ a sample be defined as:

$$\mathcal{L} = \left\{ (x_n, y_n) | n = 1, 2, \ldots, N, x_n \in \Re^d, y_n \in \{1, 2, \ldots, C\} \right\} \tag{53}$$

where N is the number of elements in the data set, C the number of distinct classes and d the number of variables defining the examples x_n in the set. Each of these examples is represented by a vector x_n, which has its corresponding class label y_n associated with it and it is defined by different variables, which can be numerical (their values are real numbers) or categorical (they take values in a finite set in which there is no ordering relationship). Sometimes, these vectors x_n are also cited as feature vectors.

A T Decision Tree is an ordered sequence of decisions, normally from questions, in which the next decision depends on the answer to the current one. These decisions are taken normally from questions that are formulated on the variables that define each x element in order to assign them a y certain class. This process, including its corresponding questions, decisions and bifurcations, is naturally represented by means of a tree.

In a Decision Tree, each node of the tree is an attribute (field) of the examples, and each branch represents a possible value of that attribute. The first node is known as the root node, which is successively connected to the other nodes until it reaches the leaf nodes, those that have no descendants, that is, the end of the branches of the tree. Each node is assigned one of the questions of the sequence, while each leaf node is assigned a class label.

In this way, the question of the root node is asked of the whole set $\mathcal{L}$, which is subdivided until reaching the last nodes (leaves), which constitute a disjoint partition of the initial feature space. This happens because, when given a node, one and only one branch will be followed by each instance of the training set.

Decision Trees perform well with large volumes of data, as it does not require loading all data into memory at once. The computation time scales well with a linearly increasing number of columns [76].

The advantages of Decision Trees are that they are easy to understand and interpret, rule generation is simple, it reduces the complexity of the problem, and training time is not very long.

One of the disadvantages is that if an error is made at a high level, successive nodes would be poorly created. In the construction of a Decision Tree, the most complicated step is to determine which attribute to base a node on, because if there are many features, the algorithm would have many options for the training data, and it would be difficult to construct a Decision Tree.

However, Decision Trees can give good results if they are combined with ensemble methods. With these methods, instead of learning a single model, several models are learned, and the estimates from each model are combined.

Ensemble methods are combinations of models. In these techniques, it is necessary both to define how different models are to be created and how the results of each model will be combined to generate the final prediction. The aim of ensemble methods is to produce a better prediction than individual models (individual members of the ensemble).

The most common ensemble methods are Bagging, Boosting and Random Forest. In all of these methods, the training set is manipulated, but in each case with a different strategy [77].

In Bagging, different samples are extracted from the training set (bootstrap samples), and these bootstrap samples are used as if they were the true training set. Boosting, on the other hand, always works with the full data set, that is, the complete dataset is always used. In Boosting, we can manipulate the weights of the data in the training set to generate different models. At each iteration, Boosting learns a model that minimizes the sum of the weights of the misclassified data.

Finally, Breiman [78] presented an ensemble method called Random Forest where bagging is used together with a random selection of attributes. At each node of each tree in the forest, a subset of the available attributes at that node is randomly selected and the best of them is selected according to the splitting criteria used in the base algorithm. The number of randomly selected attributes is an input parameter.

7. Results

We have evaluated the following of the proposed models explained in the previous sections, in order to measure the accuracy: Hysteresis Band Control, Fuzzy Logic Control, and Decision Trees (DT). We propose the IEEE microgrid test system of 69-bus [79]. The 69-bus distribution network has a nominal voltage of 12.66 kV. Its base apparent power is 10 MVA. This system has 69 nodes and 73 branches, including tie-lines, as shown in Figure 1. The order of each branch is assumed to be that of the furthest node minus one unit, except for the tie-lines in the base scheme which run from 69 to 73. Thus, a total of 73 remote switches are installed in the network, 68 of which are sectioned and ready for possible reconfiguration. As mentioned, and described by Lan et al. [80], two wind units, two solar panels and some switches are located in the microgrid, and a battery storage unit is also installed in the hybrid microgrid. We have followed the same location:

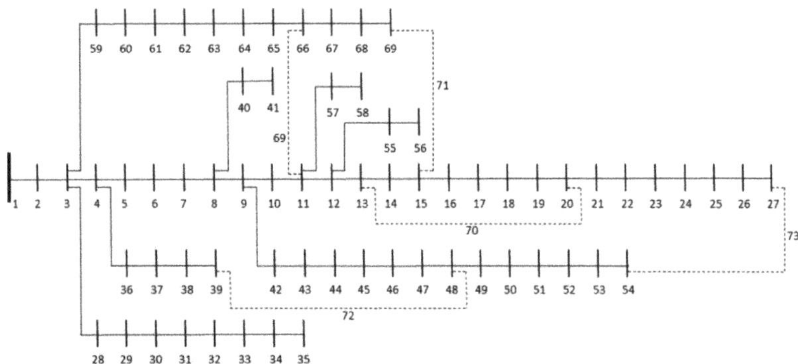

Figure 1. IEEE 69-bus test system.

In this article, we only show a first approximation of the methods to verify their viability and check primary results. In addition to achieving adequate energy management using all the tested models, regardless of the calculation time or the power and dedication of the used computers, we can highlight that for loads without large hourly differences (as it would have been the case for valley and peak periods with a large difference in the power value), the system can be controlled by applying the aforementioned modeling.

From the test results, it can be found the Machine Learning model (Decision Tree) can also recognize the states accurately for distribution systems, as we show in Table 1. We used the root medium square error for the comparison.

Table 1. Root Medium Square Error (RMSE) (voltage of the network) with different models applied to the IEE 69-bus distribution network.

	RMSE
HBC	0.1781
FL	0.1563
DT	0.1471

HBC = Hysteresis Band Control; FL = Fuzzy Logic; DT = Decision Tree.

As we can see, the three tested methods have enough viability, taking into account the characteristics of the network. We cannot conclude a better performance for the Decision Tree-based model, because, although its result is better than the others, the test is based in a single evaluation and more evaluations with different cases (for example, with different cases of distribution of solar and wind generation and different loads) should be done in the future.

In the case of the most modern method, Decision Trees, it should be noted that both solar and wind generation have followed generation patterns established in advance, according to the data of Lan et al. [80]. Obviously, in an analysis with a real system, these future generation data should be based on historical data and take forecasting into account. In Machine Learning methods, the reliability of data and predictions is very important, as they are the foundation of this type of modeling; therefore, it is advisable to use more than one database, which will usually reduce the error of calculation and analysis. These used patterns have been the same for any unit of the same generation type; that is, all solar units follow the same pattern, and all wind units follow the same pattern. This has been done to simplify the performed analysis and in accordance with Kovousi–Fard and Khodei [81].

The evaluations are based on the values shown in Figure 2 (base case), which represents the total energy consumption over a standard day.

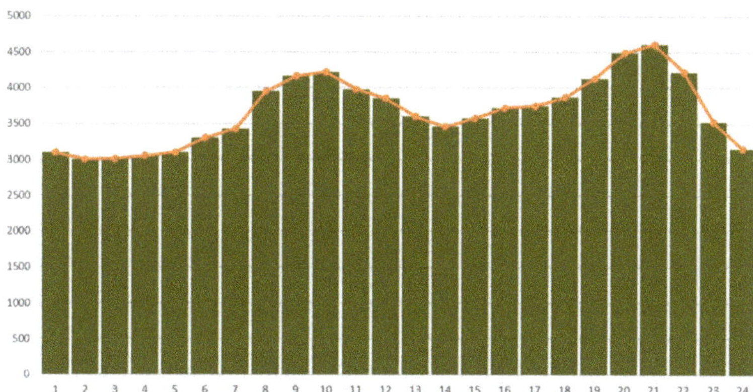

Figure 2. Total energy consumption value (kWh) over a standard day.

Figure 3 shows the values obtained in the energy storage units, including the battery allocated at node 15. It presents the 24-h values of the state of the charge (SOC) of all the devices (that is, those related to the PV systems and the aforementioned battery unit). As can be seen, the energy storage units are strongly influenced by the solar behavior and are discharged from 16–17 h onwards, as the PV power generated gradually decreases.

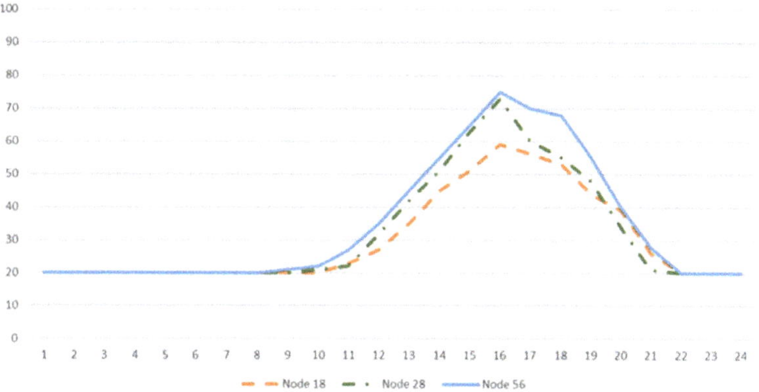

Figure 3. 24 h SOC (%) of three of the energy storage devices of the microgrid, allocated at node 18 (orange); node 28 (green); node 56 (blue).

Figures 4 and 5 show the two main cases studied: Figure 4 corresponds to the situation where there is no distributed generation, and Figure 5 where there is distributed generation. As can be seen in Figure 4, the voltage profile has been improved with the use of the three methods, not being able to conclude which of the three is the best, mainly because, although the three improve the initial case, depending on the section of the system to be analyzed, the best method is one or the other. The tie-switch distributions generated by each method are shown in Table 2.

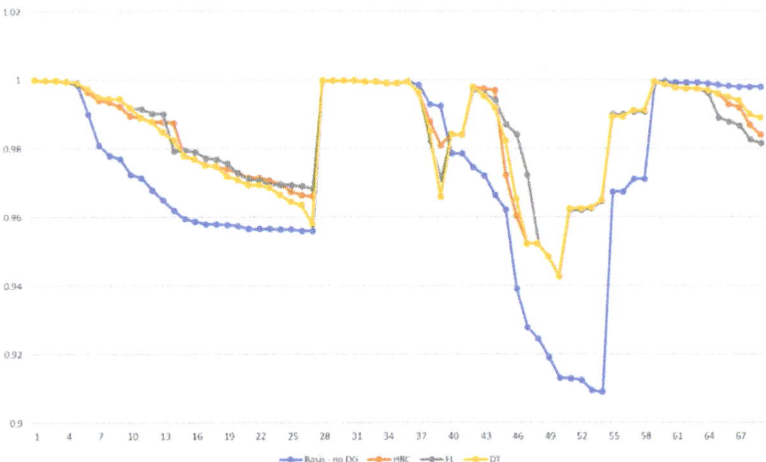

Figure 4. Case of no distributed generation. Voltage profile. Base case (blue); Hysteresis Band Control method (orange); control by means of Fuzzy Logic (grey); Decision Trees method (yellow).

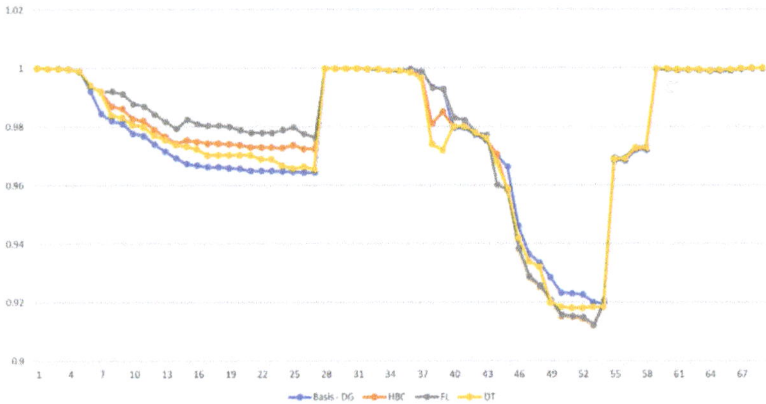

Figure 5. Case of distributed generation. Voltage profile. Base case (blue); Hysteresis Band Control method (orange); control by means of Fuzzy Logic (grey); Decision Trees method (yellow).

Table 2. Case without distributed generation: tie-switches of the configuration of the system.

	Initial	HBC	FL	DT
Tie-switches	69, 70, 71, 72, 73	14, 55, 61, 69, 70	13, 57, 61, 69, 70	13, 57, 61, 69, 70

HBC = Hysteresis Band Control; FL = Fuzzy Logic; DT = Decision Tree.

The IEEE-69 system has also been tested with distributed generation. As mentioned above, the distributions of wind and solar generation have followed the one published in the literature references, but Table 3 shows the averages of a typical day for the four nodes where they are located (6 and 68 for wind generation, and 25 and 50 for solar generation). In addition, Table 3 shows the tie-switches for each method. It can be seen in Figure 5 how, in general, the three methods improve the voltage profile of the system again after the reconfiguration of the system, being slightly different the solution (topology) found by each of them.

Table 3. Case with distributed generation: tie-switches of the configuration of the system and power (node–kW).

	Initial	HBC	FL	DT
Tie-switches	69, 70, 71, 72, 73	13, 55, 64, 69, 70	13, 58, 64, 69, 70	13, 58, 64, 69, 70
P (kW)	6–27.52 25–30.45 50–30.45 68–342.05	6–27.52 25–30.45 50–30.45 68–342.05	6–27.52 25–30.45 50–30.45 68–342.05	6–27.52 25–30.45 50–30.45 68–342.05

HBC = Hysteresis Band Control; FL = Fuzzy Logic; DT = Decision Tree.

8. Conclusions

In this article, we presented the optimization of the operation of electrical hybrid microgrids, focused particularly on the mathematical modeling. The set of loads (consume), generators (with a predominance of the renewable energies) and storage systems (at the present, particularly batteries) makes up the electrical hybrid microgrid. The management of this type of networks presents many challenges and various options to be implemented.

The mathematical modeling of the different components is a very important step in the control and management that, in recent years, is increasing with the irruption into the market of new technologies or new models for the use of energy. The main contribution is the mathematical modeling of several components of the hybrid microgrid. This modeling can be used in different methods of control and management of the network, and its feasibility has been shown in three methods, including one based on the Decision Tree method, which belongs to the Machine Learning family. The results on a test system of 69 buses show that its implementation is possible.

The three methods have been compared both in the case of existence of distributed generation and in the case of its non-existence, in order to obtain a better view of their behavior. The three methods improve the voltage profile of the system, using a different topology, although similar (the tie-switches used are topologically very close or even the same), demonstrating their effectiveness. The integration of distributed generation in the system causes small differences in the results of each method, these differences being more noticeable the longer the line (or branch) in the system.

Nevertheless, although the results indicate that the Decision Trees method is partially better than other algorithms, more tests are needed and they need to be carried out with different typologies, not only of the network itself but also of its components. As a future research line, further tests with different generation and load levels can be identified.

Author Contributions: Conceptualization, J.B. and O.G.; methodology, J.B., O.G., C.R. and C.V.; formal analysis, J.B.; investigation, J.B., E.B. and O.G.; writing—original draft preparation, J.B.; writing—review and editing, J.B., E.B. and C.R.; supervision, J.B.; project administration, J.B. All authors have read and agreed to the published version of the manuscript.

Funding: This research received no external funding.

Institutional Review Board Statement: Not applicable.

Informed Consent Statement: Not applicable.

Data Availability Statement: Not applicable.

Acknowledgments: We thank the support of the laboratory of the Applied Mathematics Department of the University of the Basque Country, UPV/EHU, for its technical support.

Conflicts of Interest: The authors declare no conflict of interest. Authors had the role in the design of the study; in the collection, analyses, or interpretation of data; in the writing of the manuscript, and in the decision to publish the results.

References

1. Lasseter, R.H. Microgrids. In Proceedings of the 2002 IEEE Power Engineering Society Winter Meeting (Conference Proceedings (Cat. No.02CH37309)), New York, NY, USA, 27–31 January 2002; pp. 305–308.
2. Piagi, P.; Lasseter, R.H. Autonomous Control of Microgrids. In Proceedings of the 2006 IEEE Power Engineering Society General Meeting, Montreal, QC, Canada, 18–22 June 2006; p. 8. [CrossRef]
3. Lopes, J.; Moreira, C.; Madureira, A. Defining Control Strategies for Microgrids Islanded Operation. *IEEE Trans. Power Syst.* **2006**, *21*, 916–924. [CrossRef]
4. Baghaee, H.R. Real-Time Verification of New Controller to Improve Small/Large-Signal Stability and Fault Ride-Through Capability of Multi-DER Microgrids. *IET Gener. Transm. Distrib.* **2016**, *10*, 3068–3084. [CrossRef]
5. Hirsch, A.; Parag, Y.; Guerrero, J. Microgrids: A Review of Technologies Key Drivers and Outstanding Issues. *Renew. Sustain. Energy Rev.* **2018**, *90*, 402–411. [CrossRef]
6. Farrokhabadi, M.; Cañizares, C.A.; Simpson-Porco, J.W.; Nasr, E.; Fan, L. Microgrid Stability Definitions, Analysis, and Examples. *IEEE Trans. Power Syst.* **2019**, *35*, 13–29. [CrossRef]
7. Ullah, S.; Khan, L.; Jamil, M.; Jafar, M.; Mumtaz, S.; Ahmad, S. A Finite-Time Robust Distributed Cooperative Secondary Control Protocol for Droop-Based Islanded AC Microgrids. *Energies* **2021**, *14*, 2936. [CrossRef]
8. Moazeni, F.; Khazaei, J. Optimal Operation of Water-Energy Microgrids; A Mixed Integer Linear Programming Formulation. *J. Clean. Prod.* **2020**, *275*, 122776. [CrossRef]
9. Bidram, A.; Lewis, F.L.; Davoudi, A. Distributed Control Systems for Small-Scale Power Networks: Using Multiagent Cooperative Control Theory. *IEEE Control Syst. Mag.* **2014**, *34*, 56–77. [CrossRef]
10. Majumder, R. A Hybrid Microgrid with DC Connection at Back to Back Converters. *IEEE Trans. Smart Grid* **2014**, *5*, 251–259. [CrossRef]
11. Tabar, V.S.; Ghassemzadeh, S.; Tohidi, S. Energy Management in Hybrid Microgrid with Considering Multiple Power Market and Real Time Demand Response. *Energy* **2019**, *174*, 10–23. [CrossRef]
12. Minchala-Avila, L.I.; Garza-Castañón, L.E.; Vargas-Martín, A.; Zhang, Y. A Review of Optimal Control Techniques Applied to the Energy Management and Control of Microgrids. *Procedia Comput. Sci.* **2015**, *52*, 780–787. [CrossRef]
13. Gómez Sánchez, M.; Macia, Y.M.; Fernández Gil, A.; Castro, C.; Nuñez González, S.M.; Pedrera Yanes, J. A Mathematical Model for the Optimization of Renewable Energy Systems. *Mathematics* **2021**, *9*, 39. [CrossRef]
14. Olivares, D.E.; Mehrizi-Sani, A.; Etemadi, A.H.; Canizares, C.A.; Iravani, R.; Kazerani, M.; Hajimiragha, A.H.; Gomis-Bellmunt, O.; Saeedifard, A.; Palma-Behnke, R.; et al. Trends in Microgrid Control. *IEEE Trans. Smart Grid* **2014**, *5*, 1905–1919. [CrossRef]
15. Nguyen, T.-L.; Guillo-Sansano, E.; Syed, M.H.; Nguyen, V.-H.; Blair, S.M.; Reguera, L.; Tran, Q.-T.; Caire, R.; Burt, G.M.; Gavriluta, C.; et al. Multi-Agent System with Plug and Play Feature for Distributed Secondary Control in Microgrid—Controller and Power Hardware-in-the-Loop Implementation. *Energies* **2018**, *11*, 3253. [CrossRef]
16. Bazmohammadi, N.; Anvari-Moghaddam, A.; Tahsiri, A.; Madary, A.; Vasquez, J.C.; Guerrero, J.M. Stochastic Predictive Energy Management of Multi-Microgrid Systems. *Appl. Sci.* **2020**, *10*, 4833. [CrossRef]
17. Díaz, N.L.; Vasquez, J.C.; Guerrero, J.M. A Communication-Less Distributed Control Architecture for Islanded Microgrids with Renewable Generation and Storage. *IEEE Trans. Power Electron.* **2018**, *33*, 1922–1939. [CrossRef]
18. Lin, P.; Jin, C.; Xiao, J.; Li, X.; Shi, D.; Tang, Y.; Wang, P. A Distributed Control Architecture for Global System Economic Operation in Autonomous Hybrid AC/DC Microgrids. *IEEE Trans. Smart Grid* **2019**, *10*, 2603–2617. [CrossRef]
19. Llaria, A.; Terrasson, G.; Curea, O.; Jiménez, J. Application of Wireless Sensor and Actuator Networks to Achieve Intelligent Microgrids: A Promising Approach towards a Global Smart Grid Deployment. *Appl. Sci.* **2016**, *6*, 61. [CrossRef]
20. Chen, C.; Duan, S.; Cai, T.; Liu, B.; Hu, G. Smart energy management system for optimal microgrid economic operation. *IET Renew. Power Gener.* **2011**, *5*, 258–267. [CrossRef]
21. Ahmad, A.; Khan, A.; Javaid, N.; Hussain, H.M.; Abdul, W.; Almogren, A.; Alamri, A.; Azim Niaz, I. An Optimized Home Energy Management System with Integrated Renewable Energy and Storage Resources. *Energies* **2017**, *10*, 549. [CrossRef]
22. Bergen, A.R.; Vittal, V. *Power Systems Analysis*, 2nd ed.; Pearson: Upper Saddle River, NJ, USA, 2000.
23. Nagrath, I.; Kothari, D. *Modern Power System Analysis*; McGraw-Hill: New Delhi, India, 1982.
24. Akbari, T.; Bina, M.T. Linear Approximated Formulation of AC Optimal Power Flow Using Binary Discretisation. *IET Gener. Transm. Distrib.* **2016**, *10*, 1117–1123. [CrossRef]
25. Zhang, H.; Vittal, V.; Heydt, G.; Quintero, J. A Relaxed AC Optimal Power Flow Model Based on a Taylor Series. In Proceedings of the 2013 IEEE Innovative Smart Grid Technologies-Asia (ISGT Asia), Bangalore, India, 10–13 November 2013. [CrossRef]
26. Yang, Z.; Zhong, H.; Xia, Q.; Kang, C. A Novel Network Model for Optimal Power Flow with Reactive Power and Network Losses. *Electr. Power Syst. Res.* **2017**, *144*, 63–71. [CrossRef]
27. Akbari, T.; Bina, M.T. A Linearized Formulation of AC Multi-Year Transmission Expansion Planning: A Mixed-Integer Linear Programming Approach. *Electr. Power Syst. Res.* **2014**, *114*, 93–100. [CrossRef]
28. Attarha, A.; Amjady, N.; Conejo, A.J. Adaptive Robust AC Optimal Power Flow Considering Load and Wind Power Uncertainties. *Int. J. Electr. Power Energy Syst.* **2018**, *96*, 132–142. [CrossRef]
29. Yang, J.; Zhang, N.; Kang, C.; Xia, Q. A State-Independent Linear Power Flow Model with Accurate Estimation of Voltage Magnitude. *Trans. Power Syst.* **2017**, *32*, 5. [CrossRef]

30. Koster, A.M.C.A.; Lemkens, S. Designing AC Power Grids Using Integer Linear Programming. In *INOC 2011, Lecture Notes in Computer Science*; Pahl, J., Reiners, T., Voß, S., Eds.; Springer: Berlin/Heidelberg, Germany, 2011; Volume 6701, pp. 478–483. [CrossRef]

31. Taylor, J.; Hover, F. Linear Relaxations for Transmission System Planning. *IEEE Trans. Power Syst.* **2011**, *26*, 4, 2533–2538. [CrossRef]

32. Yang, Z.; Zhong, H.; Bose, A.; Zheng, T.; Xia, Q.; Kang, C. A Linearized OPF Model with Reactive Power and Voltage Magnitude: A Pathway to Improve the MW-Only DC OPF. *IEEE Trans. Power Syst.* **2018**, *33*, 2, 1734–1745. [CrossRef]

33. Morvaj, B.; Evins, R.; Carmeliet, J. Optimization Framework for Distributed Energy Systems with Integrated Electrical Grid Constraints. *Appl. Energy* **2016**, *171*, 296–313. [CrossRef]

34. Kothari, D.P. Power System Optimization. In Proceedings of the 2012 2nd National Conference on Computational Intelligence and Signal Processing (CISP), Guwahati, India, 2–3 March 2012; pp. 18–21. [CrossRef]

35. Patel, N.; Porwal, D.; Bhoi, A.K.; Kothari, D.P.; Kalam, A. An Overview on Structural Advancements in Conventional Power System with Renewable Energy Integration and Role of Smart Grids in Future Power Corridors. In *Green Energy and Technology*; Bhoi, A., Sherpa, K., Kalam, A., Chae, G.S., Eds.; Springer: Singapore, 2020; pp. 1–15.

36. Barsali, S. *Benchmark Systems for Network Integration of Renewable and Distributed Energy Resources*; Technical Brochure, TF C6.04.02; CIGRE: Paris, France, 2014.

37. Rocha, R.C.C.; Berredo, R.C.; Bernis, R.A.O.; Gomes, E.M.; Nishimura, F.; Cicarelli, L.D.; Soares, M.R. New Technologies, Standards, and Maintenance Methods in Spacer Cable Systems. *IEEE Trans. Power Deliv.* **2002**, *17*, 2, 562–568. [CrossRef]

38. Van den Brom, H.E.; van Leeuwen, R.; Rietveld, G.; Houtzager, E. Voltage Dependence of the Reference System in Medium- and High-Voltage Current Transformer Calibrations. *IEEE Trans. Instrum. Meas.* **2021**, *70*, 1502908. [CrossRef]

39. Pantic, L.S.; Pavlović, T.M.; Milosavljević, D.D.; Radonjic, I.S.; Radovic, M.K.; Sazhko, G. The Assessment of Different Models to Predict Solar Module Temperature, Output Power and Efficiency for Nis, Serbia. *Energy* **2016**, *109*, 38–48. [CrossRef]

40. Sun, V.; Asanakham, A.; Deethayat, T.; Kiatsiriroat, T. A New Method for Evaluating Nominal Operating Cell Temperature (NOCT) of Unglazed Photovoltaic Thermal Module. *Energy Rep.* **2020**, *6*, 1029–1042. [CrossRef]

41. DFIG 2.1 MW—114. 2021. Available online: https://www.siemensgamesa.com/en-int/products-and-services/onshore/wind-turbine-sg-2-1-114 (accessed on 29 September 2021).

42. Bilbao, J.; Torres, E.; Saenz, J. Load Curve Estimation by Means of Prediction Intervals. In Proceedings of the 2000 10th Mediterranean Electrotechnical Conference. Information Technology and Electrotechnology for the Mediterranean Countries. Proceedings. MeleCon 2000 (Cat. No.00CH37099), Lemesos, Cyprus, 29–31 May 2000; Volume 3, pp. 970–972. [CrossRef]

43. Grandjean, A.; Adnot, J.; Binet, G. A Review and an Analysis of the Residential Electric Load Curve Models. *Renew. Sustain. Energy Rev.* **2012**, *16*, 9, 6539–6565. [CrossRef]

44. Ranaboldo, M.; Ferrer-Martí, L.; García-Villoria, A.; Pastor Moreno, R. Heuristic Indicators for the Design of Community Off-Grid Electrification Systems Based on Multiple Renewable Energies. *Energy* **2013**, *50*, 501–512. [CrossRef]

45. Borghei, M.; Ghassemi, M. Optimal Planning of Microgrids for Resilient Distribution Networks. *Int. J. Electr. Power Energy Syst.* **2021**, *128*, 106682. [CrossRef]

46. Devarapalli, R.; Sinha, N.K.; Rao, B.V.; Knypinski, Ł.; Lakshmi, N.J.N.; García Márquez, F.P. Allocation of Real Power Generation Based on Computing over All Generation Cost: An Approach of Salp Swarm Algorithm. *Arch. Electr. Eng.* **2021**, *70*, 2, 337–349. [CrossRef]

47. Muhammad, M.; Mokhlis, H.; Naidu, K.; Amin, A.; Franco, F.; Othman, M. Distribution Network Planning Enhancement via Network Reconfiguration and DG Integration Using Dataset Approach and Water Cycle Algorithm. *J. Mod. Power Syst. Clean Energy* **2020**, *8*, 86–93. [CrossRef]

48. Helmi, A.; Carli, R.; Dotoli, M.; Ramadan, H. Efficient and Sustainable Reconfiguration of Distribution Networks via Metaheuristic Optimization. *IEEE Trans. Autom. Sci. Eng.* **2021**, *19*, 82–98. [CrossRef]

49. Orosz, T.; Rassõlkin, A.; Kallaste, A.; Arsénio, P.; Pánek, D.; Kaska, J.; Karban, P. Robust Design Optimization and Emerging Technologies for Electrical Machines: Challenges and Open Problems. *Appl. Sci.* **2020**, *10*, 6653. [CrossRef]

50. Yang, X.S. Nature-Inspired Optimization Algorithms: Challenges and Open Problems. *J. Comput. Sci.* **2020**, *46*, 101104. [CrossRef]

51. Chen, S.; Peng, G.-H.; He, X.-S.; Yang, X.-S. Global Convergence Analysis of the Bat Algorithm Using a Markovian Framework and Dynamic System Theory. *Expert Syst. Appl.* **2018**, *114*, 173–182. [CrossRef]

52. Wolpert, D.H.; Macready, W.G. No Free Lunch Theorems for Optimization. *IEEE Trans. Evolut. Comput.* **1997**, *1*, 67–82. [CrossRef]

53. Capitanescu, F. Critical Review of Recent Advances and Further Developments Needed in AC Optimal Power Flow. *Electr. Power Syst. Res.* **2016**, *136*, 57–68. [CrossRef]

54. Bai, X.; Wei, H.; Fujisawa, K.; Wang, Y. Semidefinite Programming for Optimal Power Flow Problems. *Int. J. Electr. Power Energy Syst.* **2008**, *30*, 383–392. [CrossRef]

55. Yuan, Z.; Hesamzadeh, M.R. Second-Order Cone AC Optimal Power Flow: Convex Relaxations and Feasible Solutions. *J. Mod. Power Syst. Clean Energy* **2019**, *7*, 268–280. [CrossRef]

56. Garces, A. A Quadratic Approximation for the Optimal Power Flow in Power Distribution Systems. *Electr. Power Syst. Res.* **2016**, *130*, 222–229. [CrossRef]

57. Hosseini, S.M.; Carli, R.; Dotoli, M. Robust Day-Ahead Energy Scheduling of a Smart Residential User under Uncertainty. In Proceedings of the 18th European Control Conference (ECC), Naples, Italy, 25–28 June 2019; pp. 935–940. [CrossRef]

58. Barbato, A.; Capone, A. Optimization Models and Methods for Demand-Side Management of Residential Users: A Survey. *Energies* **2014**, *7*, 5787–5824. [CrossRef]
59. Ipsakis, D.; Voutetakis, S.; Seferlis, P.; Stergiopoulos, F.; Elmasides, C. Power Management Strategies for a Stand-Alone Power System Using Renewable Energy Sources and Hydrogen Storage. *Int. J. Hydrogen Energy* **2009**, *34*, 16, 7081–7095. [CrossRef]
60. Singh, S.; Zindani, D.; Kumar Roy, A.; Kumar, K. Application of Renewable Energy System with Fuzzy Logic. In *Advanced Fuzzy Logic Approaches in Engineering Science*; Ram, M., Ed.; IGI-Global: Hershey, PA, USA, 2019. [CrossRef]
61. Tao, S.; Si-jia, Y.; Guang-yi, C.; Xi-jian, Z. Modelling and Control PEMFC Using Fuzzy Neural Networks. *J. Zheijang Univ.-Sci. A* **2005**, *6*, 1084–1089. [CrossRef]
62. Parisio, A.; Rikos, E.; Glielmo, L. A Model Predictive Control Approach to Microgrid Operation Optimization. *IEEE Trans. Control Syst. Technol.* **2014**, *22*, 5, 1813–1827. [CrossRef]
63. Nelson, J.R.; Johnson, N.G. Model Predictive Control of Microgrids for Real-Time Ancillary Service Market Participation. *Appl. Energy* **2020**, *269*, 114963. [CrossRef]
64. Edwards, H. *How Machines Learn*; Koru Ventures LLC: Bend, OR, USA, 2016.
65. Krohn, J. *Deep Learning Illustrated*; Pearson Education: Boston, MA, USA, 2020.
66. Kourou, K.; Exarchos, T.P.; Exarchos, K.P.; Karamouzis, M.V.; Fotiadis, D.I. Machine Learning Applications in Cancer Prognosis and Prediction. *Comput. Struct. Biotechnol. J.* **2015**, *13*, 8–17. [CrossRef] [PubMed]
67. Malhotra, R. A Systematic Review of Machine Learning Techniques for Software Fault Prediction. *Appl. Soft Comput.* **2015**, *27*, 504–518. [CrossRef]
68. Mosavi, A.; Ozturk, P.; Chau, K.-W. Flood Prediction Using Machine Learning Models: Literature Review. *Water* **2018**, *10*, 1536. [CrossRef]
69. Johnson, A.E.W.; Ghassemi, M.M.; Nemati, S.; Niehaus, K.E.; Clifton, D.A.; Clifford, G.D. Machine Learning and Decision Support in Critical Care. *Proc. IEEE* **2016**, *104*, 444–466. [CrossRef]
70. Mitchell, T. *Machine Learning*; McGraw Hill: New York, NY, USA, 1997; p. 2.
71. Batista, G.E.A.P.A.; Prati, R.C.; Monard, M.C. A Study of the Behavior of Several Methods for Balancing Machine Learning Training Data. *SIGKDD Explor. Newsl.* **2004**, *6*, 1, 20–29. [CrossRef]
72. Bilbao, J.; Bilbao, I.; Feniser, C. Generalized Delta Rule with Entropy Error Function. *Acta Tech. Napoc. Ser. Appl. Math. Mech. Eng.* **2017**, *60*, 165–170.
73. Belkin, M.; Hsu, D.; Ma, S.; Mandal, S. Reconciling Modern Machine-Learning Practice and the Classical Bias—Variance Trade-off. *Proc. Natl. Acad. Sci. USA* **2019**, *116*, 15849–15854. [CrossRef]
74. Bilbao, I.; Bilbao, J.; Feniser, C. Adopting Some Good Practices to Avoid Overfitting in the Use of Machine Learning. *WSEAS Trans. Math.* **2018**, *17*, 274–279.
75. Sharma, A.; Bhuriya, D.; Singh, U. Survey of Stock Market Prediction Using Machine Learning Approach. In Proceedings of the 2017 International conference of Electronics, Communication and Aerospace Technology (ICECA), Coimbatore, India, 20–22 April 2017; pp. 506–509. [CrossRef]
76. Bilbao, J.; Bravo, E.; Garcia, O.; Varela, C.; Rebollar, C. Particular Case of Big Data for Wind Power Forecasting: Random Forest. *Int. J. Tech. Phys. Probl. Eng.* **2020**, *12*, 25–30.
77. Breiman, L. Bagging Predictors. *Mach. Learn.* **1996**, *24*, 123–140. [CrossRef]
78. Breiman, L. Random Forests. *Mach. Learn.* **2001**, *45*, 5–32. [CrossRef]
79. Rostami, M.-A.; Kavousi-Fard, A.; Niknam, T. Expected Cost Minimization of Smart Grids with Plug-In Hybrid Electric Vehicles Using Optimal Distribution Feeder Reconfiguration. *IEEE Trans. Ind. Inform.* **2015**, *11*, 388–397. [CrossRef]
80. Lan, T.; Jermsittiparsert, K.; Alrashood, S.T.; Rezaei, M.; Al-Ghussain, L.; Mohamed, M.A. An Advanced Machine Learning Based Energy Management of Renewable Microgrids Considering Hybrid Electric Vehicles' Charging Demand. *Energies* **2021**, *14*, 569. [CrossRef]
81. Kavousi-Fard, A.; Khodaei, A. Efficient Integration of Plug-in Electric Vehicles via Reconfigurable Microgrids. *Energy* **2016**, *111*, 653–663. [CrossRef]

mathematics

Article

Optimal Management of Energy Consumption in an Autonomous Power System Considering Alternative Energy Sources

Vadim Manusov [1], Svetlana Beryozkina [2,*], Muso Nazarov [1], Murodbek Safaraliev [3], Inga Zicmane [4], Pavel Matrenin [1] and Anvari Ghulomzoda [5]

1 Department of Industrial Power Supply Systems, Novosibirsk State Technical University, 630073 Novosibirsk, Russia; manusov@corp.nstu.ru (V.M.); musso-6556@mail.ru (M.N.); matrenin.2012@corp.nstu.ru (P.M.)
2 College of Engineering and Technology, American University of the Middle East, Kuwait
3 Department of Automated Electrical Systems, Ural Federal University, 620002 Yekaterinburg, Russia; murodbek_03@mail.ru
4 Faculty of Electrical and Environmental Engineering, Riga Technical University, LV-1048 Riga, Latvia; inga.zicmane@rtu.lv
5 Department of Automated Electric Power Systems, Novosibirsk State Technical University, 630073 Novosibirsk, Russia; anvar_4301@mail.ru
* Correspondence: svetlana.berjozkina@aum.edu.kw

Citation: Manusov, V.; Beryozkina, S.; Nazarov, M.; Safaraliev, M.; Zicmane, I.; Matrenin, P.; Ghulomzoda, A. Optimal Management of Energy Consumption in an Autonomous Power System Considering Alternative Energy Sources. *Mathematics* 2022, 10, 525. https://doi.org/10.3390/math10030525

Academic Editors: Nicu Bizon, Francisco Chiclana and Mario Versaci

Received: 4 December 2021
Accepted: 3 February 2022
Published: 8 February 2022

Publisher's Note: MDPI stays neutral with regard to jurisdictional claims in published maps and institutional affiliations.

Abstract: This work aims to analyze and manage the optimal power consumption of the autonomous power system within the Pamir region of Republic of Tajikistan, based on renewable energy sources. The task is solved through linear programming methods, production rules and mathematical modeling of power consumption modes by generating consumers. It is assumed that power consumers in the considered region have an opportunity to independently cover energy shortage by installing additional generating energy sources. The objective function is to minimize the financial expenses for own power consumption, and to maximize them from both the export and redistribution of power flows. In this study, the optimal ratio of power generation by alternative sources from daily power consumption for winter was established to be hydroelectric power plants (94.8%), wind power plant (3.8%), solar photovoltaic power plant (0.5%) and energy storage (0.8%); while it is not required in summer due to the ability to ensure the balance of energy by hydroelectric power plants. As a result, each generating consumer can independently minimize their power consumption and maximize profit from the energy exchange with other consumers, depending on the selected energy sources, thus becoming a good example of carbon-free energy usage at the micro- and mini-grid level.

Keywords: autonomous power system; generating power consumer; hydroelectric power plant; optimal power consumption; wind power plant; solar photovoltaic power plant; energy storage

1. Introduction

In the last decade, a depletion of the fossil fuel reserves and growing demand for power energy have been the most important issues in a global context. Modern energy systems are going through a period of serious changes associated with the transition from centralized, top-down structures with a heavy dependence on fossil fuels to distributed, decentralized, environmentally friendly energy solutions in accordance with the Paris Climate Agreement 2016, which was aimed at combating climate change and depletion of natural resources as well as energy security at both national and continental scale. Therefore, the use of renewable energy sources (RES) as an alternative to traditional energy sources is becoming a prioritized direction for energy policy in most countries around the world [1,2].

By making "green" decisions about power generation, researchers are doing everything possible to obtain the most reliable and efficient way of generating energy by using RES. Therefore, increased attention to technologies used for the RES on such a large scale has led to a constant cost reduction of distributed technologies for the production, storage and conversion of renewable energy [2,3]. The use of RES is especially attractive and suitable for supplying electricity to remote areas, which operate in an autonomous mode from both technical and economic points of view. For instance, the introduction of renewable energy technologies with their rational use could help in supplying power energy to areas with a weak fuel base and poor transport conditions, and solve the problem of efficient use of consumed resources and involvement in the energy balance of regions of previously unused energy sources and resources, to improve the ecological situation in the places of heat and power energy production. As result, these activities and measures will contribute to the accelerated economic development of these regions and improve the social and living conditions of their population, which corresponds to the UN's Sustainable Development Goal 7, on ensuring universal sustainable access to reliable energy sources for settlements remote from national networks by creating autonomous, low-voltage, low-inertia local networks based on renewable energy sources [2,4,5].

The mass adoption of microgrids reached 1.4 GW in 2015 and is expected to increase to 8.8 GW by 2024 [6]. Such considerable interest is due to their potential advantages in terms of facilitating the integration of the RES into both the existing and new energy systems. For example, the microgrid multiple energy carrier refers to an interconnected energy system that provides a platform for connecting various energy vectors from various sources to meet various energy needs in remote regions, and is applicable in various sectors, including commercial, industrial and military, considering the set goals, load types, geographical and climatic conditions. The model extends the concept of original microgrids focused on electricity demand, with the desire to use the interaction between different energy vectors to virtually meet all energy needs of communities, while increasing the sustainability, reliability, efficiency and availability of the RES [1].

Electrification of such isolated regions can be provided either by one type of renewable energy or by hybrid renewable energy, where solar photovoltaic coupled with wind energy sources is the most used combination today because of their complementarity. However, it is well known that the generation scheme of technologies operating on intermittent energy sources, such as solar and wind, can change greatly, quickly and unpredictably, unlike traditional technologies, for which generation can be adjusted to produce a certain amount of energy at a certain time based on fluctuations in electricity demand. As a result, network stability problems limit the possibilities of using these RES on a scale that allows them to reach their full potential in the absence of backup power at times when wind or solar photovoltaic (PV) energy suddenly becomes unavailable, and/or without providing energy management services that allow network operators to use cheaper energy generated outside the peak of consumption to meet peak electricity demand [1,7,8].

Hydropower is a type of clean renewable energy with instant power adjustment and flexibility in storage and discharge. The complementary work of hydro–wind–photovoltaic hybrid power plants ($HPP/WPP/PV$) is becoming increasingly relevant for modern energy systems. Thus, the regions with rich hydropower resources and conditions acceptable for energy generation by wind generators and photovoltaic cells have significant potential for the development of modern energy. However, there is a problem; in regions where energy resources are usually far from load centers, barriers requiring long-distance transmission of electricity remain, which undermines the economic feasibility of using such energy sources, and strong fluctuations associated with the flow and generation of PV and wind energy lead to uncertainty in additional operations and, as a result, due to the inherently unstable nature of generation of the RES, a control system of a sufficiently high level of its execution is crucial [3,9–13].

In this regard, the possible solutions are as follows:

- Energy storage (ES), for example, energy storage in HPP and compressed air, chemical batteries and active load management;
- Geographical diversification of installation sites;
- Combination of energy sources;

- Application of high-precision methods of weather and load forecasting [12].

Therefore, the optimal design of a hybrid energy system based on the RES is a complex task, which includes a feasibility study, model-based design, process modeling (simulation), as well as the integration of several hybrid RES, a hybrid energy storage system and a hybrid controller for automation to ensure the reliability of a power supply [14].

The purpose of this study is to develop a system and manage optimal power consumption in the autonomous power system within the Pamir region of the Republic of Tajikistan, based on alternative energy sources.

The work is structured as follows: Section 1 presents the introduction and relevance of the topic; Section 2 describes a problem analysis, problem statement and review of the scientific literature; Section 3 contains information about the *ES* serving as the object of research; Section 4 introduces the proposed method and algorithm of optimization of energy consumption modes; and Section 5 provides an analysis of obtained results and their discussion; Section 6 includes conclusions.

2. Problem Statement and a Brief Overview of the Proposed Approaches

2.1. Optimization of Electricity Production Based on Renewable Energy Sources

The issue of the optimal management of power energy production/consumption in modern distribution systems is becoming especially relevant in the era of Smart grid. In addition, the operation of autonomous power systems with a high share of the RES can create significant problems with system balancing. Decentralized integration of the RES based on the Smart grid (intelligent energy system) is presented as the most promising way to increase the stability and reliability of the latter, in a cost-effective way [15,16], especially since it is autonomous, intelligent and integrated renewable energy systems that underlie the "energy for all" initiatives aimed to provide modern energy services to coastal, island and mountainous regions and also in a broader sense, to rural/peripheral regions [1].

Hybrid renewable energy systems (HRES) can vary significantly in the type and number of generation sources, consumers, installed capacity, operating conditions and many other factors. Currently, conditions and opportunities have been created for power consumers to independently choose generation sources, where renewable energy sources in combination with energy storage devices can serve for autonomous and local power energy systems. Such systems are characterized by absence of a centralized energy source and high uncertainty of renewable and alternative energy sources, which implies an independent solution to the optimization problem of the most profitable combination of renewable and alternative energy sources to minimize the material, technical and financial costs of each generating electric consumer [14].

Several works are devoted to the tasks of optimizing the operation of generating power plants of power supply systems using the RES, as well as a technical and economic assessment of energy supply to isolated consumers, in which either an economic justification is given for the efficiency of connecting to centralized power supply, or the possibility of using local small energy sources is considered. Zones of expediency of centralized and decentralized power supply are given depending on electricity tariffs and the cost of diesel fuel [17–19]. This approach makes it possible at the regional level to identify those areas for which it is necessary to conduct a detailed assessment of the use of a technologically possible and economically feasible energy supply option. At the local level, specific energy supply options for each consumer are determined, the order of input of energy sources, the composition of equipment and the necessary investments. The results of research for various regions allow us to form proposals for promising areas of scientific and technological progress in the field of small-scale energy, to assess the appropriate scale of implementation and the equipment market of economically attractive projects.

2.2. Mathematical Methods of the Optimization

The authors of a number of works, using mathematical modeling methods, propose the creation of a technical and economic model to analyze the technical and economic feasibility of multi-energy complementarity. In order to solve the problem of continuous power supply at photovoltaic power plants, the authors in [11,20–22] use the theory of complementation of hydro and solar energy, which allows to solve a problem of intermittent and unstable solar energy generation. To reduce risks and increase reliability, such type of power systems is additionally supplied with the energy storage devices [23,24]. Due to the instability of wind resources, the relevant scientists have combined hydro-accumulating power plants with WPPs, striving for an optimal mode of complementary work, and maximizing profits. In [25], modeling of economic indicators of further integration of photovoltaic systems is described considering technical limitations. The economic feasibility of a large-scale hybrid hydroelectric power plant, including the transmission of energy over long distances, is presented in [10].

It should be noted that when optimizing the modes of power energy systems, the Lagrange multiplier method and gradient methods have become the most widespread. The dynamic programming method and some others are also used. Currently, alternative algorithms for optimizing modes are being developed by using fuzzy logic methods and evolutionary algorithms. For example, optimization of the modes of joint operation of solar and thermal power plants is presented in [26], where Lagrange multipliers were used to derive the optimization equation.

For isolated systems, models should be created to optimize the distributed generation management system, including the RES. Such models allow to optimize the network operation according to various parameters. Various methods for optimizing the operating modes of power energy systems and networks are discussed in [27–29] whereby an optimal method for determining the size of renewable energy parks and *ES* in a hybrid power system is proposed. The genetic algorithm is used to find optimal solutions for both renewable farms and energy storage devices. It is noted that among the traditional tasks of optimizing the modes of electric power systems and power supply systems of industrial facilities, the selection of the best configurations of electric networks, the distribution of loads between power sources of both existing and projected power supply systems as well as the rationalization of the use of energy resources are highlighted. Examples of the application of existing optimization methods are analyzed.

In general, the optimization process can be divided into single-purpose and multi-purpose methods. For example, a method for optimizing the size of renewable energy parks is proposed together with an economic analysis for a need for charging from the grid up to 50,000 plug-in electric vehicles [30]. However, the process that limits the impact of fluctuations in the capacity of a renewable energy farm on the utility network is not analyzed. Another method, which is based on a cost–benefit analysis for the optimal size of an energy storage system in a microgrid, is proposed in [31], but the corresponding size of a renewable energy farm is not specified. In addition, the energy storage system included only the use of batteries that can fail in situations of high-power fluctuations. An optimal method for determining the size of a hybrid power system with a wind solar battery is proposed for both autonomous and network mode (without an ultracapacitor) in [32].

It should be noted that most of the initial research in this area is devoted to the analysis of the short-term operation of small-scale/autonomous power systems for those cases when the requirements for the quality of electricity production seem less important. The study conducted in [13] aimed to analyze a long-term optimization model for hybrid hydro/photovoltaic systems, considering the stability of output power and total electricity generation simultaneously by creating a multi-purpose optimization model for long-term operation of a hydroelectric power plant/photovoltaic system, which is then optimized using a modified version of the non-dominated sorting genetic algorithm. In turn, to improve the long-term additional operational characteristics of a large-scale hybrid *HPP*, methods of long-term stochastic optimization were developed in [11], which simultaneously consider the uncertainty of the flow and output power of a photovoltaic installation. A multi-purpose optimization model was created to maximize the total energy production and guaranteed speed. The model was then solved using stochastic dynamic programming to obtain operational solutions. This study focused on the long-term complementary operation of hybrid power plants with *HPP*, given the uncertainty in both the flow and capacity of solar photovoltaic energy.

3. Data and Materials

3.1. An Assessment of Energy Resources of the Pamir Region of Republic of Tajikistan

Pamir is one of the richest regions of Tajikistan in terms of hydropower energy reserves. Power energy reserves are concentrated on the territory of this region, which are estimated at 32.5 billion kWh. However, approximately 0.6% of this potential is currently being used. Low development indicates a weak level of economic development and considerable potential for region growth in the future. The open hydropower resources of small rivers and watercourses of the region are so large that when the level of their use reaches 20%, the Pamirs will turn into one of the richest mining regions of the country. The hydroelectric potential of the Pamirs represents the economic efficiency of its use and commercial benefits to justify the construction of small HPPs. The main factors delaying the use of energy resources are as follows: assessment of the impact of hydropower construction on the environment, the complex nature of the use of watercourse water resources and the energy market in Central Asia. It should be noted that small and even medium-sized Pamir rivers are either insufficiently or not studied at all in terms of energy potential [17,33,34].

The network of hydrometeorological observations in Tajikistan is not dense enough; therefore, the real potential of wind energy remains not fully explored.

For example, wind speed measurements were not carried out at the level of 30 m from the earth's surface, where the potential of wind energy can be 10–20% higher than at the level of 10 m. Despite this, the wind energy potential of Tajikistan reaches 25–150 billion kWh/year according to experts' opinion [18,19]. The authors state the fact that the meteorological network of Tajikistan is still underdeveloped since a need for meteorological data on the wind and sun has not been sufficiently demanded due to the non-use of renewable energy sources. However, the authors have currently managed to collect the necessary prehistory of statistical data on wind and solar insulation from meteorological stations allocated in the main sites of industrial and domestic power consumers, which, as can be assumed, can be refined.

The total solar radiation reaches 700–800 W/m^2 or 7500–8000 MJ/m, when there are clear skies. These parameters are much higher in mountainous areas, especially in the Eastern Pamirs, where the population has limited opportunity to use the hydropower resources [35,36].

3.2. General Characteristics of the Pamir Power System

The operation and control of the modes of power stations and networks, as well as the energy generation, transmission and distribution in Tajikistan is carried out by an Open Joint-Stock Holding Company (OJSHC) "Barki Tojik", except for Gorno-Badakhshan Autonomous Region (GBAR). The power supply system of the GBAR operates based on the renewable and alternative energy sources; it was transferred from the company "Barki Tojik" to the control of the private electric company "Pamir Energy" in 2002 for a period of 25 years under a concession agreement and the system began to function in isolated (autonomous) mode from the main power system.

Currently, "Pamir Energy" manages eleven HPPs, of which the larger ones are Pamir-1 and Khorog HPPs as well as nine small HPPs with a total installed capacity of 43.5 MW, which is clearly reflected in Table 1. Three HPPs from the above list (Pamir-1 *HPP*, Khorog *HPP* and Namangut *HPP*) work for the network (for a total load), and Pamir-1 *HPP* and Khorog *HPP* are in a cascade on the Gunt River, and Namangut *HPP* is located on another river, the Panj River. The rest of these 11 stations are operating offline in separate areas.

Table 1. Information about the SHPP located on the territory of GBAR.

Substation No.	*HPP* Name	Installed Capacity, kW	Number of Hydraulic Units	Design Pressure, m	Water Flow through the Unit, m³/s
1	*HPP* Pamir-1	28,000	4	79.6	10.1
2	*HPP* Khorog	9000	5	59	3.55
3	*HPP* Namangut	2500	2	36	3.5
4	*HPP* Vanch	1200	2	21.5	3.5
5	*HPP* Ak-Su	640	2	9	5
6	*HPP* Shugnan	832	2	10	5.5
7	*HPP* Savnob	80	1	72	0.1
8	*HPP* Siponj	160	2	130	0.31
9	*HPP* Andarbek	300	1	23	1.8
10	*HPP* Techarv	360	1	110	0.59
11	*HPP* Kalai-Humb	208	2	10.8	1.55

The examined power system uses the transmission lines with a voltage of 35/10/0.4 kV and total length of 2609 km. The Pamirs' small hydroelectric power plants (SHPP) are characterized by insufficient availability of water resources in winter months, thus the *HPP*, which are working under a given load schedule, use a natural water flow without their redistribution in the daily interval. As a result, the *HPP* cannot cover the maximum loads of the daily schedule during morning and evening peaks. In winter, when the maximum demand for electricity is observed in all regions of the Pamirs, the SHPPs that do not have large-capacity reservoirs provide almost minimal power [17,33,34].

Figure 1 shows the geographical location of the stations in the Pamir power system for illustrative representation of the territorial location of the generating nodes.

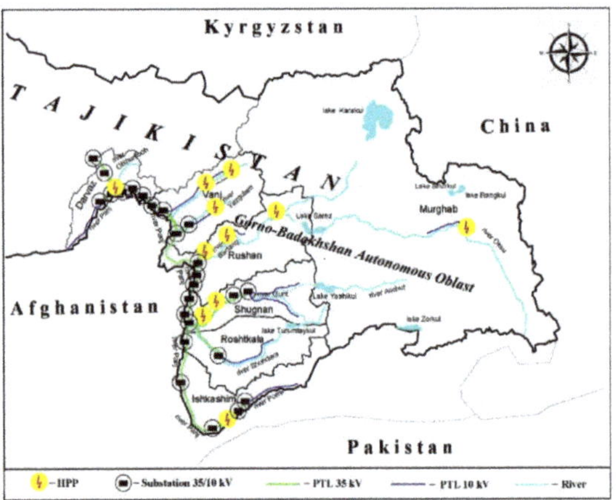

Figure 1. Geographical location of the Pamir power energy system.

Due to the efficient and economical RES, the autonomous power supply of the region can be provided by forming a hybrid power system of *HPP/WPP/PV/ES* system, in which, for an optimal combination of various renewable energy sources, methods of both technical and economic analysis are used. The development of an optimization model of the power consumption has been considered for an autonomous hybrid electric power system (EPS) based on the GBAR example (Figure 2).

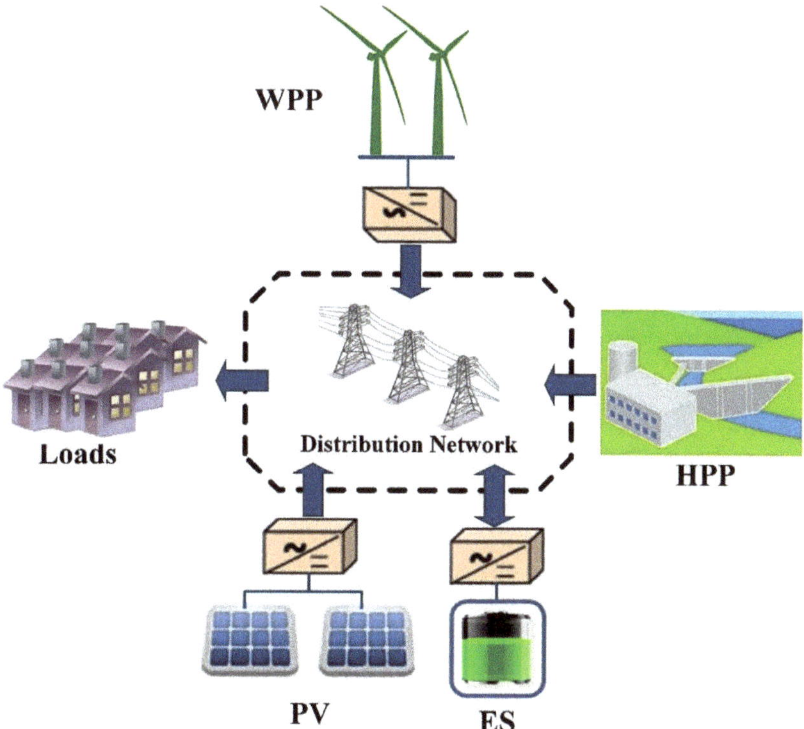

Figure 2. Overview of the proposed hybrid power system.

3.3. An Assessment of the Energy Balance of the Pamir Region

To assess the energy potential of the considered autonomous power energy system of the Pamirs, it is necessary to assess the possibility of using alternative energy sources in the daily interval. At the same time, the essential difference between the modes of this system is that it is energy-deficient in the winter and in an energy-surplus in the summer. In this regard, the most characteristic days have been chosen for both winter and summer periods. Two modes are selected characterizing the extreme points. During the summer, when there is a maximum energy of small rivers, the hydro resources are sufficient to fully cover the load, especially during the flood period. This also creates good conditions for the electricity export. During the winter, the electricity consumption for heating purposes increases, and the energy of water for activating the reservoirs can be used only by 30–40%. On these characteristic days, a power shortage is created during peak power consumption modes, exports are significantly limited, and energy storage is fully utilized. Ideally, it would be advisable to build a pumped storage station, which is a continuation of this work. To optimize power consumption modes, the statistical data of wind speed, solar insolation, power generation by *HPP* due to water resources and a daily load schedule were used for the selected characteristic days [37].

The power of the *WPP* depends on the wind speed, which varies greatly with time, weather conditions and terrain surface [16]. The dependence of the power and speed of the wind passing through a swept area of wind turbine is expressed as follows:

$$P = \frac{1}{2}\rho A V^3 C_p(\lambda), \tag{1}$$

where ρ is the air flow area (kg/m^3), depending on the temperature and air pressure; A is the surface area swept by blades (m^2); V-wind speed (m/s); C_p is the efficiency coefficient of the wind turbine; and λ-speed coefficient.

For the power system, the total installed capacity of the wind park of 10 MW was selected, which consists of 20 wind turbines with a capacity of 500 kW each. According to the datasheet, power generation begins with a wind speed of 3 m/s. When the speed reaches 12 m/s, the rated power is generated. In the range of 12–25 m/s, the rated power of wind turbines is maintained. A detailed location selection of wind turbines considering the terrain and wind speed could be carried out in accordance with the recommendations provided in [14].

Figure 3 shows wind speed and solar insolation data for a typical winter day in the examined region. The highest values of wind speed coincide approximately with the morning load peak leading to a wind energy use in the morning. It is obvious that the greatest power due to solar energy can be obtained from 8 to 18 h in a daily profile. This roughly corresponds to the duration of the electrical load during a working day. The total power of the solar panels was selected as 5 MW with an efficiency of 22.5%.

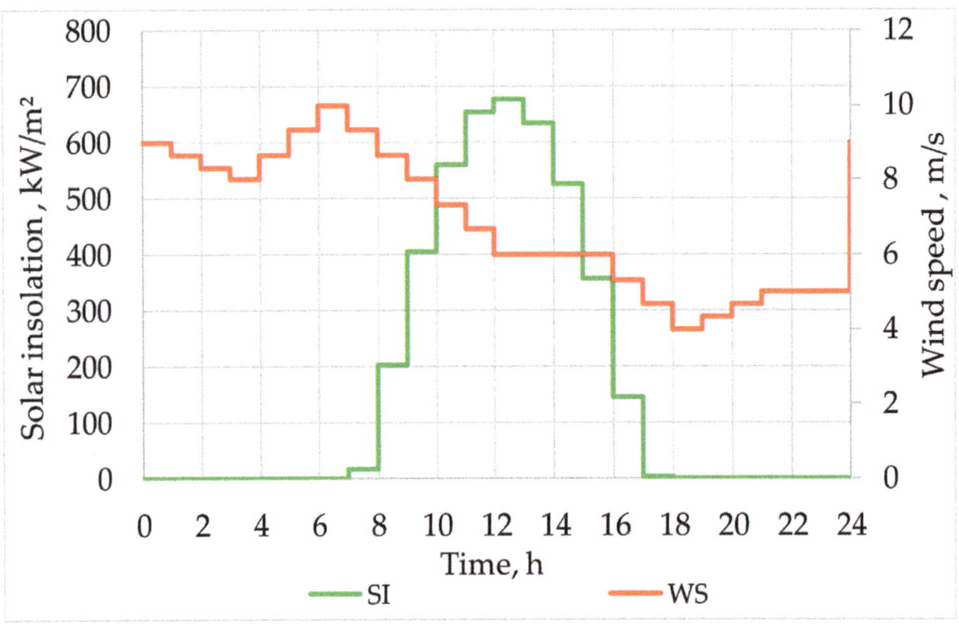

Figure 3. Wind speed and solar insolation data of a typical winter day.

For a typical winter day, the power generation by *HPP* at the expense of water resources, the daily load schedule, the power generated by the wind power plant, as well as the power generation by solar panels are selected as initial information and given in Figure 4. Additionally, the initial power of the energy storage at the beginning of the day is 2000 kW, which is optimally consumed or accumulated in accordance with the optimization algorithm in order to minimize the financial costs of electric consumers.

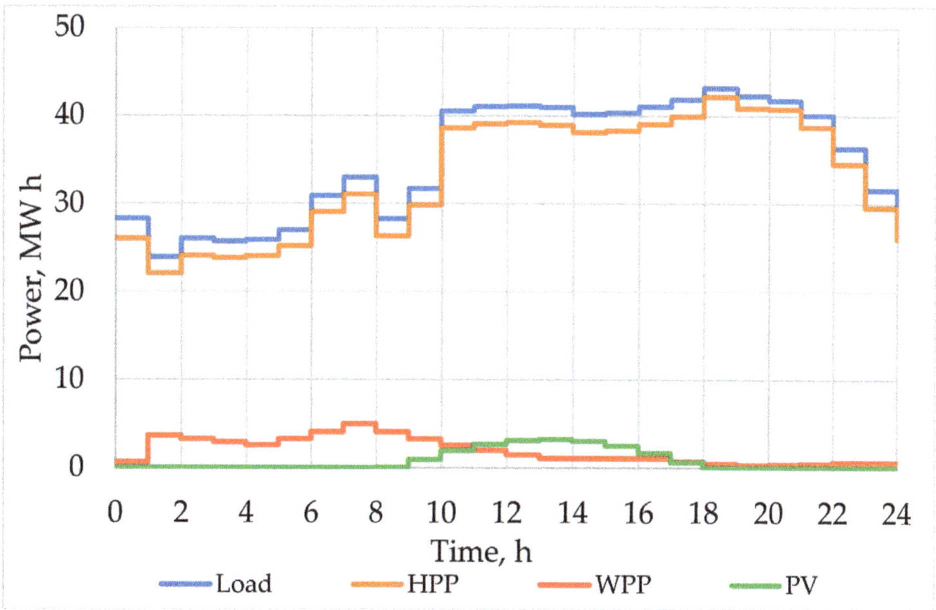

Figure 4. The initial information of the energy balance for a typical winter day.

4. Methodology

4.1. A Formation of a Mathematical Model of the Optimal Power Consumption

The main objective of this work is to offer an optimal connection of 11 potentials of large and small hydropower together with wind and photovoltaic systems in a form of a hybrid system, which will increase the reliability of the system and decrease the investment costs.

Consider an example of optimizing the generation structure of the GBAR power system, as an isolated power supply system, from the cost minimization point of view in terms of electricity generation. The power balance equation for a such autonomous system will have the following form:

$$P_{HPP} + P_{WPP} + P_{PV} \pm P_{ES} = P_{LOAD} + \Delta P, \tag{2}$$

where P_{HPP} is capacity of HPP; P_{WPP} is capacity of WPP; P_{PV} is capacity of PV; P_{ES} is capacity of energy storage (ES); P_{LOAD} is load power consumption; and ΔP is power energy losses during both the transmission and distribution.

The balance equation in integral form is given as follows:

$$\int_0^{24} P_{HPP}(t)dt + \int_0^{24} P_{WPP}(t)dt + \int_0^{24} P_{PV}(t) \pm \int_0^{24} P_{ES}(t) = \int_0^{24} P_{LOAD}(t) + \int_0^{24} \Delta P(t). \tag{3}$$

The task of minimizing the financial costs of an individual generating power consumer is solved based on the minimization of the objective function, the hourly measurements of power and energy of which presumably numerically coincide as follows:

$$\sum_{i=0}^{24} R_i \cdot P_{HPP} + \sum_{i=0}^{24} R_i \cdot P_{WPP} + \sum_{i=0}^{24} R_i \cdot P_{PV} \pm \sum_{i=0}^{24} R_i \cdot P_{ES} = \sum_{i=0}^{24} P_{i\,LOAD} + \sum_{i=0}^{24} \Delta P_i, \tag{4}$$

where R_i is hourly power energy consumption from this type of generation source; P_{HPP} is power consumption from *HPP* in the *i*-th hour; P_{WPP} is power consumption from *WPP* in the *i*-th hour; P_{PV} is power consumption from *PV* in the *i*-th hour; and P_{ES} is power consumption from *ES* in the *i*-th hour.

Due to the fact that it is advisable to minimize the financial costs of each individual consumer, Equation (4) must be written taking into account the individual cost of each alternative or renewable energy source:

$$\left(C_{HPP} \cdot \sum_{i=0}^{24} R_i \cdot P_{HPP} + C_{WPP} \cdot \sum_{i=0}^{24} R_i \cdot P_{WPP} + C_{PV} \cdot \sum_{i=0}^{24} R_i \cdot P_{PV} \pm C_{ES} \cdot \sum_{i=0}^{24} R_i \cdot P_{ES} \right) \rightarrow min, \tag{5}$$

where C_{HPP} = 0.02 EUR/kW·h, the cost of power energy generated at the *HPP*; C_{WPP} = 0.03 EUR/kW·h, the cost of power energy generated at the *WPP*; C_{PV} = 0.036 EUR/kW·h, the cost of power energy generated at the *PV*; and C_{ES} = 0.043 EUR/kW·h, the cost of power energy accumulated on *ES*.

In the case of an energy surplus, a similar following equation can be derived to maximize the income:

$$\left(k \cdot C_{HPP} \cdot \sum_{i=0}^{24} G_i \cdot W_{HPP} + k \cdot C_{WPP} \cdot \sum_{i=0}^{24} G_i \cdot W_{WPP} + k \cdot C_{PV} \cdot \sum_{i=0}^{24} G_i \cdot W_{PV} \pm k \cdot C_{ES} \cdot \sum_{i=0}^{24} G_i \cdot W_{ES} \right) \rightarrow max, \tag{6}$$

where G_i is an excess of the power energy in the *i*-th hour for each energy source; and k is the coefficient of profitability from the sale of the power energy.

Retrospective studies of this autonomous power system allows us to establish some specific values of possible power generation by energy sources, which, based on long-term observations, can be obtained as the following values for a working winter day, namely: $P_{MAX\,HPP}$ = 43.5 MW, $P_{MAX\,WPP}$ = 10 MW (rounded) and $P_{MAX\,PV}$ = 6 MW. Based on the power balance during the maximum hours of the daily load schedule, energy storage devices have been selected, taking into account the unpredictability of the generation of "green energy". For this reason, the above-given equations must be solved under constraints in the form of inequalities, namely:

$$0 \leq P_{HPP} \leq 43.5\,\text{MW}; \; 0 \leq P_{WPP} \leq 10\,\text{MW};$$

$$0 \leq P_{PV} \leq 6\,\text{MW}$$

Moreover, the possibility of transferring power energy to Afghanistan could be considered as an additional power consumer, which buys excess energy in case of GBAR autonomous power system surplus. An additional important circumstance should be noted, in particular, the balance of power and energy in the system under consideration can also be balanced by limiting the transmitted power to Afghanistan, but it cannot be a source of generation in the foreseeable future.

4.2. Solution Method and Production Rules

Software was developed in order to perform calculations. The program algorithm is based on the linear programming method by using the following conditions in the form of product rules. The "production rule" means the expression of some cause-and-effect relationships between events, phenomena or changes, which are expressed in the form of "IF, … THEN, …", where the initial condition (cause) is the antecedent, then the result (consequence) is the consequent. These rules are limited to the form of equalities and inequalities. For example, if the total capacity of an *HPP* fully covers the electricity demand of an autonomous system, then other sources are not involved. If the total capacity of the *HPP* is insufficient to cover the required load, then the next energy source is, as a priority, attracted in terms of the electricity cost; in our case, it is the *WPP*. The production rules for *HPP*, *WPP*, *PV* and *ES* were established as stated below.

1. Production rules for *HPP*:
IF $(P_{HPP} < P_{LOAD})$ THEN
$CP_{HPP} = P_{HPP}$ and $RP_{HPP} = 0$
OTHERWISE
$CP_{HPP} = P_{LOAD}$ and $RP_{HPP} = P_{HPP} - P_{LOAD}$
2. Production rules for *WPP*:
IF $(CP_{HPP} < P_{LOAD})$ THEN
IF $(P_{WPP} > 0)$ THEN
IF $(CP_{HPP} + P_{WPP} > P_{LOAD})$ THEN
$CP_{WPP} = P_{LOAD} - CP_{HPP}$ and $RP_{WPP} = P_{WPP} - CP_{WPP}$
OTHERWISE
$CP_{WPP} = P_{WPP}$ and $RP_{WPP} = 0$
OTHERWISE
$CP_{WPP} = 0$ and $RP_{WPP} = 0$
OTHERWISE
$CP_{WPP} = 0$ and $RP_{WPP} = P_{WPP}$
3. Production rules for *PV*:
IF $(CP_{HPP} + CP_{WPP} < P_{LOAD})$ THEN
IF $(P_{PV} > 0)$ THEN
IF $(CP_{HPP} + CP_{WPP} + P_{PV} > P_{LOAD})$ THEN
$CP_{PV} = P_{LOAD} - (CP_{HPP} + CP_{WPP})$ and $RP_{PV} = P_{PV} - CP_{PV}$
OTHERWISE
$CP_{PV} = P_{PV}$ and $RP_{PV} = 0$
OTHERWISE
$CP_{PV} = 0$ and $RP_{PV} = 0$
OTHERWISE
$CP_{PV} = 0$ and $RP_{PV} = P_{PV}$
4. Production rules for *ES*:
IF $(CP_{HPP} + CP_{WPP} + CP_{PV} < P_{LOAD})$ THEN
IF $(P_{ES} > 0)$ THEN
IF $(CP_{HPP} + CP_{WPP} + CP_{PV} + P_{ES} > P_{LOAD})$ THEN
$CP_{ES} = P_{LOAD} - (CP_{HPP} + CP_{WPP} + CP_{PV})$ and $RP_{ES} = P_{ES} - CP_{ES}$
OTHERWISE
$CP_{ES} = P_{ES}$ and $RP_{ES} = 0$
OTHERWISE
$CP_{ES} = 0$ and $RP_{ES} = 0$
OTHERWISE
$CP_{ES} = 0$ and $RP_{ES} = P_{ES}$

For a power energy storage device, the rules for the production of its charge from various alternative energy sources are formulated separately, taking into account restrictions in the form of equalities and inequalities. At the same time, the following priority of energy storage was established: initially from the *HPP*, then from *WPP* and further from *PV*.

Accurate losses' computation in the electrical network requires calculation of the steady state at each hourly interval. The advantage of the proposed algorithm of minimizing financial costs of electricity consumption does not imply such complex calculation. It is assumed that the losses are included in the load and conditionally range from 5 to 10%.

Figure 5 presents the UML (Unified Modeling Language) class diagram of the developed software.

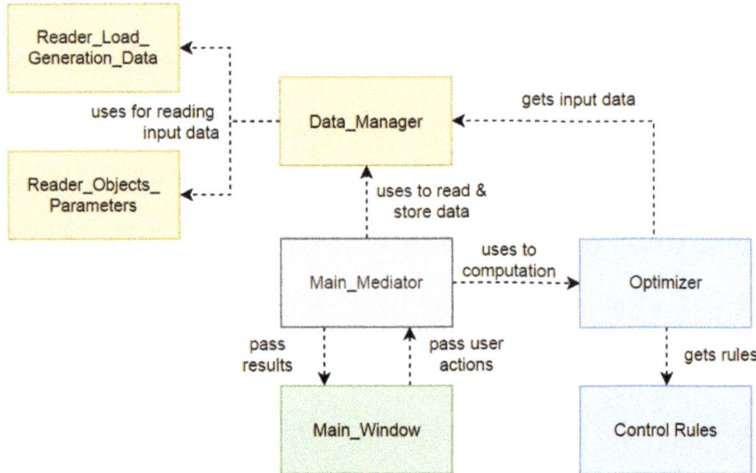

Figure 5. The UML class diagram of the developed software.

There are three main modules: Data (pale orange color), Graphical User Interface (light green), Computation (light blue), and the Mediator synchronizing all modules with each other. The modular architecture makes it easy to change individual parts in the system, for example, to add a new data format, change the optimization algorithm or modify the base of control rules.

5. Results and Discussion

The examined autonomous power system, the main generation sources of which are the HPPs of small rivers, has been adopted as a renewable energy source, and the WPPs and solar *PV* power plants have been adopted as alternative sources. An energy storage device is considered as a balancing source. Energy conservation in the storage device is necessary due to some unpredictability of power generation by alternative sources and is structurally performed based on lithium-ion batteries. The electrical diagram of the considered autonomous power system, including the installed alternative sources, is shown in Figure 6.

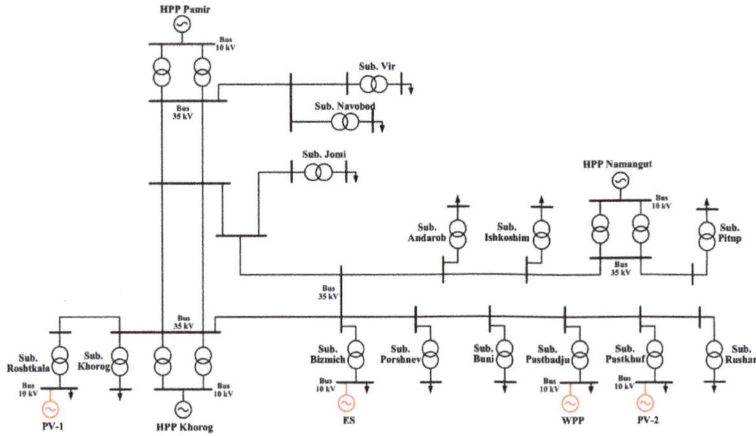

Figure 6. An electrical scheme of the considered autonomous power system.

According to the proposed algorithm, computations can be performed for each day in a certain seasonal interval of the year based on the product rules and the developed program. The calculation results of the optimal energy consumption for the examined case are shown in Figure 7.

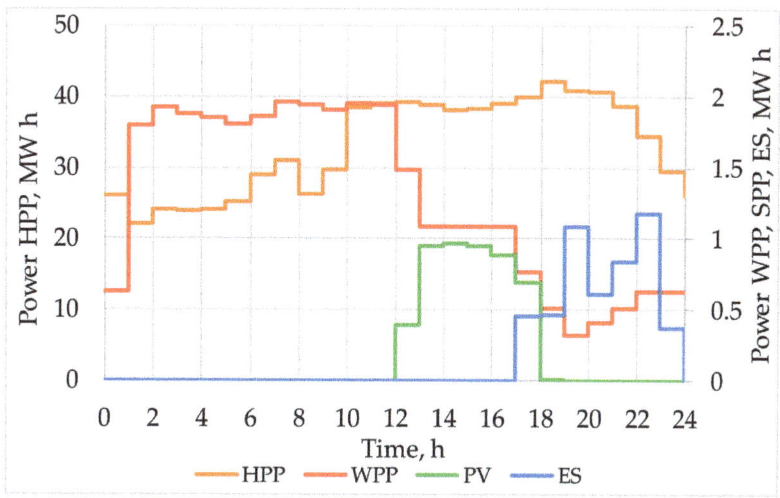

Figure 7. The optimal composition of alternative sources and energy storage to minimize the cost of power energy consumption.

The study shows that most of the daily interval, except for the periods of morning and evening peaks, renewable sources can provide coverage of electric energy consumption. However, during periods of peak load, there is a shortage of active power, which currently leads to the disconnection of power consumers or their restriction. The problem will be solved by installing an energy storage device, which should provide additional generation of stored power energy from 17 to 23 h. In the morning hours from 6 a.m. to 12 noon, there is a high speed of the wind flow, which allows to save the energy of the storage device.

The energy storage mode in terms of the optimal power consumption in the daily interval is presented in Table 2, from which it is advisable to accumulate energy at night considering the discharge of lithium-ion batteries, which should not be lower than 20% in order to extend their service life.

Table 2. An accumulation of a power energy from various generation sources.

Time, hours	P_{HPP}, kW	P_{WPP}, kW	P_{PV}, kW
1	0	1000	0
2	0	1000	0
3	0	1000	0
4	0	722	0
5	0	1000	0
6	0	278	0
7	0	0	0
8	0	0	0
9	0	0	0
...	...	...	...
24	0	0	0

In general, the minimum financial costs of power consumers in the daily interval of a winter day will be achieved with the following ratio of alternative sources of generation and energy storage reflected in Figure 8.

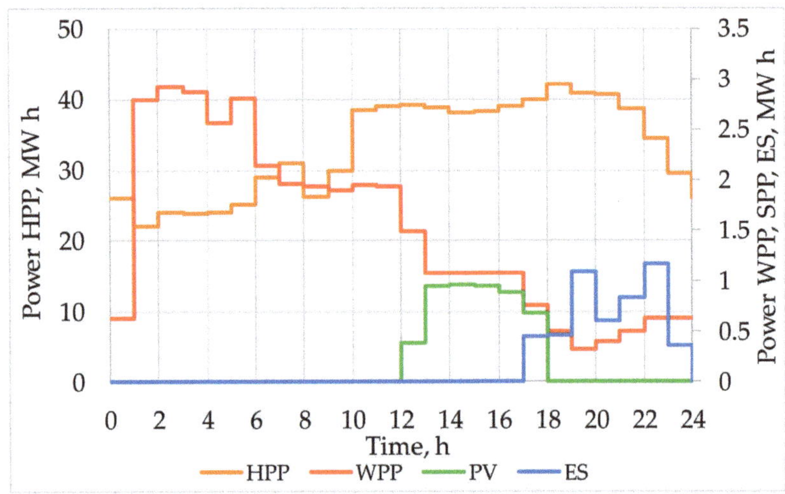

Figure 8. Selection of generating sources and energy storage devices to minimize the financial costs of electricity consumption. During this study, the most economical and efficient way of renewable and alternative energy sources joint functioning has been identified as part of the hybrid EPS aimed to achieve minimal financial costs and increase energy efficiency during such operation.

According to the obtained computation results, the electricity surplus was observed at 1–11 h for wind farms and at 8–16 h for *PV* power plant. The results are summarized in Table 3.

Table 3. The electricity surplus.

Time, h Power, kW	1	2	3	4	5	6	7	8	9	10	11	12
W_{HPP}	0	0	0	0	0	0	0	0	0	0	0	0
W_{WPP}	861	340	30	0	463	1939	3058	2140	1364	622	34	0
W_{PV}	0	0	0	0	0	0	0	81	969	1949	2688	2749
Time, h Power, kW	13	14	15	16	17	18	19	20	21	22	23	24
W_{HPP}	0	0	0	0	0	0	0	0	0	0	0	0
W_{WPP}	0	0	0	0	0	0	0	0	0	0	0	0
W_{PV}	2310	2088	1572	833	0	0	0	0	0	0	0	0

It is shown that due to the optimal management of the energy consumption in each hour of the daily time interval, minimization of the financial costs for energy consumption can be achieved, both for the region as a whole and for each electric power consumer if it has their own sources of alternative energy generation. This approach is new and corresponds to an innovative proposal for the short-term development of the examined region.

An analysis of the energy consumption modes of the presented autonomous region of the Pamirs of Republic of Tajikistan shows that for seasonal management of the power consumption, it is necessary to select the busiest working days for the winter, spring, summer and fall seasons. It showed that the most problematic daily regimes relate to the winter season, while in the summer period, all loads can be covered at the expense of water resources, including the export of excess power energy. This is caused by the mountainous terrain and the abundance of small rivers. The picture changes radically in winter, due to the need for heating, water heating and a higher level of lighting. The power energy export is not topical anymore.

This work is devoted solving the problems during the winter period, and it shows that they can be successfully solved in the short term taking into account the prospects for development of the mining industry, which plays a key role for the well-being of the population and solving the environmental issues based on carbon-free energy.

The steady-state mode and accordantly the voltage at the nodes of the system can be calculated after the optimal balancing of the mode has been achieved, which is the subject of this mathematical model. If this condition is satisfied, the frequency in the system will be ensured and correspond to the industrial frequency of 50 Hz in normal and post-emergency modes. In balanced mode, voltage and frequency will comply with the permissible deviations.

The proposed method and algorithm for optimization of energy consumption makes it possible to minimize the financial costs of several electric power consumers that have the opportunity to choose a power source from renewable (*HPP* energy) and alternative (*WPP* and *PV* energy) sources. The considered model includes an energy storage device, which plays the main balancing role in the EPS and thereby allows minimizing the risks of underutilization of the *WPP* energy, dictated by the unpredictable nature of the generation of the latter.

The advantages of the proposed method for managing energy consumption in the power system are as follows: (1) relative simplicity in comparison with more complex models; (2) versatility, which allows to replace power sources, add criteria and restrictions; (3) scalability to which the algorithm can be applied for both small generating consumers and large autonomous power systems; (4) high transparency of the control algorithm for the interpretation of each step of its work and the obtained results; and (5) low risk of error at the stage of software implementation, integration and operation of the algorithm.

6. Conclusions

Evaluation of the reliability of hybrid power generation systems using RES, which have significant differences in structure, control systems, types of converters used, etc., is of considerable importance in terms of optimizing energy distribution. The study has been motivated by the need to increase the economic viability of integrated renewable energy systems, and in particular, autonomous multiple energy carrier microgrids, which not only play a central role in integrating renewable energy variables as part of global efforts to address climate change and decentralize energy, but are also important for ensuring universal reliable access to affordable, reliable, sustainable and modern energy sources.

For the considered autonomous power system, hydroelectric power plants of small rivers are accepted as a renewable energy source, and wind power plants and solar power plants are accepted as alternative sources; energy storage is a balancing source.

As a result, the optimal ratio of power generation by alternative sources from daily power consumption for winter was established to be hydroelectric power plants (94.8%), wind power plant (3.8), solar photovoltaic power plant (0.5%) and energy storage (0.8%). In the summer, there is no need for an optimal choice of generation sources since the entire electrical load can be covered with the energy produced by hydroelectric power plants in this region.

The paper has proposed a mathematical model to minimize the financial costs of individual generating consumers. The proposed method and algorithm for optimizing energy consumption makes it possible to minimize the financial costs of several power consumers. Optimization is based on the linear programming method with variable constraints on the daily interval. The proposed hybrid energy system and optimization of its energy consumption management is implemented based on a real and practical study for the Pamir power system; computational experiments and comparison of conditions were carried out to obtain optimal results at different energy storage capacities. The research results and computational experiments are presented in detailed manner, where the analysis shows that each generating consumer can independently minimize their power energy costs and maximize the benefits of the exchange.

This study considers the methodology applied to the daily interval of the busiest day. However, the optimal growth in the daily interval will not necessarily correspond to the optimum in the annual interval (8,760 h). Preliminary calculations show that it is important to carry out calculations every day, thereby approaching the optimal solution at longer time intervals (month, year). Optimizing financial and electricity costs at monthly and longer time intervals requires introducing additional conditions, including consideration of the possible transmission and export of the power energy system, which, in turn, is determined by the load schedule of neighboring electric power consumers as well as the selling and buying prices of electricity. Further work is planned on carrying out computational experiments based on several years' data of the considered power system; to improve the optimal control approach, including the accounting of the forecast values of consumption and generation of the power system; and to select the optimal equipment of the power plants based on renewable energy, considering economic efficiency and their impact on decarbonization in the examined power system.

Author Contributions: All authors contributed extensively to the work presented in this paper. Conceptualization, M.N., V.M., M.S., I.Z., S.B., P.M. and A.G.; methodology, M.N., V.M., M.S., I.Z., S.B., P.M. and A.G.; software, P.M. and A.G.; validation, M.N., I.Z. and S.B.; formal analysis, M.N., V.M., M.S., I.Z., S.B., P.M. and A.G.; investigation, V.M., M.S., I.Z., S.B., P.M. and A.G.; writing—original draft preparation, M.N., V.M. and A.G.; writing—review and editing, I.Z., S.B. and M.S. visualization, M.N., P.M. and A.G.; supervision, M.S., P.M., I.Z. and S.B.; project administration, M.S., I.Z. and S.B.; funding acquisition, M.N., V.M., M.S., I.Z., S.B., P.M. and A.G. All authors have read and agreed to the published version of the manuscript.

Funding: This research received no external funding.

Institutional Review Board Statement: Not applicable.

Conflicts of Interest: The authors declare no conflict of interest.

References

1. Mohseni, S.; Brent, A.C.; Burmester, D. Off-Grid Multi-Carrier Microgrid Design Optimisation: The Case of Rakiura–Stewart Island, Aotearoa–New Zealand. *Energies* **2021**, *14*, 6522. [CrossRef]
2. Climate Action. Available online: https://ec.europa.eu/clima/eu-action/international-action-climate-change/climate-negotiations/paris-agreement_en (accessed on 18 October 2021).
3. Tailor, R.; Beňa, L.; Čonka, Z.; Kolcun, M. Design of Management Systems for Smart Grid. In Proceedings of the 2021 Selected Issues of Electrical Engineering and Electronics (WZEE), Rzeszow, Poland, 13–15 September 2021.
4. Min, H.S.; Wagh, S.; Kadier, A.; Gondal, I.A.; Azim, N.A.P.B.A.; Mishra, M.K. *Edition: 7 Chapter: Renewable Energy Technologies*; Min, H.S., Ed.; Ideal International E-Publication Pvt. Ltd.: Indore, India, 2018.
5. Sustainable Developments. Available online: https://www.un.org/sustainabledevelopment/energy/ (accessed on 18 October 2021).
6. Zahraoui, Y.; Alhamrouni, I.; Mekhilef, S.; Basir Khan, M.R.; Seyedmahmoudian, M.; Stojcevski, A.; Horan, B. Energy Management System in Microgrids: A Comprehensive Review. *Sustainability* **2021**, *13*, 10492. [CrossRef]
7. Jurasz, J.; Canales, F.A.; Kies, A.; Guezgouz, M.; Beluco, A. A review on the complementarity of renewable energy sources: Concept, metrics, application and future research directions. *Sol. Energy* **2020**, *195*, 703–724. [CrossRef]
8. Feddaoui, O.; Toufouti, R.; Labed, D.; Meziane, S. Control of an Isolated Microgrid Including Renewable Energy Resources. *Serb. J. Electr. Eng.* **2020**, *17*, 297–312. [CrossRef]
9. John Bhatti, H.; Danilovic, M. Making the World More Sustainable: Enabling Localized Energy Generation and Distribution on Decentralized Smart Grid Systems. *World J. Eng. Technol.* **2018**, *6*, 350–382. [CrossRef]
10. Deng, Z.; Xiao, J.; Zhang, S.; Xie, Y.; Rong, Y.; Zhou, Y. Economic feasibility of large-scale hydro-solar hybrid power including long distance transmission. *Glob. Energy Interconnect.* **2019**, *2*, 290–299.
11. Li, H.; Liu, P.; Guo, S.; Ming, B.; Cheng, L.; Yang, Z. Long-term complementary operation of a large-scale hydro-photovoltaic hybrid power plant using explicit stochastic optimization. *Appl. Energy* **2019**, *238*, 863–875. [CrossRef]
12. Ausfelder, F.; Beilmann, C.; Bertau, M.; Bräuninger, S.; Heinzel, A.; Hoer, R.; Koch, W.; Mahlendorf, F.; Metzelthin, A.; Peuckert, M.; et al. Energy Storage as Part of a SecureEnergy Supply. *ChemBioEng Rev.* **2017**, *4*, 144–210. [CrossRef]
13. Li, F.F.; Qiu, J. Multi-objective optimization for integrated hydro–photovoltaic power system. *Appl. Energy* **2016**, *167*, 377–384. [CrossRef]

14. Lawan, S.M.; Abidin, W.A.W.Z. *A Review of Hybrid Renewable Energy Systems Based on Wind and Solar Energy: Modeling, Design and Optimization*; Wind Solar Hybrid Renewable Energy System: London, UK, 2020; 21p. [CrossRef]
15. Burger, C.; Froggatt, A.; Mitchell, C.; Weinmann, J. *Decentralised Energy: A Global Game Changer*; Ubiquity Press: Berkeley, CA, USA, 2020.
16. Mohseni, S.; Brent, A.C.; Burmester, D. Community Resilience-Oriented Optimal Micro-Grid Capacity Expansion Planning: The Case of Totarabank Eco-Village, New Zealand. *Energies* **2020**, *13*, 3970. [CrossRef]
17. Matrenin, P.; Safaraliev, M.; Dmitriev, S.; Kokin, S.; Ghulomzoda, A.; Mitrofanov, S. Medium-term load forecasting in isolated power systems based on ensemble machine learning models. *Energy Rep.* **2022**, *8*, 612–618. [CrossRef]
18. Kirgizov, A.K.; Dmitriev, S.A.; Safaraliev, M.K.; Pavlyuchenko, D.A.; Ghulomzoda, A.H.; Ahyoev, J.S. Expert system application for reactive power compensation in isolated electric power systems. *Int. J. Electr. Comput. Eng.* **2021**, *11*, 3682–3691. [CrossRef]
19. Asanov, M.S.; Safaraliev, M.K.; Zhabudaev, T.Z.; Asanova, S.M.; Kokin, S.E.; Dmitriev, S.A.; Obozov, A.J.; Ghulomzoda, A.H. Algorithm for calculation and selection of micro hydropower plant taking into account hydro-logical parameters of small watercourses mountain rivers of Cen-tral Asia. *Int. J. Hydrog. Energy* **2021**, *46*, 37109–37119. [CrossRef]
20. Jurasz, J.; Dabek, P.B.; Kazmierczak, B.; Kies, A.; Wdowikowski, M. Large scale complementary solar and wind energy sources coupled with pumped-storage hydroelectricity for Lower Silesia (Poland). *Energy* **2018**, *161*, 183–192. [CrossRef]
21. Yang, Z.; Liu, P.; Cheng, L.; Wang, H.; Ming, B.; Gong, W. Deriving operating rules for a large-scale hydro-photovoltaic power system using implicit stochastic optimization. *J. Clean. Prod.* **2018**, *195*, 562–572. [CrossRef]
22. Kougias, I.; Szabo, S.; Monforti-Ferrario, F.; Huld, T.; Bódis, K. A methodology for optimization of the complementarity between small-hydropower plants and solar PV systems. *Renew. Energy* **2016**, *87*, 1023–1030. [CrossRef]
23. Parastegari, M.; Hooshmand, R.A.; Khodabakhshian, A.; Zare, A.H. Joint operation of wind farm, photovoltaic, pump-storage and energy storage devices in energy and reserve markets. *Int. J. Electr. Power Energy Syst.* **2015**, *64*, 275–284. [CrossRef]
24. Jure, M.; Zvonimir, G. Feasibility of the green energy production by hybrid solar + hydro power system in Europe and similar climate areas. *Renew. Sustain. Energy Rev.* **2010**, *14*, 1580–1590. [CrossRef]
25. Dinglin, L.; Yingjie, C.; Kun, Z.; Ming, Z. Economic evaluation of wind-powered pumped storage system. *Syst. Eng. Procedia* **2012**, *4*, 107–115. [CrossRef]
26. Bekirov, E.A.; Strizhakov, K. Optimization of load distribution modes in a combined system with renewable energy sources. *Motrol* **2012**, *1*, 146–150.
27. Ma, T.; Lashway, C.R.; Song, Y.; Mohammed, O. Optimal renewable energy farm and energy storage sizing method for future hybrid power system. In Proceedings of the 2014 17th International Conference on Electrical Machines and Systems (ICEMS), Hangzhou, China, 22–25 October 2014; pp. 2827–2832. [CrossRef]
28. Shahirinia, A.H.; Tafreshi, S.M.M.; Gastaj, A.H.; Moghaddomjoo, A.R. Optimal sizing of hybrid power system using genetic algorithm. In Proceedings of the 2005 International Conference on Future Power Systems, Amsterdam, The Netherlands, 18 November 2005; pp. 1–6. [CrossRef]
29. Gang, L.; Heqing, S.; Dragan, R. Power generation cost minimization of the grid-connected hybrid renewable energy system through optimal sizing using the modified seagull optimization technique. *Energy Rep.* **2020**, *6*, 3365–3376. [CrossRef]
30. Ma, T.; Mohammed, O. Economic analysis of real-time large scale PEVs network power flow control algorithm with the consideration of V2G services. In Proceedings of the 2013 IEEE Industry Applications Society Annual Meeting, Lake Buena Vista, FL, USA, 6–11 October 2013; pp. 1–8. [CrossRef]
31. Chen, S.X.; Gooi, H.B.; Wang, M.Q. Sizing of Energy Storage for Microgrids. *IEEE Trans. Smart Grid* **2012**, *3*, 142–151. [CrossRef]
32. Xu, L.; Ruan, X.; Mao, C.; Zhang, B.; Luo, Y. An Improved Optimal Sizing Method for Wind-Solar-Battery Hybrid Power System. *IEEE Trans. Sustain. Energy* **2013**, *4*, 774–785. [CrossRef]
33. Manusov, V.; Nazarov, M. Energy Consumption Conditions Optimization of the Autonomous System Based on Carbon-Free Energy. In Proceedings of the 2020 Ural Smart Energy Conference (USEC), Ekaterinburg, Russia, 13–15 November 2020; pp. 93–96. [CrossRef]
34. Matrenin, P.; Safaraliev, M.; Dmitriev, S.; Kokin, S.; Eshchanov, B.; Rusina, A. Adaptive ensemble models for medium-term forecasting of water inflow when planning electricity generation under climate change. *Energy Rep.* **2022**, *8*, 439–447. [CrossRef]
35. Ghulomzoda, A.; Gulakhmadov, A.; Fishov, A.; Safaraliev, M.; Chen, X.; Rasulzoda, K.; Gulyamov, K.; Ahyoev, J. Recloser-Based Decentralized Control of the Grid with Distributed Generation in the Lahsh District of the Rasht Grid in Tajikistan, Central Asia. *Energies* **2020**, *13*, 3673. [CrossRef]
36. Safaraliev, M.K.; Odinaev, I.N.; Ahyoev, J.S.; Rasulzoda, K.N.; Otashbekov, R.A. Energy Potential Estimation of the Region's Solar Radiation Using a Solar Tracker. *Appl. Sol. Energy* **2020**, *56*, 270–275. [CrossRef]
37. Shakirov, V.A.; Artemyev, A.Y. The choice of a site for the placement of a wind power plant using computer modeling of terrain and wind flow. *Bull. Irkutsk State Tech. Univ.* **2017**, *21*, 133–143. [CrossRef]

 *mathematics*

MDPI

Article

An Optimized Triggering Algorithm for Event-Triggered Control of Networked Control Systems

Sunil Kumar Mishra [1], Amitkumar V. Jha [1], Vijay Kumar Verma [2], Bhargav Appasani [1], Almoataz Y. Abdelaziz [3] and Nicu Bizon [4,5,6,*]

1   School of Electrical Engineering, Kalinga Institute of Industrial Technology, Bhubaneswar 751024, India; sunil.mishrafet@kiit.ac.in (S.K.M.); amit.jhafet@kiit.ac.in (A.V.J.); bhargav.appasanifet@kiit.ac.in (B.A.)
2   Control and Digital Electronics Group, U R Rao (ISRO) Satellite Centre Department of Space, Government of India, Bengaluru 560017, India; vijay@ursc.gov.in
3   Faculty of Engineering and Technology, Future University in Egypt, 90th St, First New Cairo, Cairo Governorate, Cairo 11835, Egypt; almoataz.abdelaziz@fue.edu.eg
4   Faculty of Electronics, Communication and Computers, University of Pitesti, 110040 Pitesti, Romania
5   Doctoral School, Polytechnic University of Bucharest, 313 Splaiul Independentei, 060042 Bucharest, Romania
6   ICSI Energy, National Research and Development Institute for Cryogenic and Isotopic Technologies, 240050 Ramnicu Valcea, Romania
*   Correspondence: nicu.bizon@upit.ro

Abstract: This paper presents an optimized algorithm for event-triggered control (ETC) of networked control systems (NCS). Initially, the traditional backstepping controller is designed for a generalized nonlinear plant in strict-feedback form that is subsequently extended to the ETC. In the NCS, the controller and the plant communicate with each other using a communication network. In order to minimize the bandwidth required, the number of samples to be sent over the communication channel should be reduced. This can be achieved using the non-uniform sampling of data. However, the implementation of non-uniform sampling without a proper event triggering rule might lead the closed-loop system towards instability. Therefore, an optimized event triggering algorithm has been designed such that the system states are always forced to remain in stable trajectory. Additionally, the effect of ETC on the stability of backstepping control has been analyzed using the Lyapunov stability theory. Two case studies on an inverted pendulum system and single-link robot system have been carried out to demonstrate the effectiveness of the proposed ETC in terms of system states, control effort and inter-event execution time.

Keywords: backstepping control; event-triggered control; Lyapunov stability theorem; networked control system; nonlinear system

Citation: Mishra, S.K.; Jha, A.V.; Verma, V.K.; Appasani, B.; Abdelaziz, A.Y.; Bizon, N. An Optimized Triggering Algorithm for Event-Triggered Control of Networked Control Systems. *Mathematics* 2021, 9, 1262. https://doi.org/10.3390/math9111262

Academic Editor: M. Victoria Otero-Espinar

Received: 5 May 2021
Accepted: 27 May 2021
Published: 31 May 2021

1. Introduction

In recent years, event-triggered control (ETC) has become a prominent topic of research due to the benefits offered by it as compared to the networked control system (NCS). The analog-to-digital (A/D) and digital-to-analog (D/A) conversion of the plant data is essential in the NCS to make the plant and controller signals compatible with the communication channel. This further needs the sampling and hold circuit for the plant data. Periodic sampling utilizes a fixed bandwidth on the communication channel and was preferred in the past for implementing the NCS. This, however, increases the cost of the system. In several control applications (e.g., chemical processes which take a longer time to settle), it is not needed to send the output data of the plant to the controller at fixed time intervals. The time interval between the samples can be varied (also called the aperiodic sampling of data) depending on the specific system events. Whenever the quantization error reaches beyond an acceptable limit, an event is triggered to sample the output data. Thereafter, a new control command is computed and the plant control input is updated. This control command is kept on hold until a new event is triggered and this process repeats. In this manner, the number of samples can be reduced significantly without disturbing the desired plant performance.

1.1. Literature Review

A plethora of research works have been published on this topic. Some of the original ETC studies were focused on the PID controller, which due to its basic configuration is still one of the commonly used controllers in industrial applications. In [1], a simple event-based PID control scheme was proposed, where the aim was to control the level of the upper or lower tank with the pump signal as the control signal. Another event-based PID controller was proposed in [2], where the objective was to control the angular velocity of a DC-motor. However, the emphasis has moved in the past 10 years toward ETC for optimized, adaptive and nonlinear controllers. That is because of the nonlinear nature of the control schemes. The nonlinear system has to be linearized to construct the linear controllers, which may lead to optimum performance. In [3], a critic neural network (NN) was used to approximate the cost and an actor neural network was used to approximate the optimal event-triggered controller. The controller stability was ensured by Lyapunov stability analysis. Again in [4], an approximation-based event-triggered control of multi-input multi-output uncertain nonlinear continuous-time systems was presented that included weight update law for aperiodic tuning of the NN weights at triggered instants to reduce the computation. In [5], the effects of bounded disturbances on the integral-based event-triggered control systems with observer-based output feedbacks were studied. In [6], a robust adaptive fuzzy control for a class of uncertain nonlinear systems via an event-triggered control strategy to reduce communication burden had been proposed, and results were demonstrated on a robot manipulator. In [7], the event-triggered function under the conditions of limited network bandwidth resources and the incomplete observability of the state of the system was considered. Then, denial-of-service (DoS) attacks that occur on the network transmission channel were applied using observer-based event-triggered control.

It is well known that the ETC can provide efficient utilization of resources by decreasing the sample size, which in turn reduces the overall bandwidth requirement of the communication channel. However, the ETC technique may lead the system towards instability if it is not driven by a suitable event triggering algorithm. Many triggering algorithms for the ETC scheme have been proposed in recent times, with impractical applications in wireless networks [8], robotics [9] and power systems [10].

A detailed review of these earlier techniques in the area of ETC has been published in [11], whereas a detailed review of several control techniques developed till recently could be found in [12]. Recently, ETC schemes that use a sliding mode controller (SMC) have been proposed for linear and nonlinear systems. In [13], a robust stabilization of a linear time-invariant system using an SMC with the self-triggering strategy was proposed. The self-triggering technique does not require additional dedicated hardware for continuous state measurement to determine the next possible triggering instant. In [14], robust stabilization for a class of nonlinear systems subject to external disturbances using an ETC-based SMC was proposed. Similarly, in [15], a global event-triggering realization of an SMC was proposed for linear systems under the effect of uncertainties. While the ETC–SMC scheme reduces the control updates considerably, the SMC itself is plagued by chattering.

The ETC schemes other than SMC have also been demonstrated in many studies [16–22]. In [16], it was explained how self-triggering rules could be deduced from the developed event-triggered strategies. The stabilization of nonlinear systems using event-triggered output feedback controllers was presented by [17], which guarantees an asymptotic stability property and enforces a minimum amount of time between two consecutive transmission instants. In [18], an event-driven tracking control algorithm for a marine vessel, based on the backstepping method, was proposed. In [19], adaptive event-triggered control of nonlinear continuous time systems in the strict feedback form is presented. An adaptive model and an associated event-triggered controller were designed by using the backstepping method. The backstepping method proposed by [18] was limited to marine vessels only, whereas [19] focused upon a neural network-based adaptive event-triggered backstepping controller scheme. In [20], both the parameter estimator and the controller were aperiodically updated only at the event-sampled instants, and an adaptive event sampling condition was designed to determine the event sampling instants. The system in [20] was not considered in a very generalized form of nonlinear system in strict-feedback form. An event-triggered nonlinear control based on the backstepping controller of an oscillating water column ocean wave energy plant was proposed by [21,22]. The second order nonlinear plant dynamics were considered in [21], whereas several wave energy plants were considered in [22] for forming an array of ocean energy plants, which also included an NCS.

In recent years, there has been a main focus on nonlinear systems while designing ETC [23–32]. In [23], a user-adjustable event-triggered mechanism based on the sampled state vectors and backstepping techniques for a nonlinear system was developed to determine the sampling state instants using the negative definite property of the derivatives of Lyapunov functions. In [24], event-triggered adaptive control for a class of nonlinear systems in Brunovsky form was considered. In [25], the considered nonlinear system contained not only unknown system parameters, but the nonlinear functions with no requirement to be globally Lipschitz. It was claimed to be in contrast to most of the existing results in the area. In [26], time-varying external disturbances were considered apart from the conditions given in [25], and the ETC was designed which can dynamically compensate for both errors caused by disturbances and the sampled-data implementation of the controller. In [27], the feedback linearization approach was applied to design the ETC for a nonlinear system, whereas [28] considered both the disturbances and transmission delays while designing the ETC scheme. In [29], an ETC scheme with nonlinear model predictive control was designed for an unmanned aerial vehicle. In [30], a self-learning robust ETC scheme was proposed for nonlinear interconnected systems subject to uncertainty. A fuzzy logic-based ETC–SMC was designed for a nonlinear system in [31]. In [32], a fractional order system was controlled using a fractional order controller where the controller interacted using a communication network. The above-mentioned studies have maintained the closed-loop stability of the systems but not considered the optimized event triggering parameters. The non-optimized event triggering parameters affect the closed-loop performance of the system significantly. Additionally, some classes of nonlinear systems have been considered in the studies discussed above. Most of the systems considered have restricted the system order. Hence, an optimized event-triggered control for a generalized nonlinear system is to be designed to accommodate different practical nonlinear systems. It is also to be noted that the backstepping controller is free from the chattering problem which predominantly exists in the SMC [20–22].

1.2. Main Contributions

Based on the literature review, this paper contributes two novel concepts to the NCS: (i) the design of an ETC scheme for an NCS wherein the NCS consists of a *nth* order nonlinear plant (or generalized nonlinear system) in strict-feedback form, a backstepping controller, and a communication channel; and (ii) the selection of the optimum value of event triggering parameter α to attain the best possible outcome of ETC as compared to the conventional approach. Firstly, a backstepping controller has been designed using the conventional approach and then it has been extended to ETC. The closed-loop asymptotic stability of the system is ensured while deriving the triggering rule for the backstepping controller. Next, the particle swarm optimization (PSO) technique [33] has been applied to obtain the optimum value of event triggering parameter α. In this regard, a fitness function consisting of error in system states and the number of triggering pulses are defined, wherein α has been chosen as an unknown parameter.

Two case studies on the inverted pendulum system [34] and a single-link robot system [20] are considered for demonstration purposes. Simulations are performed in MAT-

LAB software for these two nonlinear systems in strict-feedback form. Additionally, the analysis of a control system with different values and optimum values of event triggering parameter α is presented to show the effectiveness of the proposed ETC scheme.

The remaining sections are arranged in the following order: Section 2 gives an overview of the ETC scheme for the NCS and defines the control problem. In Section 3, the ETC using a backstepping controller has been developed. Section 4 discusses the simulation results, followed by the conclusion in Section 5.

2. Description of ETC for NCS and Control Problem Statement

The block diagram of an NCS with a backstepping controller, *nth* order nonlinear plant, communication network, and event triggering section is shown in Figure 1. The output of the plant first passes through a triggering section where non-uniform sampling is performed. Next, it is sent to the backstepping controller using the communication channel and a control command is computed. The control command then passes through the communication channel. It is further processed using zero-order-hold (ZOH) and given to the plant. It is obvious that there will be a quantization error in system states due to the sampling and hold of plant output data. This quantization error is taken as the criterion for event triggering, which then governs the sample and hold switch function.

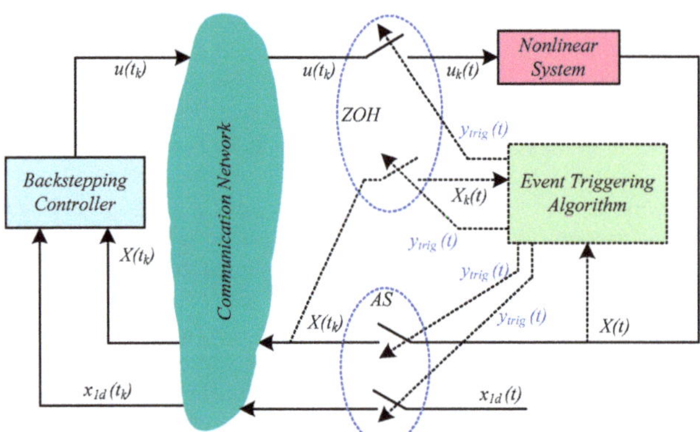

Figure 1. Block diagram of ETC for NCS.

Consider a generalized *nth* order nonlinear system in strict-feedback form as:

$$\left.\begin{aligned}\dot{x}_i &= f_i(x_1, x_2, \ldots, x_i) + g_i(x_1, x_2, \ldots, x_i)\cdot x_{i+1} \\ \dot{x}_n &= f_n(x_1, x_2, \ldots, x_n) + g_n(x_1, x_2, \ldots, x_n)\cdot u\end{aligned}\right\} \tag{1}$$

where $n \geq 2$, x_i corresponds to system states and $i = 1, 2, \ldots, n - 1$; u is the control input to the plant. The $f_i(x_1, x_2, \ldots, x_i)$, $f_n(x_1, x_2, \ldots, x_n)$ and $g_i(x_1, x_2, \ldots, x_i)$, ..., $g_n(x_1, x_2, \ldots, x_n)$ are the nonlinear functions dependent upon the system states. The control input $u(t)$ is represented by u in the continuous time domain, but with ETC where sample and hold interface is involved, u is represented as u_k, where k represents the kth triggering instant for $k = 1, 2, 3, \ldots, \infty$.

For simplicity of representation, the system in Equation (1) can be written as:

$$\left.\begin{aligned}\dot{x}_i &= f_i + g_i\cdot x_{i+1} \\ \dot{x}_n &= f_n + g_n\cdot u\end{aligned}\right\} \tag{2}$$

where $g_i \neq 0$; for $i = 1, 2, \ldots, n$. Here, it is also assumed that the time derivatives of f_i and g_i do exist for all time, t. The system states in the vector form are defined as:

$$X = [x_1 \quad x_2 \quad \ldots \quad x_n] \tag{3}$$

3. Design of Backstepping Controller and ETC Scheme

3.1. Backstepping Controller Design

Initially, we derive a backstepping control law using the conventional approach and then a triggering rule for implementing ETC is designed. In the backstepping controller, a step by step design method is followed. The error components for system states can be defined as:

$$\left.\begin{array}{l} \widetilde{x}_i = x_{id} - x_i \\ \widetilde{x}_{n+1} = x_{(n+1)d} - u \end{array}\right\} \quad \text{for } i = 1, 2, \ldots, n \tag{4}$$

where x_{1d} is the reference trajectory for system state x_1 and $x_{2d}, x_{3d}, \ldots, x_{(n-1)d}, x_{nd}$ are the virtual control inputs for corresponding system states, and:

$$u = x_{(n+1)d} - \widetilde{x}_{n+1} \tag{5}$$

Here, we define the error variable $\widetilde{x}_{n+1}$ additionally to maintain the uniformity of all the error variables of system states. This can also be called the virtual control input for actual control input u, as given in Equation (5).

The first order derivative of ith error variable $\widetilde{x}_i$ is given by:

$$\dot{\widetilde{x}}_i = \dot{x}_{id} - \dot{x}_i = \dot{x}_{id} - f_i - g_i \cdot x_{i+1} \tag{6}$$

Now, we add and subtract the $g_i \cdot x_{(i+1)d}$ term into Equation (7) as:

$$\left.\begin{array}{l} \dot{\widetilde{x}}_i = \dot{x}_{id} - f_i - g_i \cdot x_{i+1} + g_i \cdot x_{(i+1)d} - g_i \cdot x_{(i+1)d} \\ \quad = \dot{x}_{id} - f_i + g_i \cdot \widetilde{x}_{i+1} - g_i \cdot x_{(i+1)d} \end{array}\right\} \tag{7}$$

Define the ith virtual control law $x_{(i+1)d}$,

$$x_{(i+1)d} = g_i^{-1}\left[\dot{x}_{id} - f_i + k_i \cdot \widetilde{x}_i\right] \tag{8}$$

Such that,

$$\dot{\widetilde{x}}_i = -k_i \cdot \widetilde{x}_i + g_i \cdot \widetilde{x}_{i+1} \tag{9}$$

where $k_i > 0$ for $i = 1, 2, \ldots, n$.

Next, define the control law as:

$$u = g_n^{-1}\left[\dot{x}_{nd} - f_n + k_n \cdot \widetilde{x}_n\right] + (\phi_n)^{-1}\left\{\sum_{i=1}^{n-1} \phi_i\right\} + \eta \phi_n \tag{10}$$

where η is a positive constant and $\eta > 0$. Additionally, $\phi_i = g_i \cdot \widetilde{x}_i \cdot \widetilde{x}_{i+1}$ and $\phi_n = g_n \cdot \widetilde{x}_n$.

After substituting the $x_{(n+1)d}$ and u in Equation (5), the error variable $\widetilde{x}_{n+1}$ can be written as:

$$\widetilde{x}_{n+1} = -(\phi_n)^{-1}\left\{\sum_{i=1}^{n-1} \phi_i\right\} - \eta \phi_n \tag{11}$$

Theorem 1. *Consider a nonlinear system in strict-feedback form (Equation (2)) and the error dynamics (Equation (9)). Then, the virtual control law (Equation (8)) and backstepping control law (Equation (10)) ensure that all the equilibrium points, $\widetilde{x}_i$ for $i = 1, 2, \ldots, n$, are globally asymptotically stable. Therefore, the closed-loop system states, x_i, are bounded.*

Proof. In order to prove the global asymptotic stability of a nonlinear system in strict-feedback form (Equation (2)) and the error dynamics (Equation (9)), we define a positive definite function $V(t)$ as a Lyapunov function candidate as follows:

$$V(t) = \frac{1}{2}\sum_{i=1}^{n} \widetilde{x}_i^2 \tag{12}$$

As per Lyapunov stability theory, if the $V(t)$ is a positive definite function and its first order derivative $\dot{V}(t)$ is a negative definite, then $\widetilde{x}_i$ will be of converging nature and will be globally asymptotically stable. The Lyapunov function candidate in Equation (13) will then become a Lyapunov function. Hereafter, in the next few steps, it will be proved that $\dot{V}(t)$ is negative definite function.

The first order time derivative of Equation (12) is given as:

$$\dot{V}(t) = \sum_{i=1}^{n} \widetilde{x}_i \cdot \dot{\widetilde{x}}_i \tag{13}$$

Substituting Equation (9) into Equation (13) as:

$$\dot{V}(t) = -\sum_{i=1}^{n} k_i \cdot \widetilde{x}_i^2 + \sum_{i=1}^{n} g_i \cdot \widetilde{x}_i \cdot \widetilde{x}_{i+1} \tag{14}$$

$$\Rightarrow \dot{V}(t) = \dot{V}_1(t) + \dot{V}_2(t) \tag{15}$$

Here, we choose,

$$\dot{V}_1(t) = -\sum_{i=1}^{n} k_i \cdot \widetilde{x}_i^2 \quad \text{for } k_i > 0 \tag{16}$$

And

$$\dot{V}_2(t) = \sum_{i=1}^{n} g_i \cdot \widetilde{x}_i \cdot \widetilde{x}_{i+1} \tag{17}$$

$$\Rightarrow \dot{V}_2(t) = \sum_{i=1}^{n-1} g_i \cdot \widetilde{x}_i \cdot \widetilde{x}_{i+1} + g_n \cdot \widetilde{x}_n \cdot \widetilde{x}_{n+1} \tag{18}$$

$$\Rightarrow \dot{V}_2(t) = \sum_{i=1}^{n-1} \phi_i + \phi_n \cdot \widetilde{x}_{n+1} \tag{19}$$

In Equation (19), substituting the error variable $\widetilde{x}_{n+1}$ from Equation (11), we then have:

$$\dot{V}_2(t) = -\eta \phi_n^2 \Rightarrow \dot{V}_2(t) \leq 0 \tag{20}$$

where $\eta > 0$.

Therefore, after combining the results of Equations (16) and (20) into Equation (15), we have:

$$\dot{V}(t) = -\sum_{i=1}^{n} k_i \cdot \widetilde{x}_i^2 - \eta \phi_n^2 \leq 0 \tag{21}$$

Equation (21) is negative definite and, hence, the Lyapunov function candidate defined in Equation (12) has now become the Lyapunov function and Theorem 1 is proved. $\square$

Remark 1. *Theorem 1 implies that the control law in Equation (10) will result in $\lim\limits_{t\to\infty} \widetilde{x}_i \to 0$, for $i = 1, 2, \ldots, n$. Next, from Equation (4), if $\lim\limits_{t\to\infty} \widetilde{x}_i \to 0$ then $\lim\limits_{t\to\infty} x_i \to x_{id}$, for $i = 1, 2, \ldots, n$. Hence, it can be concluded that if $\widetilde{x}_i$ is globally asymptotically stable around the origin as $t \to \infty$, then x_i would be globally asymptotically stable around x_{id} as $t \to \infty$. Therefore, as per Theorem 1, the nonlinear system (Equation (2)) would always lead towards global asymptotic stability with the control input defined by Equation (10).*

3.2. ETC Scheme with Backstepping Controller

As shown in Figure 1, the nonlinear system state vector, $X(t)$, is first sent to the sampling block where it is discretized and is represented by $X(t_k)$. Additionally, the reference variable, x_{1d}, is discretized into $x_{1d}(t_k)$. Then, $X(t_k)$ and $x_{1d}(t_k)$ are sent to the controller via the communication channel and the controller output, $u(t_k)$, which is in discrete form, is calculated. Next, the controller output, $u(t_k)$, is sent back to the nonlinear system again via the same communication channel, and after passing through the ZOH block, it is converted into continuous form, i.e., $u_k(t)$. Hence, $u_k(t)$ given by Equation (23) is not a discrete signal, but it is a continuous signal. However, $u_k(t)$ would not be exactly similar to the original continuous control signal $u(t)$ due to the quantization error created by the sample and hold circuit. Due to ETC implementation, the system dynamics given by Equation (2) can be rewritten as:

$$\left.\begin{array}{l} \dot{x}_i = f_i + g_i{\cdot}x_{i+1} \\ \dot{x}_n = f_n + g_n{\cdot}u_k \end{array}\right\} \tag{22}$$

The designed control law in Equation (10) ensures the asymptotic stability of the closed-loop system in continuous time. Now, in this section, the backstepping control law (in Equation (10)) would be converted into ETC law, which is given as:

$$u_k = g_{nk}^{-1}\left[\dot{x}_{ndk} - f_{nk} + k_n{\cdot}\tilde{x}_{nk}\right] + (\phi_{nk})^{-1}\left\{\textstyle\sum_{i=1}^{n-1}\phi_{ik}\right\} + \eta\phi_{nk} \tag{23}$$

where k represents the sample number.

Therefore, all time dependent variables in the control law of Equation (10) are replaced by suffix k to represent ETC. Next, the virtual control law from Equation (8) can be recalled as:

$$x_{(i+1)d} = g_i^{-1}\left[\dot{x}_{id} - f_i + k_i{\cdot}\tilde{x}_i\right] \quad \text{for } i = 1,\, 2,\, \ldots,\, n-1 \tag{24}$$

and because of the effect of the ETC implementation, the virtual control law can be written as:

$$x_{(n+1)dk} = g_{nk}^{-1}\left[\dot{x}_{ndk} - f_{nk} + k_n{\cdot}\tilde{x}_{nk}\right] \tag{25}$$

The error dynamics of Equation (9) due to ETC implementation can now be rewritten as:

$$\left.\begin{array}{l} \dot{\tilde{x}}_i = -k_i{\cdot}\tilde{x}_i + g_i{\cdot}\tilde{x}_{i+1} \\ \dot{\tilde{x}}_n = -k_n{\cdot}\tilde{x}_n + g_n{\cdot}\tilde{x}_{(n+1)k} \end{array}\right\} \quad \text{for } i = 1,\, 2,\ldots,\, n-1 \tag{26}$$

Additionally, Equation (11), due to ETC implementation, can be written as:

$$\tilde{x}_{(n+1)k} = -(\phi_{nk})^{-1}\left\{\textstyle\sum_{i=1}^{n-1}\phi_{ik}\right\} - \eta\phi_{nk} \tag{27}$$

The virtual control law in Equation (27) creates a quantization error in the system states and subsequently affects its stability. Therefore, we define a quantization error due to ETC as follows:

$$\xi_i = \phi_i - \gamma\phi_{ik} \quad \text{for } i = 1,\, 2,\, \ldots,\, n-1 \tag{28}$$

where $\gamma = \phi_n\phi_{nk}^{-1}$.

The ξ_i in Equation (28) is the quantization error created during the sample and hold process of the closed-loop system. Due to non-uniform sampling, there might be an undesired time interval between two sampling instants. It may lead the closed-loop system towards instability unless there is a mechanism to limit the quantization error within the permissible range of operation. Hence, we place an upper bound on $\sum_{i=1}^{n-1}|\xi_i|$ such that,

$$\textstyle\sum_{i=1}^{n-1}|\xi_i| = \textstyle\sum_{i=1}^{n-1}|\phi_i - \gamma\phi_{ik}| \le \alpha,\ \forall\, t \ge 0 \tag{29}$$

or,

$$\left(\textstyle\sum_{i=1}^{n-1}|\xi_i|\right)_{max} = \alpha \tag{30}$$

where the upper bound α is a non-zero positive constant and, hence, we refer to it as a triggering parameter.

Next, we also place a lower bound on $\phi_n\phi_{nk}$ such that,

$$\phi_n\phi_{nk} \ge \frac{\alpha}{\eta},\quad \forall\, t \ge 0 \tag{31}$$

Remark 2. *During the implementation of the event triggering algorithm, if $\phi_n\phi_{nk} < \alpha/\eta$ then the virtual control law (Equations (24) and (25) and backstepping control law (Equation (23)) would fail to ensure the semiglobal asymptotic stability of the error variables, $\tilde{x}_i$ and closed-loop system states, x_i for $i = 1, 2, \ldots, n$. It is therefore necessary to maintain $\phi_n\phi_{nk} \geq \alpha/\eta$, $\forall\, t \geq 0$. The region of attraction exists as long as the conditions: (i) $\sum_{i=1}^{n-1}|\xi_i| \leq \alpha$, $\forall\, t \geq 0$ and (ii) $\phi_n\phi_{nk} \geq \alpha/\eta$, $\forall\, t \geq 0$, are satisfied.*

In order to ensure the stability of the closed-loop system with ETC, the upper boundedness on $\sum_{i=1}^{n-1}\xi_i$ as defined by Equations (29) and (30) and the necessary condition, $\phi_n\phi_{nk} \geq \alpha/\eta \,\forall\, t \geq 0$, must be maintained. Therefore, a triggering algorithm has been designed and is described in Algorithm 1. If the condition given in Algorithm 1 is satisfied, then the system will trigger an event which will send the new sample to the controller. The control command will then be computed and sent back to the nonlinear system after passing through the communication network and ZOH block. If the condition given in Algorithm 1 is not satisfied, then the previous control command will remain in the nonlinear system input.

Algorithm 1. Event Triggering Algorithm

---

if $\sum_{i=1}^{n-1}|\xi_i| \geq \alpha$ *or* $\phi_n\phi_{nk} \leq \alpha/\eta$
   (i) an event is triggered ($y_{trig}(t) = 1$) and
   (ii) control command is updated
else
   (i) no event triggering ($y_{trig}(t) = 0$) and
   (ii) control command is on hold at its previous value
end

Next, the stability analysis of the closed-loop system with ETC law is again performed using the Lyapunov stability theory. In this regard, Theorem 2 is described next, followed by its proof.

Theorem 2. *Consider a nonlinear system in strict-feedback form (Equation (2)), and the error dynamics due to ETC (Equation (26)). Then, the virtual control law (Equations (24) and (25), backstepping control law (Equation (23)) and event triggering algorithm (Algorithm 1) ensure that all the equilibrium points, $\tilde{x}_i$ for $i = 1, 2, \ldots, n$, are semiglobally asymptotically stable with region of attraction-$\sum_{i=1}^{n-1}|\xi_i| \leq \alpha$ and $\phi_n\phi_{nk} \geq \alpha/\eta$. Therefore, the closed-loop system states, x_i, are bounded.*

Proof. It is again required to revisit the Lyapunov function candidate (Equation (12)) to analyze the effect of non-uniform sampling and ETC implementation on the system stability. The derivative of the Lyapunov function candidate from Equation (13) is recalled here as:

$$\dot{V}(t) = \sum_{i=1}^{n} \tilde{x}_i \cdot \dot{\tilde{x}}_i = \sum_{i=1}^{n-1} \tilde{x}_i \cdot \dot{\tilde{x}}_i + \tilde{x}_n \cdot \dot{\tilde{x}}_n \tag{32}$$

Substituting Equation (26) into Equation (32), we have:

$$\dot{V}(t) = -\sum_{i=1}^{n} k_i \cdot \tilde{x}_i^2 + \sum_{i=1}^{n-1} \phi_i + \phi_n \cdot \tilde{x}_{(n+1)k} \tag{33}$$

Here, Equation (33) can be divided into two parts as $\dot{V}_1(t)$ and $\dot{V}_2(t)$. $\dot{V}_1(t)$ is independent of ETC law and reduces to Equation (16). $\dot{V}_2(t)$ is dependent on $\tilde{x}_{(n+1)k}$ and can be written as:

$$\dot{V}_2(t) = \sum_{i=1}^{n-1} \phi_i + \phi_n \cdot \tilde{x}_{(n+1)k} \tag{34}$$

Putting the expression of $\widetilde{x}_{(n+1)k}$ from Equation (27) into Equation (34), we have:

$$\dot{V}_2(t) = \sum_{i=1}^{n-1} \phi_i + \phi_n \cdot \left[-(\phi_{nk})^{-1} \left\{ \sum_{i=1}^{n-1} \phi_{ik} \right\} - \eta \phi_{nk} \right] \tag{35}$$

$$\Rightarrow \dot{V}_2(t) = -\eta \phi_n \phi_{nk} + \sum_{i=1}^{n-1} (\phi_i - \gamma \phi_{ik}) \tag{36}$$

where $\gamma = \phi_n \cdot \phi_{nk}^{-1}$.

Equation (36) can further be simplified by substituting Equation (28) as:

$$\dot{V}_2(t) = -\eta \phi_n \phi_{nk} + \sum_{i=1}^{n-1} \xi_i \tag{37}$$

While implementing the ETC scheme, two conditions arise due to sampling.

Condition 1. At sampling instants, e.g., at $t = t_k$ and $t = t_{k+1}$, we have:

$$\phi_i = \phi_{ik} \quad \text{for } i = 1, 2, \ldots, n \tag{38}$$

Therefore, Equation (38) implies that:

$$\sum_{i=1}^{n-1} \xi_i = 0 \text{ and } \gamma = 1 \tag{39}$$

Therefore, Equation (37) reduces to the following form:

$$\dot{V}_2(t) = -\eta \phi_n^2 \Rightarrow \dot{V}_2(t) \leq 0 \tag{40}$$

which is same as Equation (20) and further results in Equation (21), and implies that the system states would always lead towards asymptotic stability.

Condition 2. Between any two consecutive sampling instants, e.g., $t_k < T_k < t_{k+1}$, we have:

$$\phi_i \neq \phi_{ik} \quad \text{for } i = 1, 2, \ldots, n \tag{41}$$

Therefore, Equation (41) implies that:

$$\sum_{i=1}^{n-1} \xi_i \neq 0 \text{ or } \sum_{i=1}^{n-1} |\xi_i| > 0 \tag{42}$$

and $\gamma \neq 1$.

Next, we can further simplify Equation (37) as:

$$\dot{V}_2(t) \leq -\eta \phi_n \phi_{nk} + \left(\sum_{i=1}^{n-1} |\xi_i| \right)_{max} \tag{43}$$

Substituting Equation (30) into Equation (43) as:

$$\dot{V}_2(t) \leq -\eta \phi_n \phi_{nk} + \alpha \tag{44}$$

$$\Rightarrow \dot{V}_2(t) \leq -\eta \left(\phi_n \phi_{nk} - \frac{\alpha}{\eta} \right) \tag{45}$$

Substituting Equation (31) in Equation (45), we get:

$$\dot{V}_2(t) \leq 0 \tag{46}$$

Now, Equation (46) along with Equation (16) is converted to the following form:

$$\dot{V}(t) \leq -\sum_{i=1}^{n} k_i \cdot \widetilde{x}_i^2 \leq 0 \tag{47}$$

From Equations (40) and (47), it can be concluded that the Lyapunov function candidate defined by Equation (12) is now a Lyapunov function and the closed-loop system with the ETC scheme is semi-globally asymptotically stable. Hence, Theorem 2 is proved for both Condition 1 and Condition 2. $\square$

3.3. Design of Optimized Event Triggering Algorithm

As given in Algorithm 1, it is also very important to choose triggering parameter α very carefully so that both the objectives of attaining desired system performance and of sending a minimum number of samples over the communication network are achieved successfully. A numerical analysis of the need for an optimum value of triggering parameter α is presented in Section 4. In order to optimize α, a fitness function consisting of error in system states and the number of triggering pulses is defined wherein α has been chosen as an unknown parameter. This can be seen in Figure 2 wherein the output of the nonlinear system, i.e., $X(t)$, and the output of the triggering block, i.e., $y_{trig}(t)$, are given as an input to the fitness function block. The fitness function is given as:

$$J_{fit}(t) = \int_0^{T_s} \{t \cdot \|X(t)\| + |y_{trig}(t)|\} dt \tag{48}$$

where $\|X(t)\|$ is the Euclidean norm, and its mathematical expression is given as:

$$\|X(t)\| = \sqrt{x_1^2 + x_2^2 + \ldots + x_n^2} \tag{49}$$

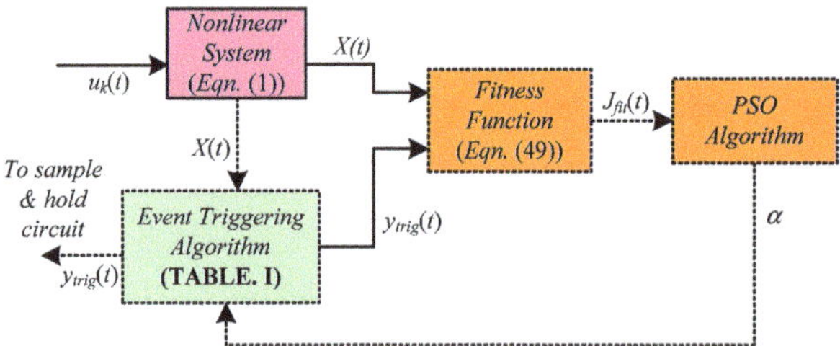

Figure 2. Block diagram for optimizing triggering parameter α based on PSO (offline tuning).

The $y_{trig}(t)$, represents the triggering pulses which actually drive the sample and hold circuit. The y_{trig} always toggles between 0 and 1. The $y_{trig}(t) = 1$ means that the switches in *AS* and *ZOH* blocks (in Figure 1) are closed, whereas the switches are opened for $y_{trig}(t) = 0$.

$J_{fit}(t)$ in Equation (48) is the summation of two fitness functions as: (i) integral time norm of system states, i.e., $\|X(t)\|$ and (ii) integral absolute of triggering function, i.e., $y_{trig}(t)$. The reason for multiplying time with $\|X(t)\|$ is that in the transient phase of the simulation, the errors in the system state are high, whereas in the steady state stage the errors become very low. Hence, in the final stage of simulation, if the time is multiplied with $\|X(t)\|$, then the errors become significant for optimization purposes. However, the time is not multiplied to $|y_{trig}(t)|$ because the triggering pulses will always jump between 0 and 1 for the whole simulation period. The total simulation run-time has been chosen to be T_s.

Next, an optimization algorithm is required to minimize $J_{fit}(t)$ for obtaining the best possible value of triggering parameter α. In this regard, the PSO algorithm [33] has been used due to its simplicity and popularity. It should be noted that any other optimization algorithm could also be used to optimize α. However, optimization in itself is a very big research area and a large number of optimization tools have been developed. Hence, the focus of this paper is not on developing optimization techniques, but on applying them for an event-triggered control.

Next, we consider two case studies on a single-link robot system and an inverted pendulum system to demonstrate the proposed ETC scheme for an NCS.

3.4. Case Study 1: Stabilization of Inverted Pendulum System

The inverted pendulum system, a nonlinear and unstable system, is widely used in laboratories to implement and validate new ideas emerging in control engineering. In this case study, we have considered the balancing control of the inverted pendulum system to demonstrate the proposed control scheme of event triggering.

The dynamics of the inverted pendulum system [34] can be written as:

$$(M+m)\ddot{x} + ml\cos\theta\,\ddot{\theta} - ml\sin\theta\,\dot{\theta}^2 = F_x \tag{50}$$

$$\cos\theta\,\ddot{x} - g\sin\theta = 0 \tag{51}$$

where θ is the pendulum angle (rad) and $|\theta| < \pi/2$; M is the mass of the cart (kg); m is the mass of the pendulum rod (kg); x is the cart position (m); l is the distance from the pivot to the mass center of the pendulum; F_x is the force applied on the cart (N); and g is the gravitational constant (N/kg).

From Equations (50) and (51), it can be concluded that the angle dynamics are independent of cart position and can be written as:

$$\ddot{\theta} = \frac{(M+m)g\sin\theta - ml\sin\theta\cos\theta\,\dot{\theta}^2}{Ml + ml\sin^2\theta} - \frac{\cos\theta\,F_x}{Ml + ml\sin^2\theta} \tag{52}$$

Equation (52) can be written in state space form as follows:

$$\left.\begin{aligned}
\dot{x}_1 &= x_2 \\
\dot{x}_2 &= \frac{(M+m)g\,\sin x_1 - ml\,\sin x_1\,\cos x_1\,x_2^2}{Ml + ml\,\sin^2 x_1} - \frac{\cos x_1}{Ml + ml\,\sin^2 x_1}\cdot u
\end{aligned}\right\} \tag{53}$$

where $x_1 = \theta$; $u = F_x$. Additionally, comparing Equation (53) with Equation (2) we get: $n = 2; f_1 = 0; g_1 = 1; f_2 = \frac{(M+m)g\,\sin x_1 - ml\,\sin x_1\,\cos x_1\,x_2^2}{Ml + ml\,\sin^2 x_1}$; and $g_2 = -\frac{\cos x_1}{Ml + ml\,\sin^2 x_1}$.

Now, the backstepping control law u from Equation (11) can be written as:

$$u = g_2^{-1}\left[\dot{x}_{2d} - f_2 + k_2\cdot\tilde{x}_2\right] + (\phi_2)^{-1}\left\{\phi_1 + \eta\phi_2^2\right\} \tag{54}$$

where $\phi_1 = g_1.\tilde{x}_1.\tilde{x}_2$ and $\phi_2 = g_2.\tilde{x}_2$, and after solving for $x_{1d} = 0$, we obtain:

$$\left.\begin{aligned}
\phi_1 &= x_1(k_1 x_1 + x_2) \\
\phi_2 &= \frac{\cos x_1}{Ml + ml\,\sin^2 x_1}(k_1 x_1 + x_2)
\end{aligned}\right\} \tag{55}$$

Therefore, the final expression for u can be written as:

$$\left.\begin{aligned}
u = {}& \frac{Ml + ml\,\sin^2 x_1}{\cos x_1}\left[(1 + k_1 k_2)x_1 + (k_1 + k_2)x_2\right] \\
& + \frac{(M+m)g\,\sin x_1 - ml\,\sin x_1\,\cos x_1\,x_2^2}{\cos x_1} \\
& + \frac{\eta\,\cos x_1}{Ml + ml\,\sin^2 x_1}(k_1 x_1 + x_2)
\end{aligned}\right\} \tag{56}$$

Next, for implementing the event-triggered control, Equation (56) can be expressed as:

$$\left.\begin{aligned}
u_k = {}& \frac{Ml + ml\,\sin^2 x_{1k}}{\cos x_{1k}}\left[(1 + k_1 k_2)x_{1k} + (k_1 + k_2)x_{2k}\right] \\
& + \frac{(M+m)g\,\sin x_{1k} - ml\,\sin x_{1k}\,\cos x_{1k}\,x_{2k}^2}{\cos x_{1k}} \\
& + \frac{\eta\,\cos x_{1k}}{Ml + ml\,\sin^2 x_{1k}}(k_1 x_{1k} + x_{2k})
\end{aligned}\right\} \tag{57}$$

Additionally, the error due to event-triggered control can be expressed as:

$$\sum_{i=1}^{n-1}|\xi_i| = |\xi_1| = |\phi_1 - \gamma\phi_{1k}| \tag{58}$$

$$\Rightarrow |\xi_1| = |\phi_1 - \gamma\phi_{1k}| \tag{59}$$

$$\Rightarrow |\xi_1| = \left|\phi_1 - \frac{\phi_2}{\phi_{2k}}\phi_{1k}\right| \tag{60}$$

Substituting Equation (55) into Equation (60), we obtain:

$$|\xi_1| = |k_1x_1 + x_2|\cdot|x_1 - \Delta\cdot x_{1k}| \tag{61}$$

where $\Delta = \frac{cosx_1}{cosx_{1k}}\cdot\frac{Ml+ml\ sin^2x_{1k}}{Ml+ml\ sin^2x_1}$.

In Equation (61), at triggering instants, $|\xi_1| = 0$ and the closed-loop system becomes asymptotically stable in the same manner as it is without ETC. Otherwise, $|\xi_1| \neq 0$ and the closed-loop system leads towards instability. In order to avoid this, the condition given in Algorithm 1 is forced on the control law of Equation (57).

3.5. Case Study 2: Stabilization of Single-Link Robot System

The dynamics of a single-link robot system have been adopted from [20] and are given as follows:

$$J\ddot{y} + 0.5mgl\sin(y) = u \tag{62}$$

where J is the inertia, y is the angle position of the link, $\dot{y}$ is the angle velocity of the link, $\ddot{y}$ is the angle acceleration of the link, $g = 9.8$ m/s^2 is the acceleration due to gravity, l is the length of the link, m is the mass of the link, and u is the control force of the link.

We denote $x_1 = y$ and $x_2 = \ddot{y}$ and, hence, Equation (62) can be rewritten as:

$$\left.\begin{array}{l} \dot{x}_1 = x_2 \\ \dot{x}_2 = -\frac{0.5mgl}{J}\sin(x_1) + \frac{1}{J}u \end{array}\right\} \tag{63}$$

Comparing Equation (63) with Equation (2), we have: $n = 2$; $f_1 = 0$; $g_1 = 1$; $f_2 = -\frac{0.5mgl}{J}\sin(x_1)$; $g_2 = \frac{1}{J}$.

Now, the backstepping control law u from Equation (11) can be rewritten as:

$$u = g_2^{-1}\left[\dot{x}_{2d} - f_2 + k_2\cdot\tilde{x}_2\right] + (\phi_2)^{-1}\left\{\phi_1 + \eta\phi_2^2\right\} \tag{64}$$

where $\phi_1 = g_1\cdot\tilde{x}_1\cdot\tilde{x}_2$ and $\phi_2 = g_2\cdot\tilde{x}_2$, and after solving for $x_{1d} = 0$, we obtain:

$$\left.\begin{array}{l} \phi_1 = x_1(k_1x_1 + x_2) \\ \phi_2 = -\frac{1}{J}(k_1x_1 + x_2) \end{array}\right\} \tag{65}$$

Therefore, the final expression for u can be written as:

$$u = -J[(1 + k_1k_2)x_1 + (k_1 + k_2)x_2] + 0.5mgl\ sinx_1 + \frac{\eta}{J}(k_1x_1 + x_2) \tag{66}$$

Next, for implementing the event-triggered control, Equation (66) can be expressed as:

$$u_k = -J[(1 + k_1k_2)x_{1k} + (k_1 + k_2)x_{2k}] + 0.5mgl\ sinx_{1k} + \frac{\eta}{J}(k_1x_{1k} + x_{2k}) \tag{67}$$

Additionally, the error due to event-triggered control can be expressed as:

$$\sum_{i=1}^{n-1}|\xi_i| = |\xi_1| = |\phi_1 - \gamma\phi_{1k}| = \left|\phi_1 - \frac{\phi_2}{\phi_{2k}}\phi_{1k}\right| \tag{68}$$

Substituting Equation (65) into Equation (68), we obtain:

$$|\xi_1| = |k_1x_1 + x_2|\cdot|x_1 - x_{1k}| \tag{69}$$

Here, Equation (69) will be utilized for implementing the event triggering algorithm as given in Algorithm 1.

In order to study the robustness of the proposed control scheme for a single-link robot system, we have applied an external disturbance signal in the actuator/control signal u_k. Therefore, the new control signal is represented as:

$$u_k = -J[(1 + k_1 k_2)x_{1k} + (k_1 + k_2)x_{2k}] + 0.5mgl \, sinx_{1k} + \frac{\eta}{J}(k_1 x_{1k} + x_{2k}) + \delta \qquad (70)$$

where δ is the external disturbance signal and is represented as:

$$\delta = 0.5\sin(10t) \qquad (71)$$

4. Simulation Results

In this section, we have performed simulations using MATALB software for implementing the ETC scheme on an inverted pendulum and single-link robot system. In ETC, suppose that the kth triggering instant is t_k and the next triggering instant is defined by t_{k+1}, then the inter event execution time, T_k, is represented as:

$$T_k = t_{k+1} - t_k \qquad (72)$$

where $T_k \geq T_{min}$. The T_{min} is the lower bound on T_k and $T_{min} = 1ms$. The $T_{min} = 1ms$ is also the sampling time set for running the simulations in MATLAB software. If $T_k < T_{min}$, then many sampling instants will not be captured by the MATLAB processor. Hence, the minimum inter-event execution time, T_k, will always be greater than or equal to T_{min}. If the T_{min} is reduced below 1 ms, then the computation time increases significantly. However, with high-speed computers, the T_{min} can be reduced to values below 1 ms as well. The upper bound on T_k is governed by an event triggering mechanism which has been developed in Section 3.

4.1. Case Study 1: Stabilization of Inverted Pendulum System

The simulation results for the stabilization of the inverted pendulum system have been obtained using parameters given in Table 1. Table 2 presents an analysis of the inverted pendulum system for different values of triggering parameter α. It is observed that for lower values of the α, the value of fitness function $J_{fit}(t)$ is very high, whereas the lowest number of samples is needed to be sent to the controller. On the increase in α value, the $J_{fit}(t)$ starts decreasing, whereas the number of samples goes on increasing. However, after a certain value of α, both the $J_{fit}(t)$ and number of samples start increasing. Hence, it is required to find the optimum value of α so that both the fitness function $J_{fit}(t)$ and number of samples are at the minimum level.

The $J_{fit}(t)$ is further minimized using the PSO algorithm and an optimized value of α is obtained. The simulation parameters taken for running the PSO algorithm are as follows: (i) number of iterations = 50, (ii) number of particles = 20 and (iii) total simulation period $T_s = 20s$. The other PSO parameters have been taken from Mishra and Chandra et al. (2014). Now, the optimized value of α is given in the bottom row of Table 2. The value of fitness function $J_{fit}(t)$ is 0.8369, which is the minimum among all fitness values given in Table 2. The number of samples needed is 501, which is very low if we compare it with a periodic sampling of 1 ms. If $T_s = 20s$ then it would require 20,001 samples in case of periodic sampling.

Table 1. Inverted Pendulum System and Controller Parameters.

M (Kg)	m (Kg)	l (m)	g (m/s^2)	k_1	k_2	η
1	0.1	0.3	9.8	5	5	1

Table 2. Fitness Function and Number of Samples for Inverted Pendulum System.

Value of α	Fitness Function J_{fit} (t)	Number of Samples
10^{-1}	72.13	147
10^{-2}	22.92	211
10^{-3}	8.15	276
10^{-4}	2.70	335
10^{-5}	1.21	394
10^{-6}	0.85	508
10^{-7}	0.86	630
10^{-8}	0.91	736
10^{-9}	1.01	866
10^{-10}	1.12	969
0.998×10^{-6}	0.8369	501

Next, the inverted pendulum system performance has been evaluated by analyzing the behavior of its system states, control effort and inter-event execution time waveforms, which are shown Figures 3–5. Two situations of norm of system states, $\|X(t)\|$, for $\alpha = 10^{-2}$ (very high) and $\alpha = 0.998 \times 10^{-6}$ (optimized), are shown in Figure 3. For a relatively higher value of α as shown in Figure 3a, the system states have very poor performance, while the number of samples is 211. For the optimized case as shown in Figure 3b, the system states performance is very satisfactory, whereas the number of samples needed is just 501.

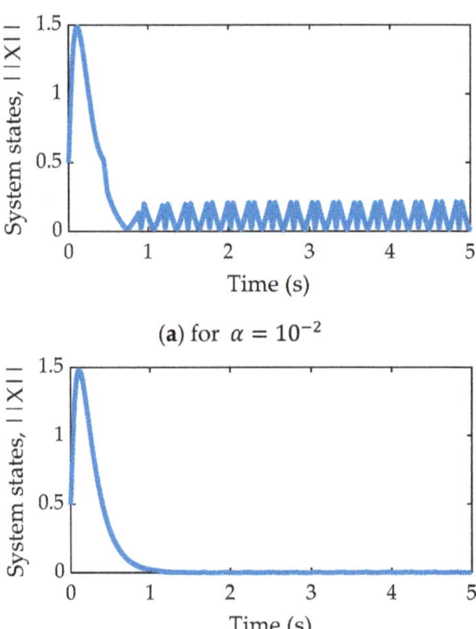

(**a**) for $\alpha = 10^{-2}$

(**b**) for $\alpha = 0.998 \times 10^{-6}$ (optimized)

Figure 3. Inverted pendulum system states, $\|X\|$.

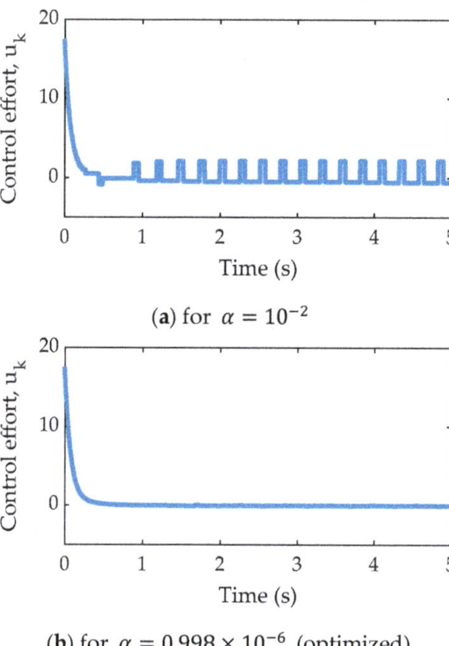

(**a**) for $\alpha = 10^{-2}$

(**b**) for $\alpha = 0.998 \times 10^{-6}$ (optimized)

Figure 4. Control effort, u_k.

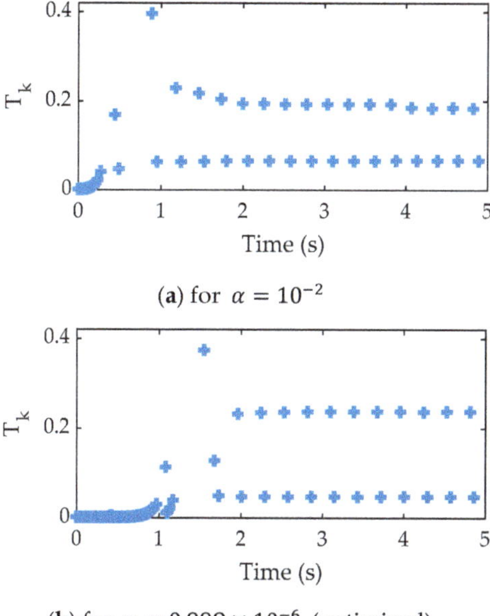

(**a**) for $\alpha = 10^{-2}$

(**b**) for $\alpha = 0.998 \times 10^{-6}$ (optimized)

Figure 5. Inter event execution time, T_k.

Similarly, in Figure 4, the control effort waveforms are shown for $\alpha = 10^{-2}$ (very high) and $\alpha = 0.998 \times 10^{-6}$ (optimized). Again, the performance of the optimized case is superior to the case of Figure 4a. Hence, it can be said that minimum control effort could be maintained with a very low number of samples as well. The inter-event execution time, T_k, waveforms are shown in Figure 5. It is already explained that the lowest number of triggering samples is required in the optimized case.

4.2. Case Study 2: Stabilization of Single-Link Robot System

We will now discuss the simulation results of the single-link robot system. The simulation results for the stabilization of the single-link robot system have been obtained using parameters given in Table 3. Table 4 presents an analysis of the single-link robot system for different values of triggering parameter α. It is observed that for lower values of the α, the value of fitness function $J_{fit}(t)$ is very high, whereas the lowest number of samples needed to be sent to the controller. On the increase in α value, the $J_{fit}(t)$ starts decreasing, whereas the number of samples goes on increasing. However, after a certain value of α, both the $J_{fit}(t)$ and number of samples start increasing. Hence, it is required to find the optimum value of α so that both the fitness function $J_{fit}(t)$ and number of samples are at the minimum level.

Table 3. Single-Link Robot and Controller Parameters.

J	m (Kg)	l (m)	g (m/s^2)	k_1	k_2	η
0.5	0.1	0.3	9.8	2	10	1

Table 4. Fitness Function and Number of Samples for Single-Link Robot.

Value of α	Fitness Function J_{fit} (t)	Number of Samples
10^{-1}	78.33	509
10^{-2}	25.47	521
10^{-3}	8.56	538
10^{-4}	3.287	556
10^{-5}	1.71	693
10^{-6}	1.49	971
10^{-7}	1.60	1240
10^{-8}	1.82	1511
10^{-9}	2.09	1782
10^{-10}	2.35	2057
1.001×10^{-6}	1.488	968

The $J_{fit}(t)$ is further minimized using the PSO algorithm, and an optimized value of α is obtained. The optimized value of α is given in the bottom row of Table 4. The value of fitness function $J_{fit}(t)$ is 1.488, which is the minimum among all fitness values given in Table 4. The number of samples needed is 968, which is very low if we compare it with a periodic sampling of 1 ms. If $T_s = 20s$ then it would require 20,001 samples in case of periodic sampling.

The simulation results for single-link robot system states, control effort and inter-event execution time waveforms are shown in Figures 6–8. Three situations of norm of system states, $\|X(t)\|$, for $\alpha = 10^{-2}$ (very high), $\alpha = 1.001 \times 10^{-6}$ (optimized) and $\alpha = 1.001 \times 10^{-6}$ (optimized) with actuator disturbance, are shown in Figure 6. For a relatively higher value of α as in Figure 6a, the system states have very poor performance, while the number of samples is 521. For the optimized case in Figure 6b, the system states performance is better compared to the case of Figure 6a, and the number of samples needed is just 968. Next, in Figure 6c, the performance of the system states is slightly affected by the actuator disturbance. However, the states always remain in the stability region provided that the disturbance signal is bounded within a small range.

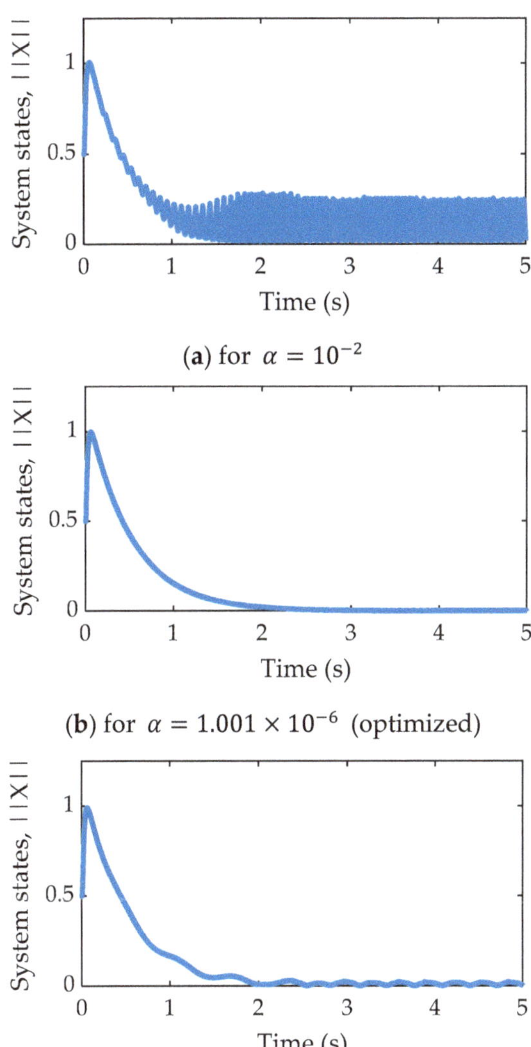

(**a**) for $\alpha = 10^{-2}$

(**b**) for $\alpha = 1.001 \times 10^{-6}$ (optimized)

(**c**) for $\alpha = 1.001 \times 10^{-6}$ (optimized) with actuator disturbance

Figure 6. Single-link robot system states, $\|X\|$.

In Figure 7, the control effort waveforms are shown for $\alpha = 10^{-2}$ (very high), $\alpha = 1.001 \times 10^{-6}$ (optimized) and $\alpha = 1.001 \times 10^{-6}$ (optimized) with actuator disturbance. Again, the performance of the optimized case (Figure 7b) is much better than that of Figure 7a. Hence, it can be said that minimum control effort could be maintained with an optimized triggering parameter. As shown in Figure 7c, a slightly high control effort is required to keep the system states in stable trajectory due to disturbance added in the actuator signal. However, it is important to notice that the proposed ETC scheme works absolutely fine for small disturbances. Figure 8 presents the inter-event execution time, T_k. The range of T_k is between 0.001 and 0.15 s for $\alpha = 10^{-2}$ (very high), as shown in Figure 8a, whereas for $\alpha = 1.001 \times 10^{-6}$ (optimized) it is between 0.001 and 0.28 s, as shown in Figure 8b. For the third case with actuator disturbance as shown in Figure 8c, the range of

T_k lies between 0.001 and 0.045 s. This is due to the fact that the number of control updates would increase when a continuous disturbance signal affects the actuator/control signal.

The chattering can also be observed in non-optimized event-triggered control cases (e.g., Figure 3a, Figure 4a, Figure 6a, and Figure 7a). The optimized event-triggered control cases do not show any chattering (e.g., Figure 3b, Figure 4b, Figure 6b, and Figure 7b).

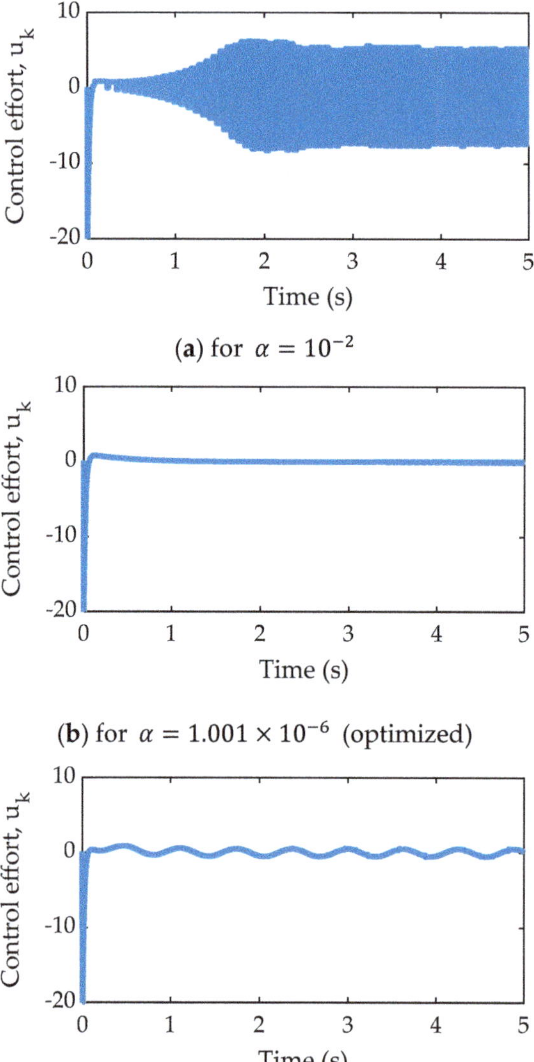

(**a**) for $\alpha = 10^{-2}$

(**b**) for $\alpha = 1.001 \times 10^{-6}$ (optimized)

(**c**) for $\alpha = 1.001 \times 10^{-6}$ (optimized) with actuator disturbance

Figure 7. Control effort, u_k.

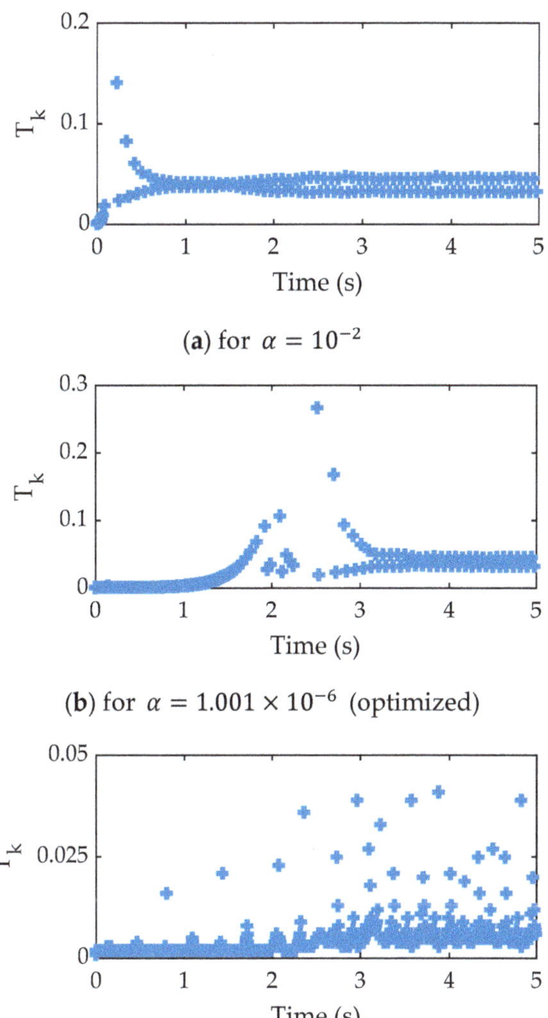

(**a**) for $\alpha = 10^{-2}$

(**b**) for $\alpha = 1.001 \times 10^{-6}$ (optimized)

(**c**) for $\alpha = 1.001 \times 10^{-6}$ (optimized) with actuator disturbance

Figure 8. Inter event execution time, T_k.

Finally, we consider the non-zero value of x_{1d}, i.e., trajectory tracking control of the proposed scheme. The reference trajectory for this objective is considered as $x_{1d} = 0.1\sin(0.2\pi t)$. Next, the reference trajectory for the second system state is evaluated using the expression of Equation (8) for $i = 1$. Hence, the x_{2d} is given as:

$$x_{2d} = g_1^{-1}\left[\dot{x}_{1d} - f_1 + k_1 \cdot \tilde{x}_1\right] \tag{73}$$

Based on the above reference trajectories, x_{1d} and x_{2d}, the closed-loop system has been simulated and the results are shown in Figure 9. It should be noted that the closed-loop response shown in Figure 9 is obtained only for the optimized event-triggered control for $\alpha = 1.001 \times 10^{-6}$. All other cases have been described in Figures 7–9. As shown in the figure, the proposed controller forces the system states (in Figure 9a,b) to follow the desired trajectory very satisfactorily. Similarly, the control effort does not vary after 1 s duration (in Figure 9c). However, the inter-event time (in Figure 9d) has been reduced as compared to the previous cases. It is due to variable reference trajectories where a continuous control command update is required. In cases where the control command does not need regular updates, the number of data samples is minimized significantly. This has been demonstrated in the case of $x_{1d} = 0$.

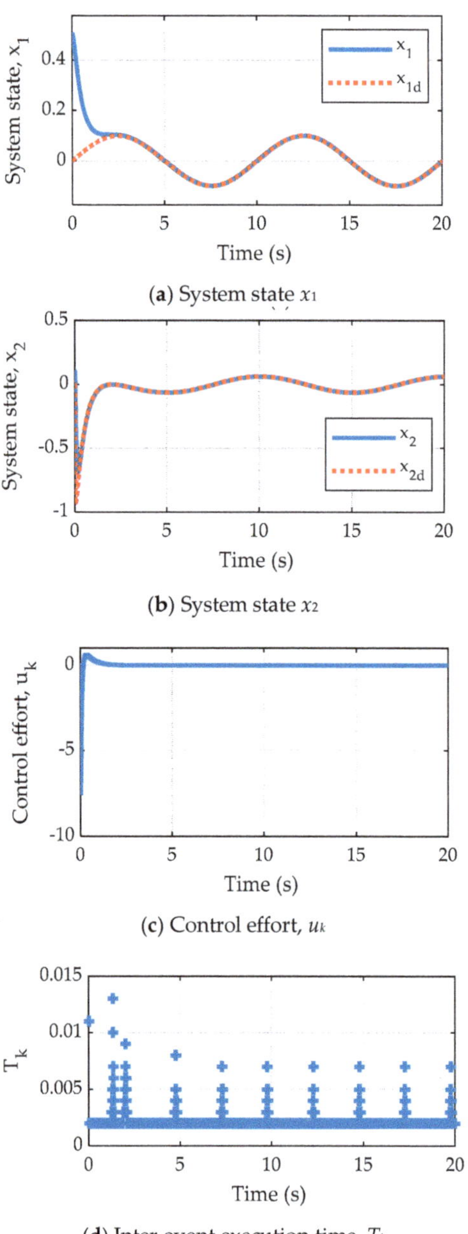

(**a**) System state x_1

(**b**) System state x_2

(**c**) Control effort, u_k

(**d**) Inter event execution time, T_k

Figure 9. Closed-loop response for trajectory tracking control using optimized event-triggered control.

5. Conclusions

This paper presented an ETC scheme for an NCS. Initially, a backstepping control law was designed for a nonlinear system in strict-feedback form. Thereafter, an optimized event triggering algorithm by assuring the closed-loop stability of the system was developed. Later, the proposed ETC scheme was implemented on two case studies of an inverted pendulum and single-link robot system. Next, the simulations were performed to verify the effectiveness of the proposed ETC scheme. The performance of the system states, control efforts, state error due to non-uniform sampling and event execution time interval was analyzed for both the nonlinear systems. The proposed ETC scheme can be applied to any generalized nonlinear system provided it is in the strict-feedback form. Another important contribution of this study is the optimized event triggering algorithm that helps the ETC to maintain the optimum performance of the nonlinear system with a reduced number of samples required to be sent over the communication channel. As compared to the non-optimized event triggering algorithm, the optimized event triggering algorithm takes more triggering pulses but provides very efficient tracking of state trajectories. The non-optimized event triggering parameter reduces the triggering samples but leads to the degradation of system performance, as shown in simulation results. The effects of a communication network for data transmission have not been taken into account in the present work, which might be the focus of future work.

Author Contributions: Conceptualization, S.K.M. and B.A.; methodology, S.K.M. and A.V.J.; software, S.K.M. and V.K.V.; validation, B.A. and A.V.J.; investigation, A.Y.A.; resources, V.K.V.; data curation, A.V.J.; writing—original draft preparation, S.K.M.; supervision, N.B.; project administration, B.A.; formal analysis, N.B.; funding acquisition, N.B.; visualization, N.B.; writing—review and editing, N.B.; figure and table, S.K.M. All authors have read and agreed to the published version of the manuscript.

Funding: There is no funding available for this research.

Institutional Review Board Statement: Not applicable.

Informed Consent Statement: Not applicable.

Data Availability Statement: Not applicable.

Conflicts of Interest: The authors declare no conflict of interest.

References

1. Åarzén, K.-E. A Simple Event-Based PID Controller. *IFAC Proc. Vol.* **1999**, *32*, 8687–8692. [CrossRef]
2. Heemels, W.P.M.H.; Sandee, J.H.; Van Den Bosch, P.P.J. Analysis of Event-Driven Controllers for Linear Systems. *Int. J. Control* **2008**, *81*, 571–590. [CrossRef]
3. Vamvoudakis, K.G. Event-Triggered Optimal Adaptive Control Algorithm for Continuous-Time Nonlinear Systems. *IEEE/CAA J. Autom. Sin.* **2014**, *1*, 282–293. [CrossRef]
4. Sahoo, A.; Xu, H.; Jagannathan, S. Neural Network-Based Event-Triggered State Feedback Control of Nonlinear Continuous-Time Systems. *IEEE Trans. Neural Netw. Learn. Syst.* **2016**, *27*, 497–509. [CrossRef]
5. Yu, H.; Hao, F. Input-to-State Stability of Integral-Based Event-Triggered Control for Linear Plants. *Automatica* **2017**, *85*, 248–255. [CrossRef]
6. Wang, A.; Liu, L.; Qiu, J.; Feng, G. Event-Triggered Robust Adaptive Fuzzy Control for a Class of Nonlinear Systems. *IEEE Trans. Fuzzy Syst.* **2019**, *27*, 1648–1658. [CrossRef]
7. Lu, W.; Yin, X.; Fu, Y.; Gao, Z. Observer-Based Event-Triggered Predictive Control for Networked Control Systems under Dos Attacks. *Sensors* **2020**, *20*, 6866. [CrossRef] [PubMed]
8. Mazo, M.; Tabuada, P. Special Issue Technical Notes and Correspondence: Decentralized Event-Triggered Control over Wireless Sensor/Actuator Networks. *IEEE Trans. Automat. Control* **2011**, *56*, 2456–2461. [CrossRef]
9. Tripathy, N.S.; Kar, I.N.; Paul, K. An Event-Triggered Based Robust Control of Robot Manipulator. In Proceedings of the 2014 13th International Conference on Control Automation Robotics and Vision, ICARCV, Singapore, 10–12 December 2014; pp. 425–430. [CrossRef]
10. Bhadu, M.; Tripathy, N.S.; Kar, I.N.; Senroy, N. Event-Triggered Communication in Wide-Area Damping Control: A Limited Output Feedbackbased Approach. *IET Gener. Transm. Distrib.* **2016**, *10*, 4094–4104. [CrossRef]

11. Mahmoud, M.S.; Sabih, M. Networked Event-Triggered Control: An Introduction and Research Trends. *Int. J. Gen. Syst.* **2014**, *43*, 810–827. [CrossRef]

12. Gautam, M.K.; Pati, A.; Mishra, S.K.; Appasani, B.; Kabalci, E.; Bizon, N.; Thounthong, P. A Comprehensive Review of the Evolution of Networked Control System Technology and Its Future Potentials. *Sustainability* **2021**, *13*, 2962. [CrossRef]

13. Behera, A.K.; Bandyopadhyay, B. Self-Triggering-Based Sliding-Mode Control for Linear Systems. *IET Control Theory Appl.* **2015**, *9*, 2541–2547. [CrossRef]

14. Behera, A.K.; Bandyopadhyay, B. Event-Triggered Sliding Mode Control for a Class of Nonlinear Systems. *Int. J. Control* **2016**, *89*, 1916–1931. [CrossRef]

15. Behera, A.K.; Bandyopadhyay, B. Robust Sliding Mode Control: An Event-Triggering Approach. *IEEE Trans. Circuits Syst. II Express Briefs* **2017**, *64*, 146–150. [CrossRef]

16. Postoyan, R.; Tabuada, P.; Nesic, D.; Anta, A. Event-Triggered and Self-Triggered Stabilization of Distributed Networked Control Systems. In Proceedings of the 2011 50th IEEE Conference on Decision and Control and European Control Conference, Orlando, FL, USA, 12–15 December 2011; pp. 2565–2570. [CrossRef]

17. Abdelrahim, M.; Postoyan, R.; Daafouz, J.; Nešić, D. Stabilization of Nonlinear Systems Using Event-Triggered Output Feedback Controllers. *IEEE Trans. Automat. Control* **2016**, *61*, 2682–2687. [CrossRef]

18. Jiao, J.; Wang, G. Event Driven Tracking Control Algorithm for Marine Vessel Based on Backstepping Method. *Neurocomputing* **2016**, *207*, 669–675. [CrossRef]

19. Li, Y.X.; Yang, G.H. Model-Based Adaptive Event-Triggered Control of Strict-Feedback Nonlinear Systems. *IEEE Trans. Neural Networks Learn. Syst.* **2018**, *29*, 1033–1045. [CrossRef] [PubMed]

20. Li, Y.X.; Yang, G.H. Event-Triggered Adaptive Backstepping Control for Parametric Strict-Feedback Nonlinear Systems. *Int. J. Robust Nonlinear Control* **2018**, *28*, 976–1000. [CrossRef]

21. Mishra, S.K.; Purwar, S.; Kishor, N. Event-Triggered Nonlinear Control of OWC Ocean Wave Energy Plant. *IEEE Trans. Sustain. Energy* **2018**, *9*, 1750–1760. [CrossRef]

22. Mishra, S.K.; Appasani, B.; Jha, A.V.; Garrido, I.; Garrido, A.J.; Costa Castelló, R. Centralized Airflow Control to Reduce Output Power Variation in a Complex OWC Ocean Energy Network. *Complexity* **2020**. [CrossRef]

23. Zhang, C.H.; Yang, G.H. Event-Triggered Control for a Class of Strict-Feedback Nonlinear Systems. *Int. J. Robust Nonlinear Control* **2019**, *29*, 2112–2124. [CrossRef]

24. Huang, J.; Wang, Q.G. Event-Triggered Adaptive Control of a Class of Nonlinear Systems. *ISA Trans.* **2019**, *94*, 10–16. [CrossRef] [PubMed]

25. Xing, L.; Wen, C.; Liu, Z.; Su, H.; Cai, J. Event-Triggered Output Feedback Control for a Class of Uncertain Nonlinear Systems. *IEEE Trans. Automat. Control* **2019**, *64*, 290–297. [CrossRef]

26. Li, T.; Wen, C.; Yang, J.; Li, S.; Guo, L. Event-Triggered Tracking Control for Nonlinear Systems Subject to Time-Varying External Disturbances. *Automatica* **2020**, *119*. [CrossRef]

27. Xu, B.; Liu, X.; Wang, H.; Zhou, Y. Event-Triggered Control for Nonlinear Systems via Feedback Linearisation. *Int. J. Control* **2020**, 1–11. [CrossRef]

28. Hu, X.; Yu, H.; Hao, F. Lyapunov-Based Event-Triggered Control for Nonlinear Plants Subject to Disturbances and Transmission Delays. *Sci. China Inf. Sci.* **2020**, *63*. [CrossRef]

29. Zou, K.; Cai, Y.; Chen, L.; Sun, X. Event-Triggered Nonlinear Model Predictive Control for Trajectory Tracking of Unmanned Vehicles. *Proc. Inst. Mech. Eng. Part D J. Automob. Eng.* **2021**. [CrossRef]

30. Cui, L.; Zhang, Y.; Wang, X.; Xie, X. Event-Triggered Distributed Self-Learning Robust Tracking Control for Uncertain Nonlinear Interconnected Systems. *Appl. Math. Comput.* **2021**, *395*. [CrossRef]

31. Su, X.; Wen, Y.; Shi, P.; Wang, S.; Assawinchaichote, W. Event-Triggered Fuzzy Control for Nonlinear Systems via Sliding Mode Approach. *IEEE Trans. Fuzzy Syst.* **2021**, *29*, 336–344. [CrossRef]

32. Mishra, S.K.; Pati, A.; Appasani, B.; Jha, A.V.; Kumar, M.R.; Pal, V.C.; Gautam, M.K. Event-Triggered Fractional-Order PID Control of Fractional-Order Networked Control System. In *Lecture Notes in Electrical Engineering*; Springer: Berlin/Heidelberg, Germany, 2021; Volume 690, pp. 205–216. [CrossRef]

33. Clerc, M. *Particle Swarm Optimization*; ISTE Ltd.: London, UK, 2010. [CrossRef]

34. Wang, J.J. Simulation Studies of Inverted Pendulum Based on PID Controllers. *Simul. Model. Pract. Theory* **2011**, *19*, 440–449. [CrossRef]

mathematics

MDPI

Article

Dynamic Stability Performance of Autonomous Microgrid Involving High Penetration Level of Constant Power Loads

Mohamed A. Hassan [1,*], Muhammed Y. Worku [1], Abdelfattah A. Eladl [2] and Mohammed A. Abido [3,4]

1   Center for Engineering Research, Research Institute, King Fahd University of Petroleum & Minerals, Dhahran 31261, Saudi Arabia; muhammedw@kfupm.edu.sa
2   Electrical Engineering Department, Faculty of Engineering, Mansoura University, Mansoura 35516, Egypt; eladle7@mans.edu.eg
3   Electrical Engineering Department, Faculty of Engineering, King Fahd University of Petroleum & Minerals, Dhahran 31261, Saudi Arabia; mabido@kfupm.edu.sa
4   King Abdullah City for Atomic and Renewable Energy (K.A.CARE), Energy Research & Innovation Center (ERIC) at KFUPM, Dhahran 31261, Saudi Arabia
*   Correspondence: mhassan@kfupm.edu.sa; Tel.: +966-13-860-7332

Citation: Hassan, M.A.; Worku, M.Y.; Eladl, A.A.; Abido, M.A. Dynamic Stability Performance of Autonomous Microgrid Involving High Penetration Level of Constant Power Loads. *Mathematics* **2021**, *9*, 922. https://doi.org/10.3390/math9090922

Academic Editor: Nicu Bizon

Received: 12 December 2020
Accepted: 4 February 2021
Published: 21 April 2021

Abstract: Nowadays, behaving as constant power loads (CPLs), rectifiers and voltage regulators are extensively used in microgrids (MGs). The MG dynamic behavior challenges both stability and control effectiveness in the presence of CPLs. CPLs characteristics such as negative incremental resistance, synchronization, and control loop dynamic with similar frequency range of the inverter disturb severely the MG stability. Additionally, the MG stability problem will be more sophisticated with a high penetration level of CPLs in MGs. The stability analysis becomes more essential especially with high-penetrated CPLs. In this paper, the dynamic stability performance of an MG involving a high penetration level of CPLs is analyzed and investigated. An autonomous MG engaging a number of CPLs and inverter distributed generations (DGs) is modeled and designed using MATLAB. Voltage, current, and power controllers are optimally designed, controlling the inverter DGs output. A power droop controller is implemented to share the output DGs powers. Meanwhile, the current and voltage controllers are employed to control the output voltage and current of all DGs. A phase-locked loop (PLL) is essentially utilized to synchronize the CPLs with the MG. The controller gains of the inverters, CPLs, power sharing control, and PLL are optimally devised using particle swarm optimization (PSO). As a weighted objective function, the error in the DC voltage of the CPL and active power of the DG is minimized in the optimal problem based on the time-domain simulation. Under the presence of high penetrated CPLs, all controllers are coordinately tuned to ensure an enhanced dynamic stability of the MG. The impact of the highly penetrated CPLs on the MG dynamic stability is investigated. To confirm the effectiveness of the proposed technique, different disturbances are applied. The analysis shows that the MG system experiences the instability challenges due to the high penetrated CPLs. The simulation results confirm the effectiveness of the proposed method to improve the MG dynamic stability performance.

Keywords: constant power load; microgrid; dynamic stability; optimization; PLL; power-sharing control

1. Introduction

Renewable energy sources (RESs) such as fuel cells, photovoltaic arrays (PVs), and wind are recently utilized to avoid the fossil fuels environmental distresses [1]. At the distribution level, passive and active loads are engaged with RESs to perform a microgrid (MG) [2,3]. A MG can operate either on autonomous or grid-connected modes. Addressing several challenges such as technologies, management, reliability, uncertainties, control, integration, islanding, operation, power quality, protection, and stability, MGs have been considered recently as a key topic [3–8]. A comprehensive review of MG technologies and their applications was presented [4]. Solar PV, wind, hydro, biomass, and conventional energy systems were involved in this MG. Considering PV, double-fed induction generator (DFIG)-based wind, diesel generator, and critical and non-critical loads, an efficient under frequency control and the energy management of an RESs-based MG were presented [5]. Additionally, an efficient power management control for MGs with energy storage was presented [6]. The proposed control scheme increases the reliability and resiliency of the MG. Both the MG and its controller were developed in a Real-Time Digital Simulator (RTDS). Due to the uncertainties of both loads and RESs, an optimal operation of RESs was probabilistically investigated [7]. An optimal PQ control scheme was proposed to control and share predefined injected real and reactive powers of the MG [8]. An optimal design of MGs in autonomous and grid-connected modes was introduced [3]. For both modes, the controller parameters of the MG are optimally designed to make the MG more stable after getting disturbed. The dynamic operation and control strategies for a hybrid MG involving wind, PV, fuel cell, and static VAR compensator (SVC) were examined. This stand-alone MG could effectively extract the maximum power from the wind and PV energy sources [9]. A static synchronous compensator (STATCOM) was controlled using a novel intelligent controller to reduce the power fluctuations, voltage support, and damping in an MG system [10]. To improve the MG protection system and reduce the maximum transient overvoltage of a wind turbine, a grounding scheme was modified [11]. An advanced demand-side management (DSM) and control strategy were introduced for an efficient energy management system (EMS) in smart microgrid (SMG) [12]. An optimal cost-effective EMS operation was firstly proposed based on a two-level genetic algorithm (GA) optimization problem and augmented with the time-of-use pricing (ToU) principle. Secondly, the SMG voltage and frequency were optimally regulated using an improved proportional integral derivative (PID)-based mixed sensitivity H-infinity (PID-MSH∞) control scheme while operating in islanded mode. The proposed DSM and control strategy harness the immense internet of things (IoT) aptitudes to ensure an economic and secure operation of the SMG. Depending on the MG operation modes, disturbances, and time frame, MG stability was classified in [13]. Additionally, a comprehensive review presented and addressed the challenges and effects of the instability MG problems. Load perturbations, changing operating conditions, and interactions between the load dynamics and generation dynamics significantly disturbed the MG stability [14]. Intensely, the MG stability depends on several parameters such as controller parameters, load dynamics, LC filters, and controller dynamics [3]. It was reported that the low-frequency modes are dominantly affected with both load demand and outer controller parameters. Meanwhile, the medium and high-frequency damped modes are mainly influenced with the load dynamics, filter components, and inner controller parameters [14]. Electrical loads are classified as passive and active loads. Resistive space heaters and incandescent lighting are common passive load examples. They are usually modeled as a resistor or an inductor–resistor combination. Meanwhile, electronics loads such as switching power supplies, motor drive systems, and electrical vehicles are considered as active loads. Recently, power electronics applications have been utilized in different applications such as RESs, electrical vehicles, MGs, and smart grids. Nevertheless, such active loads tend to behave like constant power loads (CPLs) [15].

CPLs have been tremendously used in distributed power systems. They are extensively involved in MGs with a main concern in stability studies [14–20]. It is worth mentioned that CPLs significantly reduce the MG damping [14]. A comprehensive review on CPLs compensation techniques in the DC microgrid (DCMG) and AC microgrid (ACMG) was presented and classified in [16]. The CPLs stability problem in MG was defined and analyzed [17]. Several compensation techniques were introduced to overcome this problem. CPLs characteristics such as negative incremental resistance, synchronization, and control loop dynamic with similar frequency range of the inverter affect considerably the MG dynamic [15]. CPLs show a negative incremental impedance. It means that the input current decreases with increasing supply voltage. Meanwhile, the absorbed power does not change with any input voltage fluctuations, which moderates the MG damping causing instability or oscillatory response [12]. Modeling and analysis of IBMGs with CPLs were investigated using a new linear state-space model in [18]. A simple linear model corrected through a time-step simulation was proposed for CPLs. A sliding mode controller was developed for an MG system in the presence of CPLs to keep the output voltage constant at 480 V [19]. A robustness analysis of the sliding mode controller was performed. The performance of the proportional integral derivative (PID) was compared with the sliding mode controller to prove the superiority of the sliding mode controller over the PID controller. Popov's absolute stability theorem was utilized to analyze stability conditions for an AC MG in the presence of CPL [18]. Furthermore, the stability of the DCMG was investigated in [21–24]. A comprehensive small-signal model was derived to study the overall stability of the DCMG involving CPLs [21]. The instability issue induced by the CPLs was revealed using the impedance matching criteria. The virtual-impedance-based stabilizers were presented to enhance the damping of DCMGs with CPLs and guarantee the stable operation. Using the active power control of PV arrays, a virtual inertia control (VIC) was introduced to enhance the inertia of a hybrid PV Array-battery DCMG [22]. In this proposed VIC, there is no need for any high-power energy storage system such as supercapacitors. Impedance-based stability analysis was utilized to study the VIC impact on improving the stability margin of the DCMG in the presence of CPLs. A finite-time disturbance observer (FTDO)-based backstepping control strategy with finite-time disturbance observers was presented to ensure the large signal stability of DCMGs using high boost ratio interleaved converter interfaced energy storage systems (ESSs) [23]. The proposed controller is utilized for stabilizing interleaved double dual boost converter (IDDBC) feeding CPLs in DCMGs. It is not limited to IDDBC but also applicable to other types of interleaved converters, which achieves fast dynamics and accurate tracking with large signal stability. A robust passivity-based control (PBC) strategy was presented to solve the instability problem caused by CPLs in DCMG systems [24]. The strategy was designed to stabilize and regulate the DC-bus voltage of the DCMG.

CPL must be synchronized with MG to get the right power amount on the right time [25]. Regularly, phase-locked loop (PLL) is used to synchronize the CPL with MG, track the frequency, and extract the voltage phase angle of the MG [26]. A potential instability problem was raised because of the synchronization coupling and interaction between the synchronization unit and MG impedance [15]. It is worth mentioning that the PLL parameters should be carefully tuned to have a better MG dynamic performance [3]. Additionally, the CPLs control dynamics have a similar frequency range of the inverters. Therefore, studying the interaction between dynamics of the distributed generations (DGs) and CPLs should be essentially considered [27]. With high-level penetrations of CPLs, the MG stability will be more complicated and need more investigations. Having advantages such as flexibility, redundancy, and expandability, droop control techniques are proposed to improve the low-frequency damping in both steady-state and transient modes and overcome the stability problems related to the sudden disturbances [3,12,28,29]. A coordinated virtual impedance control strategy for DGs units was presented to overcome the mismatched line impedance and avoid inaccurate power sharing and circulating current [28]. Both virtual resistance and virtual inductance were simultaneously tuned to compensate the mismatched line impedance among DGs. The MG system stability was enhanced by increasing damping for the whole system. A generalized droop control (GDC) was proposed for a grid-supporting inverter based on a comparison between traditional droop control and virtual synchronous generator (VSG) control [30]. The proposed GDC can achieve satisfactory control performance and provide virtual inertia and damping properties in autonomous mode. The output active power of an inverter can follow the changing references quickly and accurately without large overshoot or oscillation in the grid-connected mode. The controller parameters and transient gains were designed and tuned using the trial-and-error method [27]. However, this method has significant drawbacks such as being time-consuming and failing to obtain the optimal settings [30]. Additionally, in large MGs, fixed-gain controllers cannot certainly adapt to disturbances and load changes especially with parameters variation [31]. The proper selection of both MG and CPL parameters is essential to improve the power quality and system performance against load switching and different disturbances [27]. To overcome power system problems successfully, computational intelligence techniques such artificial neural networks, genetic algorithm (GA), fuzzy logic (FL), and particle swarm optimization (PSO) have been recently introduced [3,12,32–36]. These techniques are used to improve the MG transient performance [3]. However, their applications have some disadvantages [5]. A combined sizing and energy management methodology was formulated as a leader–follower problem [32]. The leader problem was focusing on sizing and aiming at selecting the optimal size for the MG components using a genetic algorithm. A low-complexity FL controller was designed and embedded in an energy management system for a residential grid-connected MG including RESs and storage capability [33]. The system assumes that neither the renewable generation nor the load demand is controllable. The main goal of the design was to minimize the grid power profile fluctuations while keeping the battery state of charge within secure limits [34]. Compared with other optimization techniques, especially GA, PSO has several advantages such as robustness, computational efficiency, simplicity, easy implementation, effective capability memory, and greater efficiency [3,12,35,36]. Additionally, using PSO can avoid some GA disadvantages such as converging toward the local solution rather than the global solution, especially with improperly defined objective function, and difficulties with the dynamic datasets [35]. Furthermore, it is worth mentioning that the best results are obtained using the PSO technique [36].

In this paper, MG stability with high-level penetrations of CPLs is analyzed and investigated. An autonomous MG including CPLs is modeled and presented. PLL is employed to synchronize the involved CPLs with the MG and track the MG frequency. An optimal design of the power-sharing controller, CPLs controllers, inverters controllers, and PLL gains is performed. The control problem is designed based on minimizing a weighted objective function to limit the error in the DC voltage and the measured active power. The DGs, CPLs, and PLL controllers are coordinately tuned to improve the MG dynamic stability. Additionally, the dynamic stability of the autonomous MG is examined. Different disturbances are applied to verify and assess the impact of the optimal power-sharing parameters on the MG stability with high-level penetrations of CPLs.

2. Autonomous Microgrid Model

The considered MG is shown in Figure 1. Three CPLs are fed from three inverter based DGs through coupling inductances, LC filters, and transmission lines. LC filters are imposed to filter out the high frequency switching noises. Current, voltage, and power controllers are employed to control the DGs. Emulating the operation of the synchronous generator, the droop controller illustrated in Figure 2 is designed to accurately share the DGs output powers, which depend mainly on the power angle and amplitude of the output voltage, respectively. The DGs output voltage and current (v_o and i_o) are measured firstly to obtain the instantaneous real and reactive powers (P_m and Q_m) as illustrated in (1). Secondly, the average real and reactive powers (P_c and Q_c) are obtained from the instantaneous powers using a low pass filter as in (2). Finally, the average powers are used to obtain the frequency (ω) and d-axis reference voltage (v^*_{od}) as given in (3).

$$P_m = v_{od}i_{od} + v_{oq}i_{oq}, \quad Q_m = v_{od}i_{oq} - v_{oq}i_{od} \tag{1}$$

$$P_c = \frac{\omega_c}{\omega_c + s}P_m, \quad Q_c = \frac{\omega_c}{\omega_c + s}Q_m \tag{2}$$

$$\begin{aligned} \omega &= \omega_n - m_p P_c, & \dot{\theta} &= \omega \\ v^*_{od} &= V_n - n_q Q_c, & v^*_{oq} &= 0 \end{aligned} \tag{3}$$

where i_{odq} and v_{odq} are the dq components of the DGs output current and voltage respectively, ω_c is the cut-off frequency, ω_n is the nominal angular frequency of DG, V_n is the nominal magnitude of the DG voltage, and m_p and n_q are the droop controller gains.

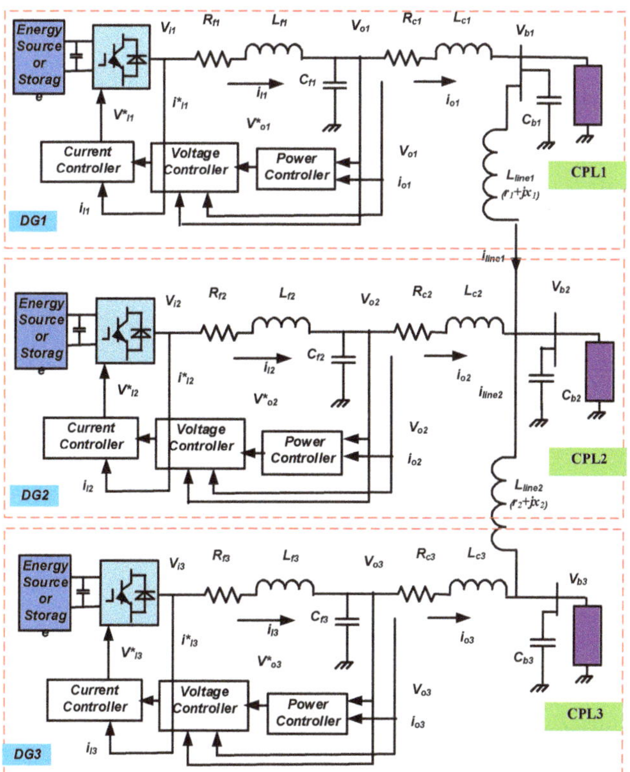

Figure 1. The considered microgrid.

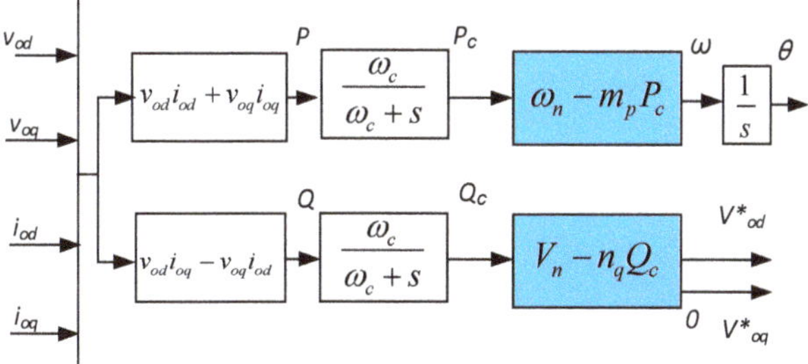

Figure 2. Distributed generations (DGs) power droop controller.

To control the DGs output voltages and currents, two controllers are employed as shown in Figure 3. The proportional integral (PI) current and voltage controllers are used to generate the reference voltage and current signals, respectively. In both controllers, the reference and measured signals are compared to generate the reference signals. To mimic the output impedance of the voltage source, the feed-forward terms are engaged [4]. On a common reference frame (DQ), the state equations of the current and voltage controllers, coupling inductance, line and load currents, LC filter, and load voltages are given respectively in Equations (4)–(11).

$$i^*_{ld} = Fi_{od} - \omega_n C_f v_{oq} + K_{pv}(v^*_{od} - v_{od}) + K_{iv}\int (v^*_{od} - v_{od})dt$$
$$i^*_{lq} = Fi_{oq} + \omega_n C_f v_{od} + K_{pv}(v^*_{oq} - v_{oq}) + K_{iv}\int (v^*_{oq} - v_{oq})dt \tag{4}$$

$$v^*_{id} = -\omega_n L_f i_{lq} + K_{pc}(i^*_{ld} - i_{ld}) + K_{ic}\int (i^*_{ld} - i_{ld})dt$$
$$v^*_{iq} = \omega_n L_f i_{ld} + K_{pc}(i^*_{lq} - i_{lq}) + K_{ic}\int (i^*_{lq} - i_{lq})dt \tag{5}$$

$$\frac{di_{ld}}{dt} = -\frac{r_f}{L_f}i_{ld} + \omega i_{lq} + \frac{1}{L_f}(v_{id} - v_{od})$$
$$\frac{di_{lq}}{dt} = -\frac{r_f}{L_f}i_{lq} - \omega i_{ld} + \frac{1}{L_f}(v_{iq} - v_{oq}) \tag{6}$$

$$\frac{dv_{od}}{dt} = \omega v_{oq} + \frac{1}{C_f}(i_{ld} - i_{od})$$
$$\frac{dv_{oq}}{dt} = -\omega v_{od} + \frac{1}{C_f}(i_{lq} - i_{oq}) \tag{7}$$

$$\frac{di_{od}}{dt} = -\frac{r_c}{L_c}i_{od} + \omega i_{oq} + \frac{1}{L_c}(v_{od} - v_{bd})$$
$$\frac{di_{oq}}{dt} = -\frac{r_c}{L_c}i_{oq} - \omega i_{od} + \frac{1}{L_c}(v_{oq} - v_{bq}) \tag{8}$$

$$\frac{di_{lineDi}}{dt} = -\frac{r_{linei}}{L_{linei}}i_{lineDi} + \omega i_{lineQi} + \frac{1}{L_{linei}}(v_{bDj} - v_{bDk})$$
$$\frac{di_{lineQi}}{dt} = -\frac{r_{linei}}{L_{linei}}i_{lineQi} - \omega i_{lineQi} + \frac{1}{L_{linei}}(v_{bQj} - v_{bQk}) \tag{9}$$

$$\frac{di_{loadDi}}{dt} = -\frac{R_{loadi}}{L_{loadi}}i_{loadDi} + \omega i_{loadQi} + \frac{1}{L_{loadi}}v_{bDi}$$
$$\frac{di_{loadQi}}{dt} = -\frac{R_{loadi}}{L_{loadi}}i_{loadQi} - \omega i_{loadDi} + \frac{1}{L_{loadi}}v_{bQi} \tag{10}$$

$$\frac{dv_{bDi}}{dt} = \omega v_{bQi} + \frac{1}{C_f}(i_{oDi} - i_{loadDi} \pm i_{lineDi,j})$$
$$\frac{dv_{bQi}}{dt} = -\omega v_{bDi} + \frac{1}{C_f}(i_{oQi} - i_{loadQi} \pm i_{lineQi,j}) \tag{11}$$

where i^*_{odq} and v^*_{odq} are the dq components of reference output current and voltage, i^*_{ldq} and v^*_{idq} are the dq components of the reference inductor current and inverter voltage, i_{ld} and i_{lq} are the dq components of the inverter current, i_{lineDQ} and i_{loadDQ} are the DQ components of the line and load currents, V_{bDQ} are the DQ load voltage components and i_{oDQ} are the DQ output current components. L_f, r_f, and C_f are the filter components. r_c, L_c, L_{line}, and r_{line} are the coupling inductor and line components, L_{load} and r_{load} are the load parameters. F is the voltage controller feed-forward gain. K_{pc}, K_{ic}, K_{pv}, and K_{iv} are the PI current and voltage controller parameters of each DG.

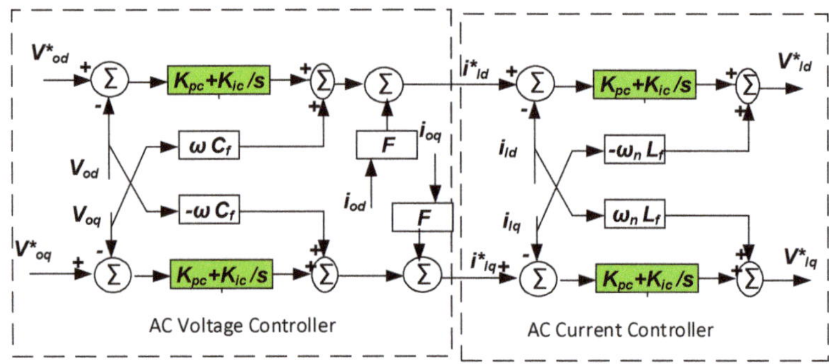

Figure 3. AC voltage and current controllers.

3. Constant Power Load (CPL) Model

Nowadays, CPLs have been recently used in MGs [14–24]. CPLs have two important characteristics: negative incremental resistance and control loop dynamic with a similar frequency range of the inverter [14]. The negative resistance property and the interaction between the DGs and CPLs reduce the system damping and lead to instability or unacceptable oscillatory responses [15]. Furthermore, the highly penetrated CPLs in an MG make the MG stability problem more sophisticated. The dynamic behavior of an MG challenges both stability and control effectiveness in the presence of CPLs. Several methods have been proposed to cope with the mentioned CPL instability [14–24]. The CPL circuit shown in Figure 4 is modeled in MATLAB. The model contains the rectifier bridge, DC load, LC filter, AC current, and DC voltage controllers, and PLL. The rectifier bridge is used to convert the AC into DC and feed the DC load. The AC current and DC voltage controllers are shown in Figure 5. The DC voltage controller is employed to control the DC voltage of the CPL and to generate the reference current of the current controller, while the current controller is utilized to control the AC inductor current and to generate the required rectifier pulses. The PLL depicted in Figure 6 is used to synchronize the CPLs with the MG, extract the phase angle, and deliver the reference voltage for the MG. To extract the grid frequency, one of the *dq* voltage components is devised to be constant. The *q*-component of the voltage of the point of common connection (PCC) and reference voltage are compared. The error signal is fed to the PI controller to generate the frequency. This frequency is compared with the reference frequency to extract the synchronized MG frequency. The synchronized frequency is integrated to obtain the inverter phase reference θ needed to convert from *abc* to *dq* and vice versa, as given in (13).

$$\omega = k_P^{PLL}(v_{oq} - v_{oq}^*) + k_I^{PLL} \int (v_{oq} - v_{oq}^*)dt \tag{12}$$

$$\theta = \int (\omega - \omega_{ref})dt + \theta(0) \tag{13}$$

where $k_p{}^{PLL}$ and $k_I{}^{PLL}$ are the PLL controller parameters.

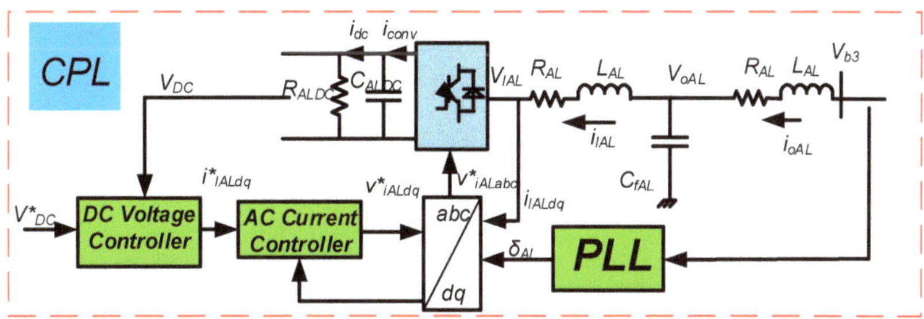

Figure 4. Constant power loads (CPL) circuit.

The common reference (*DQ*) frame of the whole MG, reference (*dq_i*) frame of each DGi, and reference (*dq_{ALj}*) frame of the jth CPL are shown in Figure 7. These frames are rotated at frequencies ω_{com}, ω_i, and ω_{ALj}, respectively. The angles δ_i and δ_{ALj} represent the angles between the DGi reference frame, CPL reference frame, and DQ frame, respectively. Each DG and CPL must be modeled on its rotating reference frame. DGs and CPLs share the MG DQ frame. The CPL current controller has the same state equations of the DG inverter. However, the CPL current has an opposite sign to the DG current because the CPL receives the current from the MG. The *dq* components of the inductor current are decoupled using the feed-forward as displayed in Figure 5. The CPLs state-space equations are given in (14)–(20).

$$v_{idAL}i_{idAL} + v_{iqAL}i_{iqAL} = i_{convAL}v_{DCAL} \tag{14}$$

$$\frac{dv_{DCAL}}{dt} = \frac{1}{C_{dcAL}}i_{convAL} - \frac{1}{R_{dcAL}C_{dcAL}}v_{DCAL} \tag{15}$$

$$i^*_{ldAL} = K_{pv\_AL}(v^*_{DCAL} - v_{DCAL}) + K_{iv\_AL}\int (v^*_{DCAL} - v_{DCAL})dt \tag{16}$$

$$\begin{aligned}
v^*_{idAL} &= \omega_n L_{fAL}i_{lqAL} - K_{pc_AL}(i^*_{ldAL} - i_{ldAL}) - K_{ic_AL}\int (i^*_{ldAL} - i_{ldAL})dt \\
v^*_{iqAL} &= -\omega_n L_{fAL}i_{ldAL} - K_{pc_AL}(i^*_{lqAL} - i_{lqAL}) - K_{ic_AL}\int (i^*_{lqAL} - i_{lqAL})dt
\end{aligned} \tag{17}$$

$$\begin{aligned}
\frac{di_{ldAL}}{dt} &= -\frac{r_{fAL}}{L_{fAL}}i_{ldAL} + \omega_{AL}i_{lqAL} + \frac{1}{L_{fAL}}(v_{idAL} - v_{odAL}) \\
\frac{di_{lqAL}}{dt} &= -\frac{r_{fAL}}{L_{fAL}}i_{lqAL} - \omega_{AL}i_{ldAL} + \frac{1}{L_{fAL}}(v_{iqAL} - v_{oqAL})
\end{aligned} \tag{18}$$

$$\begin{aligned}
\frac{dv_{odAL}}{dt} &= \omega_{AL}v_{oqAL} + \frac{1}{C_{fAL}}(i_{ldAL} - i_{odAL}) \\
\frac{dv_{oqAL}}{dt} &= -\omega_{AL}v_{odAL} + \frac{1}{C_{fAL}}(i_{lqAL} - i_{oqAL})
\end{aligned} \tag{19}$$

$$\begin{aligned}
\frac{di_{odAL}}{dt} &= -\frac{r_{cAL}}{L_{cAL}}i_{odAL} + \omega i_{oqAL} + \frac{1}{L_{cAL}}(v_{odAL} - v_{bdAL}) \\
\frac{di_{oqAL}}{dt} &= -\frac{r_{cAL}}{L_{cAL}}i_{oqAL} - \omega_{AL}i_{odAL} + \frac{1}{L_{cAL}}(v_{oqAL} - v_{bqAL})
\end{aligned} \tag{20}$$

where v_{idqAL} and i_{idqAL} are the dq components of the CPL output voltage (v_{iAL}) and current (i_{lAL}) respectively; i_{convAL} is the DC side current of the CPL; C_{dcAL} and R_{dcAL} are the DC side capacitor components of the CPL; L_{fAL} is the filter inductance of the CPL; and k_{pv_AL}, k_{iv_AL}, K_{pc_AL}, and K_{ic_AL} are the DC voltage and AC current controllers' parameters of the CPL.

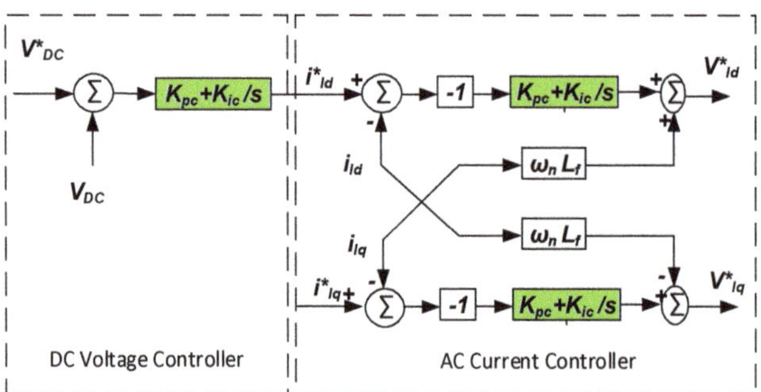

Figure 5. CPL controllers.

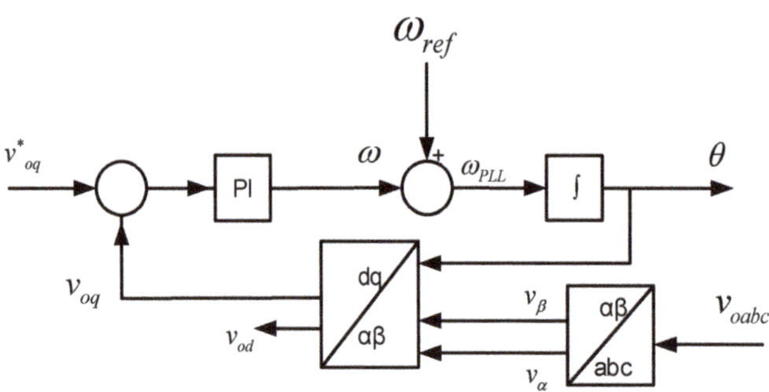

Figure 6. Phase-locked loop (PLL).

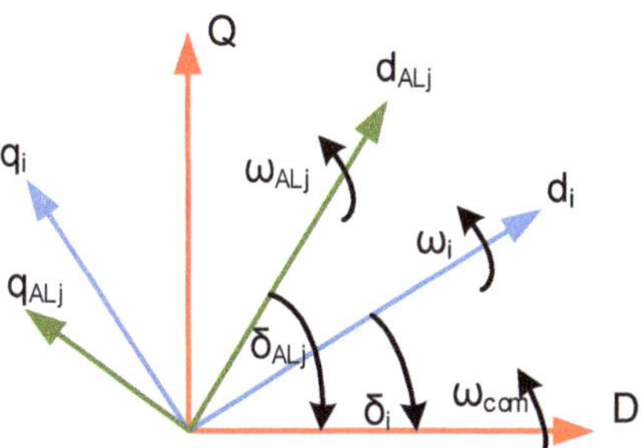

Figure 7. Common reference frame (DQ).

4. Problem Formulation

To enhance the MG stability, the controller and power-sharing parameters of both DGs and CPLs, and PLL parameters are cautiously tuned. One of the big advantages of obtaining optimal parameters is that the controller parameters can be obtained easily to get the stability of the MG. In this paper, the PSO technique is used to optimally tune these parameters. As reported in [33], compared with other optimization techniques, especially GA, the best results are obtained using the PSO method because of its easy implementation with less tuned parameters, effective memory, and greater efficiency for maintaining swarm diversity. The design problem is formulated as follows:

4.1. Objective Function and Problem Constraints

The optimal controller gains are established based on the time domain simulation. Therefore, the controller design problem has been solved using PSO. The error in the DC voltage and measured active power of the CPL has been curtailed using a time-domain weighted objective function (J). The problem is optimally designed and formulated with the parameter constraints as follows:

$$\text{Minimize } J = \int_{t=0}^{t=t_{sim}} \left[\left(P_m - P_{ref} \right)^2 + \left(V_{dc} - V_{dcref} \right)^2 \right] . t \, dt \tag{21}$$

Subject to

$$K^{min} \leq K \leq K^{max}, m_p{}^{max} \leq m_p \leq m_p{}^{max}, m_{pAL}{}^{max} \leq m_{pAL} \leq m_{pAL}{}^{max}, n_q{}^{min} \leq n_q \leq n_q{}^{max}, n_{qAL}{}^{min} \leq n_{qAL} \leq n_{qAL}{}^{max},$$
$$\text{and } k_p{}^{PLL\ min} \leq k_p{}^{PLL} \leq k_p{}^{PLL\ max}, k_I{}^{PLL\ min} \leq k_I{}^{PLL} \leq k_I{}^{PLL\ max}. \tag{22}$$

Bounded as $K^{min} \leq K \leq K^{max}$, the controller parameters are $K = [k_{pv}, k_{iv}, k_{pc}, k_{ic}, k_{pv\_AL}, k_{iv\_AL}, k_{pc\_AL}, k_{ic\_AL}]^T$. m_p, n_q, m_{p_AL}, and n_{q_AL} are the power sharing gains. $k_p{}^{PLL}$ and $k_I{}^{PLL}$ are the PLL controller gains. To guarantee the minimum settling time, t is added.

4.2. Particle Swarm Optimization

In this paper, PSO is exploited to gain the optimal parameters. Based on the swarm behavior, a bird flock is filing in a stochastic manner for food searching. PSO is developed to emulate the swarm behavior in 1995 [31]. As an efficient optimization tool, PSO could poise between the local and global search methods. Behind its robustness and flexibility, PSO can enhance the search capability and overcome the premature convergence problem [32]. Figure 8 depicts the PSO computational flow. In a PSO algorithm, the population has n particles that represent candidate solutions. Each particle is an m–dimensional real-valued vector, where m is the number of optimized parameters. Therefore, each optimized parameter represents a dimension of the problem space. The PSO technique can be summarized as follows.

- Initially, n particles and their velocities are randomly created. The time starts counting. The initial related objective functions are determined. The lowest one is nominated as a global best function J_{best}. Its associated particle is chosen as the global best particle x_{best}. The inertia weight is carefully initiated to control the current velocity.
- The time counter is updated.
- The new inertia weight is calculated as $w(t) = \alpha * w(t-1)$; α is close to 1.
- At each time step, each particle velocity is adapted depending on the distance between the particle and its personal position, the distance between the particle and the global best position, and the current velocity.

$$v^i_{n+1} = wv^i_n + c_1 r_1 \left(p_{best} - k^i_n \right) + c_2 r_2 \left(g_{best} - k^i_n \right) \tag{23}$$

where r_1 and r_2 are random numbers between 0 and 1; c_1 and c_2 are the "trust" parameters; w is the inertia weight.

At time n, g_{best} is the best position in the swarm; p_{best} is the best position for particle i.

It is worth mentioning that based on its own thinking and memory, the particle changes its velocity. The second term is the PSO cognitive part, while the third term is the PSO social part. It is based on the social–psychological adaptation of knowledge.

- At iteration $n + 1$, the new particle position is determined based on the updated velocities as follows,

$$k^i_{n+1} = k^i_n + v^i_{n+1} \tag{24}$$

where k^i_{n+1} and v^i_{n+1} are the position and velocity of particle i, respectively.

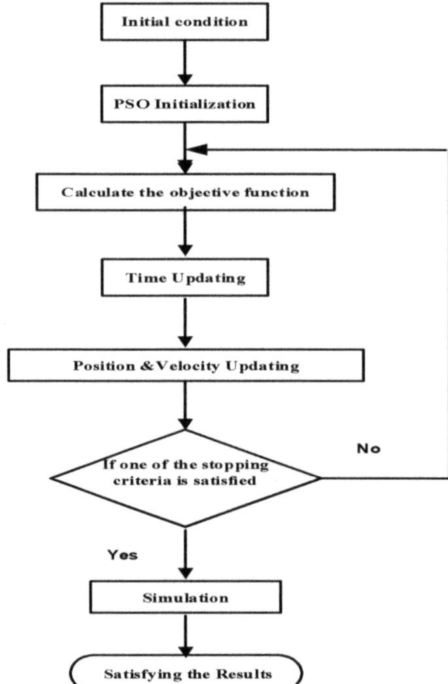

Figure 8. Particle swarm optimization (PSO) computational flow.

The cost function J of each particle is updated depending on its new position. Compared with the previous one, the lowest one will be selected to be the global best J_j^*. Additionally, the individual best will be nominated as a global best.

- From all global best values J_j^*, the minimum one will be chosen as given.
- If $J_{min} > J^{**}$, then update the global best as $X^{**} = X_{min}$ and $J^{**} = J_{min}$
- PSO stops searching when the number of iterations exceeds the pre-specified number or maximum allowable iterations.

4.3. PSO Implementation

The proposed PSO has been built in MATLAB. The maximum allowable velocity and initial inertia weight are the main parameters that affect the PSO performance. To have a good optimization performance, these parameters should be cautiously designated. In this paper, based on the authors' experiences, the selected PSO parameters are given as follows.

- Inertia weight factor =1
- Generation or iteration = 100
- Population size = 20
- Acceleration constants: $c_1 = c_2 = 2$
- Decrement constant (α) = 0.98.

5. Results and Discussion

To investigate the proposed control effectiveness, a MATALB code has been created to model the MG shown in Figure 1. The model contains three CPLs (7.4 kVA) engaged with three identical DGs (10 kVA) through filters, coupling inductances, and transmission lines. Table 1 depicts the MG parameters. The optimal parameters are given in Table 2.

5.1. Simulation Results

To examine and verify the impact of the optimal controllers' parameters of the CPL, PLL, droop, and DGs on the MG stability, different disturbances have been applied. Firstly, a three-phase fault has been applied at CPL3. Figure 9a–d illustrate the active and reactive powers, output voltage, and output current responses of all DGs, respectively, while Figure 9e depicts the DC output voltage of the CPLs. At t = 3.3 s, the measured active powers of DG1, DG2, and DG3 have been increased from 1.4 to 2.4 p.u., from 0.4 to 1.5 p.u. and from 0.4 to 2.48 p.u., respectively. While the measured reactive powers of DG1, DG2, and DG3 have been increased from 0.384 to 1.14 p.u., from −0.04 to 0.435 p.u., and from 0.0 to 0.54 p.u., respectively. The measured d-axis voltages of DG1, DG2, and DG3 have been increased from 1.0 to 1.14, 1.15, and 1.28 p.u. respectively. The d-axis of the measured output currents of DG1 has been increased from 1.5 to 2.45 p.u. while the d-axis of the measured output currents of DG2 and DG3 have been increased from 0.5 to 1.6 and 3.48 p.u., respectively. The DC voltages of the three CPLs have been increased from 1.17 to 1.47, 1.39, and 1.38 p.u., respectively. As shown in Figure 9, with the high penetration level of CPLs, the system is getting stable within 2 s after the transient has been settled. The proposed controller needs this time to return back to the stable mode because of the high number of CPLs. In terms of overshoot and settling time, the results impressively show an enhanced and adequate damping MG performance under this fault. In addition, the DGs output powers have been tightly increased after occurring the fault directly.

Table 1. System parameters.

Microgrid Parameters			
Parameter	Value	Parameter	Value
f_s	8 kHz	V_n	381 V
L_f	1.35 mH	L_c	0.35 mH
C_f	50×10^{-6} F	C_b	50×10^{-6} F
r_f	0.1 Ω	r_c	0.03 Ω
ω_n	314.16 rad/s	ω_c	31.416 rad/s
$r_1 + jx_1$	$(0.23 + j0.1)$ Ω	$r_2 + jx_2$	$(0.35 + j0.58)$ Ω
CPL Parameters			
L_f	2.3 mH	L_c	0.93 mH
C_f	8.8×10^{-6} F	r_c	0.03 Ω
r_f	0.1 Ω		
R_{dc}	72 Ω	C_{dc}	2040×10^{-6} F

Table 2. Optimal parameters.

Power Sharing Parameters of the Three DG Units			
m_p	3.79404×10^{-7}	n_q	9.36593×10^{-5}
	6.75934×10^{-7}		1.86121×10^{-5}
	1.71857×10^{-7}		3.21349×10^{-5}
m_d	0.2957×10^{-5}	n_d	0.2618×10^{-6}
	0.1572×10^{-5}		0.2374×10^{-6}
	-0.0030×10^{-5}		0.2374×10^{-6}
PLL Parameters			
$K_p{}^{PLL}$	1	$K_I{}^{PLL}$	50
Controller Parameters of the Three DG Units			
Parameter	Value	Parameter	Value
K_{pv} (Amp/Watt)	1.19585	K_{pc} (Amp/Watt)	44.1091
	1.43531		31.8037
	1.6380		40.8816
K_{iv} (Amp/Joule)	4.4568	K_{ic} (Amp/Joule)	35.8275
	6.17159		26.904
	-0.69434		13.4463
CPL Parameters			
K_{pv_AL} (Amp/Watt)	0.331792	K_{pc_AL} (Amp/Watt)	33.2732
K_{iv_AL} (Amp/Joule)	4.33114	K_{ic_AL} (Amp/Joule)	-4.61844

The three-phase fault has been relocated at CPL1 to check the controller effectiveness. Figure 10a,b illustrates the responses of the active power of the three DGs and DC output voltage of the CPLs, respectively. At t = 3.3 s, the measured active powers of DG1, DG2, and DG3 have been increased from 1.44 to 3.4 p.u., from 0.44 to 2.7 p.u., and from 0.44 to 1.5 p.u., respectively. Meanwhile, the DC voltages of the three CPLs have been increased from 1.17 to 1.45, 1.38, and 1.37 p.u., respectively. The results display satisfactory damping characteristics of the proposed controller for this disturbance. Moreover, Figure 11 illustrates the MG response when the voltage at CPL3 has been stepped down to 0.25 p.u. The measured active powers of DG1, DG2, and DG3 have been increased from 1.44 to 2.3 p.u., from 0.44 to 1.5 p.u., and from 0.44 to 2.45 p.u., respectively. Meanwhile, the DC voltages of the three

CPLs have been increased from 1.17 to 1.44, 1.37, and 1.35 p.u., respectively. The output responses display the satisfactory damping characteristics of the proposed controller. The DGs output power responses are displayed in Figure 11a, while Figure 10b depicts the DC voltage response of CPLs. It has been demonstrated that the system oscillations are highly damped, which shows the effectiveness of the proposed method.

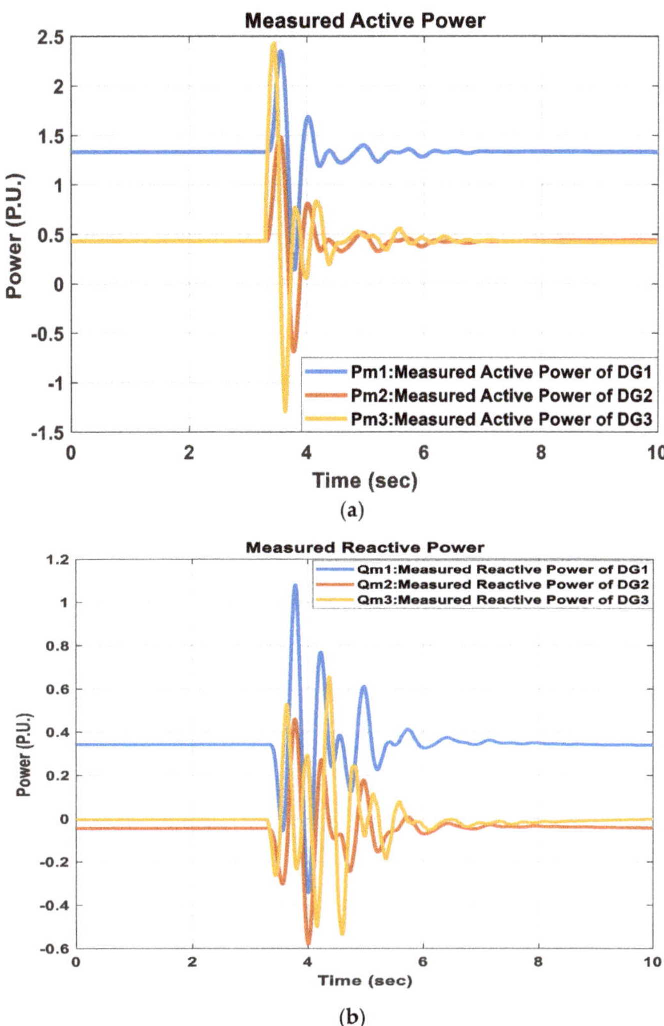

Figure 9. *Cont.*

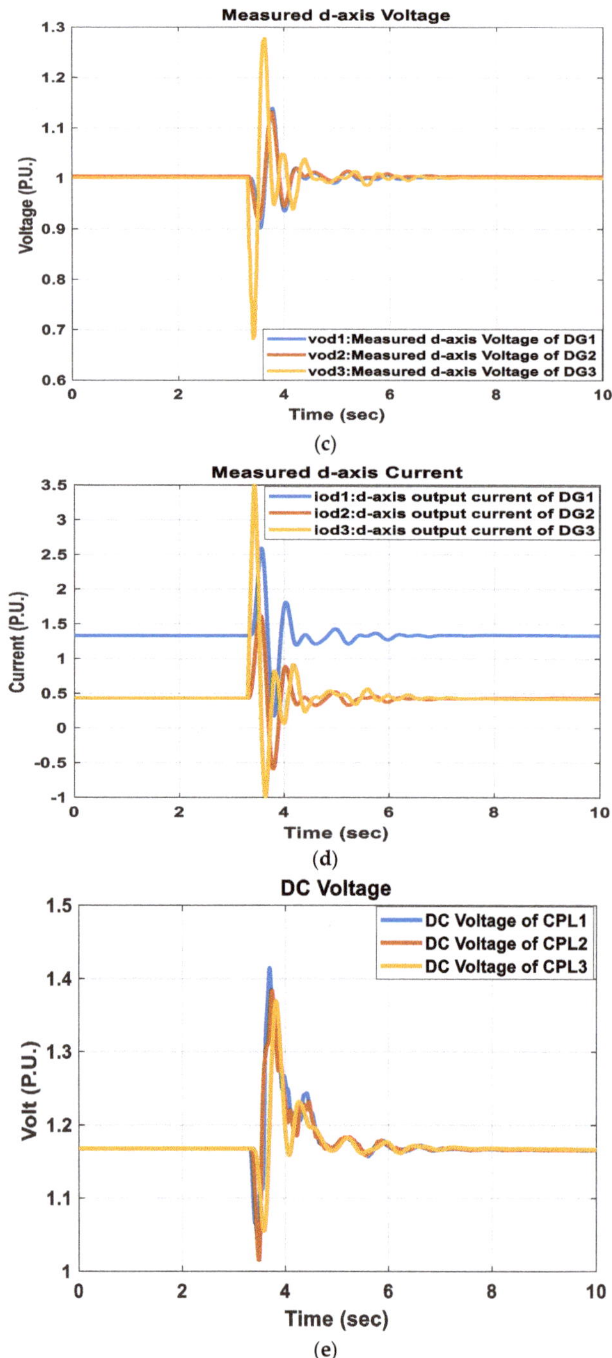

Figure 9. Microgrid response with a three-phase fault at CPL.3. (**a**) Measured active powers of DGs. (**b**) Measured reactive powers of DGs. (**c**) Measured d-axis voltages of DGs. (**d**) Measured d-axis currents of DGs. (**e**) DC voltages of CPLs.

Additionally, at t = 3.3 s, a three-phase fault is located at line 2. The line has been recovered at t = 3.6 s. During the fault, CPL1 and CPL2 have been fed from DG1 and DG2, while CPL3 has been fed from DG3. The system is getting stable after this disturbance. The DGs capability to support the loads is clearly illustrated in Figure 12. The DGs output power response is depicted in Figure 12a, while Figure 12b shows the DC voltage response of the CPLs. With the optimal settings, the DC reference voltage of CPL1 is stepped up to 1.1 p.u. The proposed controller has been investigated through the time domain simulation. The DGs output power responses are displayed in Figure 13a, while Figure 13b depicts the responses of the DGs output voltage for the considered disturbance.

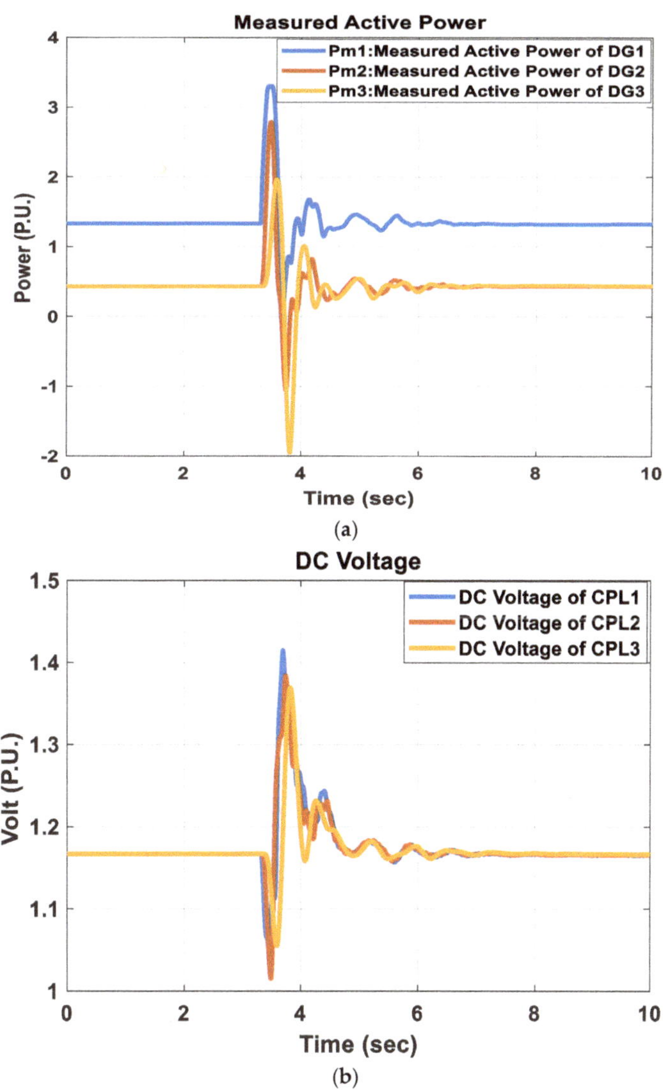

Figure 10. Microgrid response with a three-phase fault at CPL1. (**a**) Measured active powers of DGs. (**b**) DC voltages of CPLs.

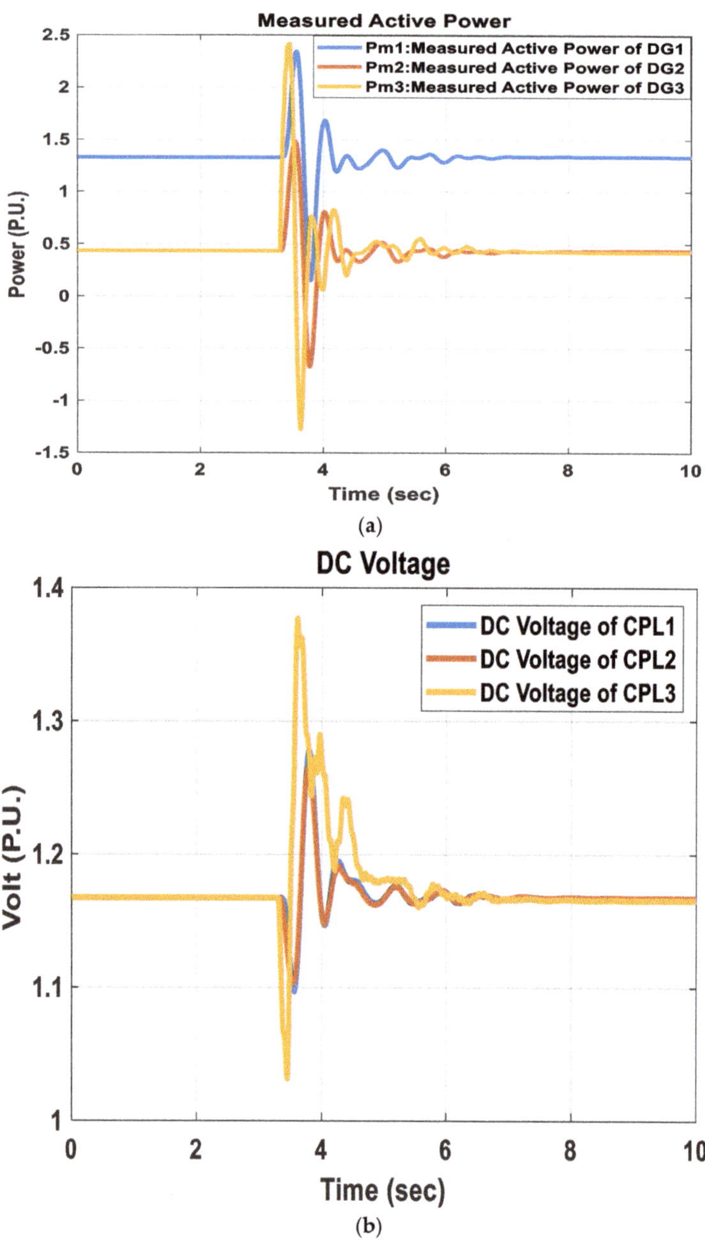

Figure 11. Microgrid response with a voltage step change at CPL.3. (**a**) Measured active powers of DGs. (**b**) DC voltages of CPLs.

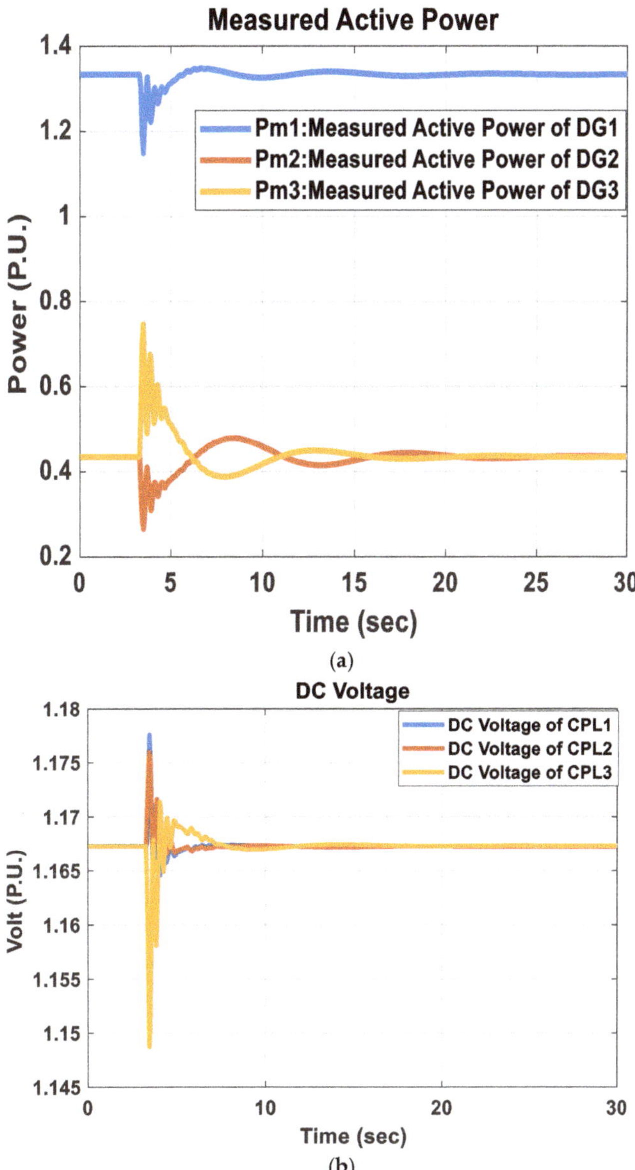

Figure 12. Microgrid response with a fault at TL2. (**a**) Measured active powers of DGs. (**b**) DC voltages of CPLs.

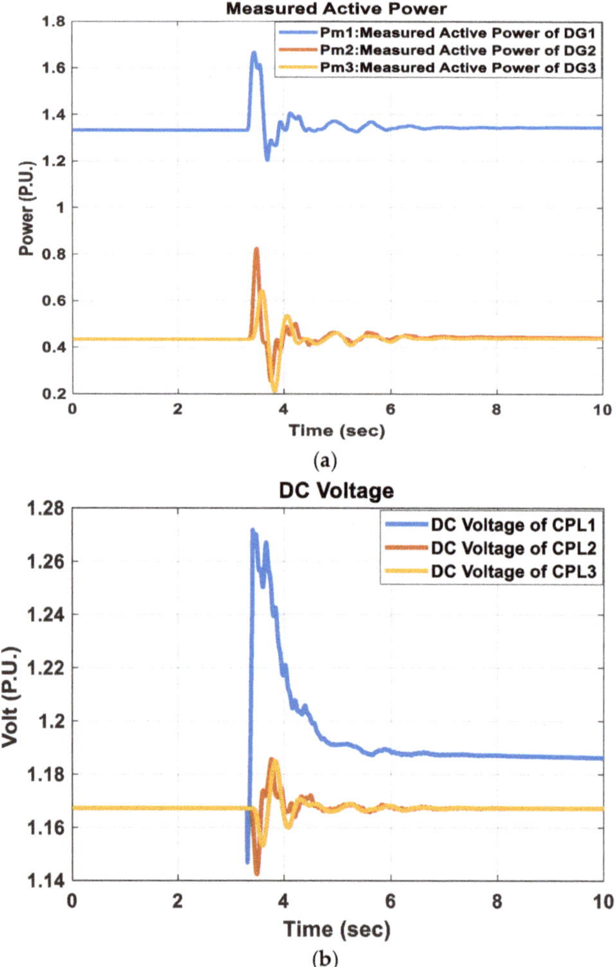

Figure 13. Microgrid response with a reference DC voltage step change at CPL3. (**a**) Measured active powers of DGs. (**b**) DC voltages of CPLs.

For more investigations, the MG performance has been tested when DG1 has been lost. In this case, all CPLs are fed from DG2 and DG3. Figure 14a,b proves the responses of the DGs output power and DC voltage of CPLs for this disturbance. The output responses display the satisfactory damping characteristics of the proposed controller. The output responses show the controller capability for making the system stable after getting disturbance in both cases. With the proposed controller, the MG became stable after getting disturbed and has better dynamic performance. It can be seen from the different applied disturbances that despite variations in the DGs or CPLs side, the system is highly stable with no oscillatory behavior. This indicates the effectiveness of the proposed controller in mitigating the AC power and DC-bus voltage oscillations in the considered MG.

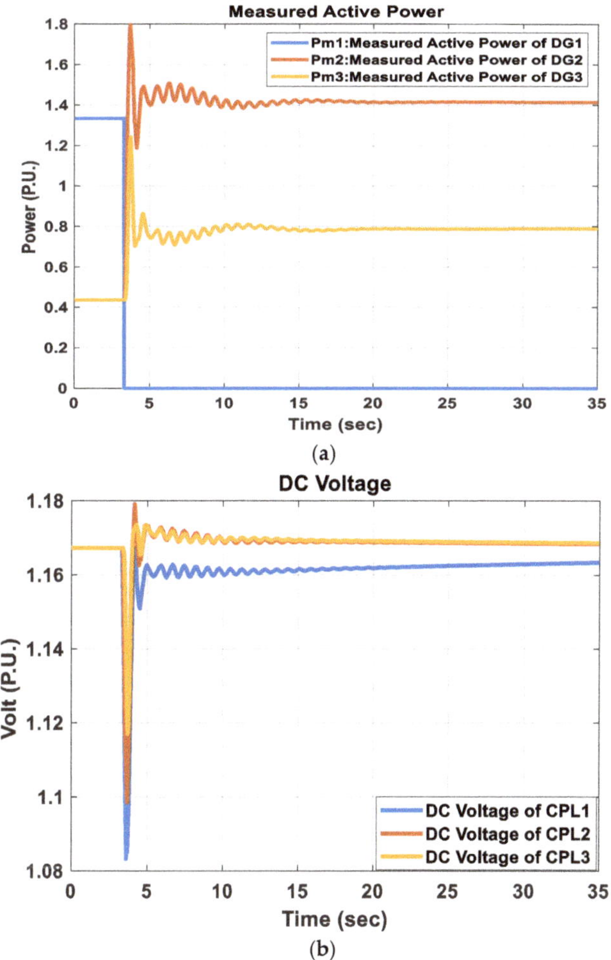

Figure 14. Microgrid response with losing DG1. (**a**) Measured active powers of DGs. (**b**) DC voltages of CPLs.

5.2. Impacts of the Uncertainties of Controller Parameters

In this section, the impacts of the uncertainties of controller parameters are investigated. A three-phase fault has been applied at CPL3 when the integral controller parameter (K_{ic_AL}) of the AC current of the CPL has been deviated from its optimal value by 10% from this value. The system dynamic response of the DGs output powers and DC voltages of the CPLs are depicted in Figure 15a,b. There is no variation in both DGs output power and DC voltages of the CPLs as compared with the results shown in Figure 9. It can be concluded that the optimal design has worked fine, since the system goes to stable mode with fast dynamics and accurate power sharing as expected.

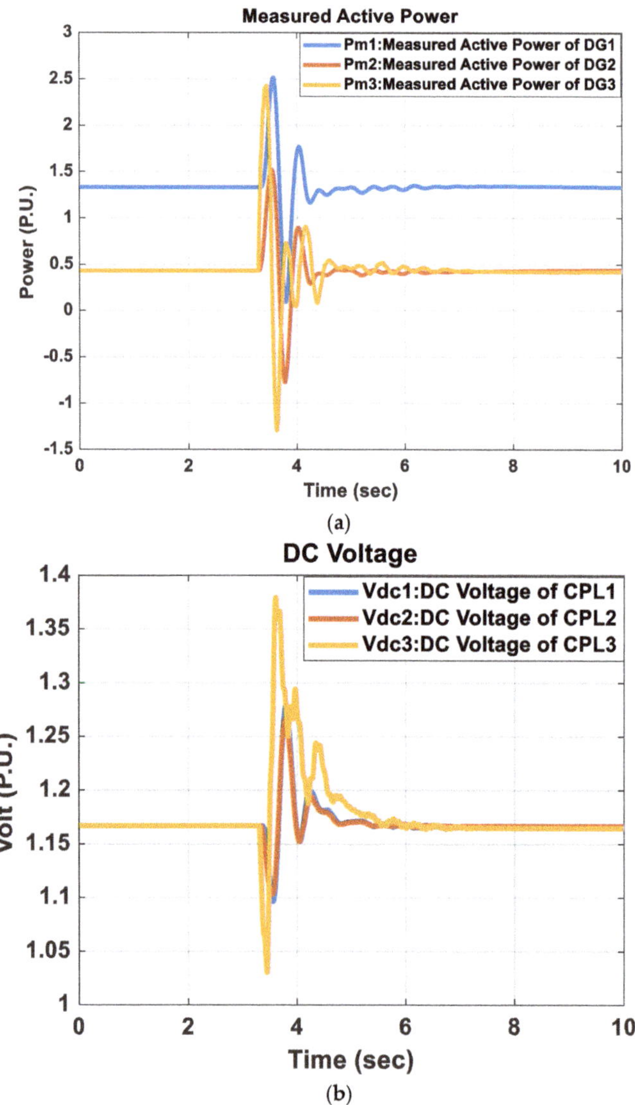

Figure 15. Microgrid response with a three-phase fault at CPL3 with different value of K_{ic_AL}. (**a**) Measured active powers of DGs. (**b**) DC voltages of CPLs.

5.3. Proposed Method Limitations

As a result of the incremental negative impedance effects of CPLs, and the non-linearity and time dependency of converters' operation, classical linear control methods such as a conventional proportional integral PI controller have stability limitations around the operating points. Of course, the problem will be more complicated with the presence of highly penetrated CPLs. To ensure large signal stability, nonlinear stabilizing control methods must be applied [17]. Different compensation techniques need to be addressed and examined to enhance the MG dynamic stability in the presence of high penetrated CPLs. Additionally, our lab capability could not help us verify the obtained results experimentally and examine the effectiveness of the proposed controller.

6. Conclusions

In this paper, MG stability with high penetration level of CPLs is examined and investigated. An autonomous MG including CPLs is modeled using MATALB. Current, voltage, and power controllers are optimally designed to enhance the transient response of the autonomous MG considering CPLs. A power droop controller is utilized to share the DGs output powers in the presence of high penetrated CPLs. PLL is employed to synchronize the involved CPLs with the MG and track the MG frequency. Based on error curtailing in the DC voltage of the CPL and measured active power of the DG, an optimal control design for the controllers of DGs, CPLs, and PLL is presented. Several disturbances are applied to assess the optimal parameters impact on the MG stability. The impact of the high penetration level of CPLs on MG stability is investigated. Additionally, the dynamic stability of the autonomous MG is examined under these disturbances. CPLs are a potential instability source for the MG system. It can be observed from the different applied disturbances in DGs or CPLs sides that the MG system is highly stable with no oscillatory behavior. The simulation results confirm the effectiveness of the proposed method to improve the performance of the MG dynamic stability in the presence of highly penetrated CPLs. The proposed controller provides a fast and accurate power control, an efficient damping characteristic, and satisfactory performance. However, the classical linear control methods such as a PI controller have stability limitations around the operating points. Future work will focus on hardware implementation, testing and verifying the proposed controller effectiveness. Additionally, different MG topologies can be used to study the destabilization effects of the high penetrated CPLs on the AC MGs.

Author Contributions: M.A.H. and M.A.A. initiated the idea, formulated the problem, performed the simulation, and analyzed the results. M.Y.W. and A.A.E. participated in the paper revision stage, contributed to enhancing the simulation results, and shared in paper writing. All authors have read and agreed to the published version of the manuscript.

Funding: The authors acknowledge the support provided by the Center for Engineering (CER), Research Institute at King Fahd University of Petroleum & Minerals. Abido would like also to acknowledge the funding support by King Abdullah City for Atomic and Renewable Energy (K.A. CARE), Energy Research & Innovation Center (ERIC) at KFUPM through the Direct Funded project #DF191004.

Institutional Review Board Statement: Not applicable.

Informed Consent Statement: Not applicable.

Data Availability Statement: Not applicable.

Conflicts of Interest: The authors declare no conflict of interest.

Nomenclature

v_{od}, v_{oq}	dq components of the inverter output voltage v_o
i_{od}, i_{oq}	dq components of the inverter output current i_o
P_m, Q_m	instantaneous active and reactive powers
P_c, Q_c	average active and reactive powers
m_p, n_q	droop controller gains
θ	phase reference
ω	nominal frequency
ω_c	cut-off frequency of the low-pass filter
ω_n	nominal angular frequency of DG
V_n	nominal magnitude of the DG voltage
F	voltage controller feed-forward gain
v^*_{od}, v^*_{oq}	dq components of the reference output voltage
i^*_{ld}, i^*_{lq}	dq components of inductor reference current
v^*_{id}, v^*_{iq}	dq components of the reference inverter voltage
i_{ld}, i_{lq}	dq components of the coupling inductor current i_L

v_{id}, v_{iq}	dq components of the inverter voltage v_i
C_f, L_f, R_f	capacitance, inductance, and resistance of the LC filter
L_c, R_c	inductance and resistance of the coupling inductor
C_{dc}, R_{dc}	capacitance and resistance of the DC load of the active load
δ_i	angle between the reference frame of each inverter (dq)and the common reference frame (DQ)
δ_{AL}	angle between the reference frame of CPL (dq_{AL})and the common reference frame (DQ)
i_{lineDQ}	DQ components of the line
i_{loadDQ}	DQ components of the load currents
v_{dc}, i_{dc}	DC voltage and DC current of the active load respectively
$v^*{}_{DC}$	DC Reference voltage of the CPL
i_{conv}	DC side current of the CPL
v_{idqAL}	dq components of the active load output voltage (v_{iAL})
i_{odqAL}	dq components of the active load output current (i_{oAL})
i_{ldqAL}	dq components of the input current to the bridge (i_{lAL})
K_{pv}, K_{iv}	PI voltage controller parameters of the DG inverter
K_{pc}, K_{ic}	PI current controller parameters of the DG inverter
K_{pv_AL}, K_{iv_A}	PI controller parameters of the DC voltage of the CPL
K_{pc_AL}, K_{ic_AL}	PI controller parameters of the AC current of the CPL

References

1. Colmenar-Santos, A.; Reino-Rio, C.; Borge-Diez, D.; Collado-Fernández, E. Distributed Generation: A Review of Factors that can Contribute Most to Achieve a Scenario of DG Units Embedded in the New Distribution Networks. *Renew. Sustain. Energy Rev.* **2016**, *59*, 1130–1148. [CrossRef]
2. IEEE Standards Association. *IEEE Guide for Design, Operation, and Integration of Distributed Resource Island Systems with Electric Power Systems*; IEEE Std. 1547.4™-2011; IEEE: New York, NY, USA, 2011; pp. 1–54.
3. Hassan, M.; Abido, M. Optimal Design of Microgrids in Autonomous and Grid-Connected Modes Using Particle Swarm Optimization. *IEEE Trans. Power Electron.* **2011**, *26*, 755–769. [CrossRef]
4. Akinyele, D.; Belikov, J.; Levron, Y. Challenges of Microgrids in Remote Communities: A STEEP Model Application. *Energies* **2018**, *11*, 432. [CrossRef]
5. Worku, M.Y.; Hassan, M.A.; Abido, M.A. Real Time-Based under Frequency Control and Energy Management of Microgrids. *Electronics* **2020**, *9*, 1487. [CrossRef]
6. Worku, M.Y.; Hassan, M.A.; Abido, M.A. Real Time Energy Management and Control of Renewable Energy based Microgrid in Grid Connected and Island Modes. *Energies* **2019**, *12*, 276. [CrossRef]
7. Pooranian, Z.; Nikmehr, N.; Najafi-Ravadanegh, S.; Mahdin, H.; Abawajy, J. Economical and environmental operation of smart networked microgrids under uncertainties using NSGA-II. In Proceedings of the 24th International Conference on Software, Telecommunications and Computer Networks (SoftCOM), Split, Croatia, 22–24 September 2016; pp. 1–6. [CrossRef]
8. Hassan, M.A.; Worku, M.Y.; Abido, M.A. Optimal Power Control of Inverter-Based Distributed Generations in Grid-Connected Microgrid. *Sustainability* **2019**, *11*, 5828. [CrossRef]
9. Ou, T.C.; Hong, C.M. Dynamic Operation and Control of Microgrid Hybrid Power Systems. *Energy* **2014**, *66*, 314–323. [CrossRef]
10. Ou, T.C.; Lu, K.H.; Huang, C.J. Improvement of Transient Stability in a Hybrid Power Multi-System Using a Designed NIDC (Novel Intelligent Damping Controller). *Energies* **2017**, *10*, 488. [CrossRef]
11. Hajeforosh, S.F.; Pooranian, Z.; Shabani, A.; Conti, M. Evaluating the High Frequency Behavior of the Modified Grounding Scheme in Wind Farms. *Appl. Sci.* **2017**, *7*, 1323. [CrossRef]
12. Sedhom, B.E.; El-Saadawi, M.M.; El Moursi, M.S.; Hassan, M.A.; Eladl, A.A. IoT-based optimal Demand Side Management and Control Scheme for Smart Microgrid. *Int. J. Electr. Power Energy Syst.* **2021**, *127*, 106674. [CrossRef]
13. Shuai, Z.; Sun, Y.; Shen, Z.J.; Tian, W.; Tu, C.; Li, Y.; Yin, X. Microgrid stability: Classification and a Review. *Renew. Sustain. Energy Rev.* **2016**, *58*, 167–179. [CrossRef]
14. Hassan, M.A. Dynamic Stability of an Autonomous Microgrid Considering Active Load Impact with a New Dedicated Synchronization Scheme. *IEEE Trans. Power Syst.* **2018**, *33*, 4994–5005. [CrossRef]
15. Bottrell, N.; Prodanovic, M.; Green, T. Dynamic Stability of a Microgrid with an Active Load. *IEEE Trans. Power Electron.* **2013**, *28*, 5107–5119. [CrossRef]
16. Hossain, E.; Perez, R.; Nasiri, A.; Padmanaban, S. A Comprehensive Review on Constant Power Loads Compensation Techniques. *IEEE Access* **2018**, *6*, 33285–33305. [CrossRef]
17. Al-Nussairi, M.; Bayindir, R.; Padmanaban, S.; Siano, L.M.P.P. Constant Power Loads (CPL) with Microgrids: Problem Definition, Stability Analysis and Compensation Techniques. *Energies* **2017**, *10*, 1656. [CrossRef]
18. Mahmoudi, A.; Hosseinian, S.; Kosari, M.; Zarabadipour, H. A New Linear Model for Active Loads in Islanded Inverter-Based Microgrid. *Int. J. Electr. Power Energy Syst.* **2016**, *81*, 104–113. [CrossRef]

19. Hossain, E.; Perez, R.; Sanjeevikumar, P.; Siano, P. Investigation on Development of Sliding Mode Controller for Constant Power Loads in Microgrids. *Energies* **2017**, *10*, 1086. [CrossRef]
20. Karimipour, D.; Salmasi, R. Stability Analysis of AC Microgrids with Constant Power Loads Based on Popov's Absolute Stability Criterion. *IEEE Trans. Circuits Syst. II* **2015**, *62*, 7, 696–700. [CrossRef]
21. Lu, X.; Sun, K.; Guerrero, J.M.; Vasquez, J.C.; Huang, L.; Wang, J. Stability Enhancement Based on Virtual Impedance for DC Microgrids with Constant Power Loads. *IEEE Trans. Smart Grid* **2015**, *6*, 2770–2783. [CrossRef]
22. Hosseinipour, A.; Hojabri, H. Virtual Inertia Control of PV Systems for Dynamic Performance and Damping Enhancement of DC Microgrids with Constant Power Loads. *IET Renew. Power Gener.* **2018**, *12*, 430–438. [CrossRef]
23. Xu, Q.; Jiang, W.; Blaabjerg, F.; Zhang, C.; Zhang, X.; Fernando, T. Backstepping Control for Large Signal Stability of High Boost Ratio Interleaved Converter Interfaced DC Microgrids with Constant Power Loads. *IEEE Trans. Power Electron.* **2020**, *35*, 5397–5407. [CrossRef]
24. Hassan, M.A.; He, Y. Constant Power Load Stabilization in DC Microgrid Systems using Passivity-Based Control with Nonlinear Disturbance Observer. *IEEE Access* **2020**, *8*, 92393–92406. [CrossRef]
25. Svensson, J. Synchronization Methods for Grid-Connected Voltage Source Converter. *IEE Proc.-Generat. Transmiss. Distrib.* **2001**, *148*, 229–235. [CrossRef]
26. Luna, A.; Rocabert, J.; Candela, I.; Hermoso, J.; Teodorescu, R.; Blaabjerg, F.; Rodriquez, P. Grid Voltage Synchronization for Distributed Generation Systems under Grid Fault Conditions. *IEEE Trans. Ind. Appl.* **2015**, *51*, 3414–3425. [CrossRef]
27. Du, W.; Zhang, J.; Zhang, Y.; Qian, Z. Stability Criterion for Cascaded System with Constant Power Load. *IEEE Trans. Power Electron.* **2013**, *28*, 1843–1851. [CrossRef]
28. Pham, M.; Lee, H. Effective Coordinated Virtual Impedance Control for Accurate Power Sharing in Islanded Microgrid. *IEEE Trans. Ind. Electron.* **2021**, *68*, 2279–2288. [CrossRef]
29. Meng, X.; Liu, J.; Liu, Z. A Generalized Droop Control for Grid-Supporting Inverter Based on Comparison between Traditional Droop Control and Virtual Synchronous Generator Control. *IEEE Trans. Power Electron.* **2019**, *34*, 5416–5438. [CrossRef]
30. Seidi Khorramabadi, S.; Bakhshai, A. Critic-Based Self-Tuning PI Structure for Active and Reactive Power Control of VSCs in Microgrid Systems. *IEEE Trans. Smart Grid* **2015**, *6*, 92–103. [CrossRef]
31. Mohamed, Y.A.R.I.; El-Saadany, E.F. Adaptive Decentralized Droop Controller to Preserve Power Sharing Stability of Paralleled Inverters in Distributed Generation Microgrids. *IEEE Trans. Power Electron.* **2008**, *23*, 2806–2816. [CrossRef]
32. Li, B.; Roche, R.; Miraoui, A. Microgrid Sizing with Combined Evolutionary Algorithm and MILP Unit Commitment. *Appl. Energy* **2017**, *188*, 547–562. [CrossRef]
33. Arcos-Aviles, D.; Pascual, J.; Marroyo, L.; Sanchis, P.; Guinjoan, F. Fuzzy Logic-Based Energy Management System Design for Residential Grid-Connected Microgrids. *IEEE Trans. Smart Grid* **2016**, *9*, 530–543. [CrossRef]
34. Kennedy, J.; Eberhart, R. Particle Swarm Optimization. In Proceedings of the 1995 IEEE International Conference on Neural Networks, Perth, Australia, 27 November–1 December 1995; pp. 1942–1948.
35. Abido, M. Optimal Design of Power-System Stabilizers Using Particle Swarm Optimization. *IEEE Trans. Energy Convers.* **2002**, *17*, 406–413. [CrossRef]
36. Al-Saedi, W.; Lachowicz, S.; Habibi, D.; Bass, O. PSO Algorithm for an Optimal Power Controller in a Microgrid. *IOP Conf. Ser. Earth Environ. Sci.* **2017**, *73*, 12–28. [CrossRef]

MDPI

Article

Smart Campus Microgrids towards a Sustainable Energy Transition—The Case Study of the Hellenic Mediterranean University in Crete

Alexandros Paspatis, Konstantinos Fiorentzis, Yiannis Katsigiannis and Emmanuel Karapidakis *

Department of Electrical and Computer Engineering, Hellenic Mediterranean University, GR-71410 Heraklion, Greece; agpaspatis@hmu.gr (A.P.); kfiorentzis@hmu.gr (K.F.); katsigiannis@hmu.gr (Y.K.)
* Correspondence: karapidakis@hmu.gr

Abstract: Smart campus microgrids are considered in this paper, with the aim of highlighting their applicability in the framework of the sustainable energy transition. In particular, the campus of the Hellenic Mediterranean University (HMU) in Heraklion, Crete, Greece, is selected as a case study to highlight the multiple campus microgrids' advantages. Crete represents an interesting insular power system case, due to the high renewable energy sources capacity and the large summer tourism industry. There is also a high density of university and research campuses, making the campus microgrid concept a promising solution for the energy transition and decarbonization of the island. In this sense, policy directions that could facilitate the development of the smart campus microgrid are also given, to motivate areas with similar characteristics. For the performed case study, the HMU microgrid is assumed to consist of PV systems, wind turbines, battery energy storage systems and EV chargers. The analysis explores the financial feasibility and environmental impact of such an investment through the optimal sizing of the systems under investigation, while a sensitivity analysis regarding the battery system cost is also performed. Apart from the financial benefits of the investment, it is evident that the main grid experiences a significant load reduction, with the microgrid acting as a RES producer for many hours, hence improving system adequacy. Moreover, it is shown that the location of HMU makes the investment more sustainable compared to other locations in northern Europe, such as Stockholm and London. The methodology and the derived results are expected to motivate such investments, especially in areas with high RES capacity and a high density of university and research campuses.

Keywords: microgrids; university campus; battery energy storage; renewable energy; optimization

MSC: 90C31

Citation: Paspatis, A.; Fiorentzis, K.; Katsigiannis, Y.; Karapidakis, E. Smart Campus Microgrids towards a Sustainable Energy Transition—The Case Study of the Hellenic Mediterranean University in Crete. *Mathematics* **2022**, *10*, 1065. https://doi.org/10.3390/math10071065

Academic Editors: Denis N. Sidorov and Mario Versaci

Received: 28 January 2022
Accepted: 19 March 2022
Published: 25 March 2022

Publisher's Note: MDPI stays neutral with regard to jurisdictional claims in published maps and institutional affiliations.

1. Introduction

From 2021, the European Union (EU) requires its member states to guarantee that all new buildings will be nearly zero-energy buildings [1]. This is expected to be a decisive movement towards a sustainable energy transition [2]. To meet this goal, the use of proper building materials and renewable energy sources (RES) are encouraged [3]. Similarly, in the research community, the adoption of such actions in areas with multiple buildings is also under investigation, through the smart city concept, while from an electric power perspective, such a structure could be regarded as a microgrid [4,5].

Microgrids are usually based on RES and correspond to a set of closely located distributed energy resources (DERs), including renewable energy sources, battery energy storage systems (BESS) and loads. Microgrids can operate either connected to the main electrical grid or in islanded mode. In the latter case (i.e., islanded mode), the BESS system is of crucial importance since it is responsible for maintaining a balance between the production and consumption of electricity [6]. All these features require advanced control and protection devices, which contain the "smartness" of the microgrid operation.

University campuses are a promising application for the establishment of smart microgrids [7,8], while they could be also characterized as small-scale smart cities [8], since they are formed by several adjacent buildings [1]. Apart from the smart microgrid concept, the smart energy system concept has recently been introduced, with emphasis on HVAC needs and water consumption, further from the electrical energy vector [1]. A thorough review of smart energy systems in university campuses is performed in [1]. Nevertheless, the emphasis in this paper will be on the electrical energy vector and the concept of microgrids [9], which can play a crucial role towards the energy transition of power systems [2].

As is common in power system sizing studies, the first factor to be considered during a microgrid sizing study is the expected load to be supplied. The electrical load profile of university campuses is investigated in [10], using the Democritus University of Thrace, Greece, as an example. Based on that case study, a simulation tool is developed for smart microgrids, which aims to minimize electricity cost and achieve optimal components sizing, in [11]. The financial feasibility of campus microgrids is extensively investigated in [7], where an economic model for the analysis and planning of microgrids and hybrid energy systems is proposed, while in [8], a communication framework for the smart campus operation is presented. In fact, many case studies for the transition of campuses into smart campus microgrids have been reported, either focusing on the system sizing [12], the energy management system [13] or the different components under consideration [14]. As discussed in [15], many of the tools employed when a microgrid sizing or operational characterization has to be performed are based on optimization techniques.

Indeed, from an optimization perspective, the desired functionality of a microgrid is usually either to minimize the cost of electricity [16], which can be achieved by a load curve optimization alleviating the peaks in load demand and hence avoiding the spikes in electricity price in hours of high system stress (load shifting), or possibly to govern the power flow through the microgrid/main grid tie-line for multiple objectives. In a microgrid under operation, these functionalities are usually governed through an energy management system (EMS) that gathers all the required measurements and data to perform the optimization-based control action [17]. Note that similar objectives may also be considered during the preliminary sizing study of a microgrid, which is also usually performed through optimization techniques [15].

To highlight the benefits of smart campus microgrids while aiming towards a sustainable energy transition, this paper will consider as a case study a campus on the island of Crete, Greece. Crete, in general, is an interesting case with a significant number of universities and research institutions. In particular, the island hosts four universities (Hellenic Mediterranean University, University of Crete, Technical University of Crete, Patriarchal University Ecclesiastical Academy of Crete), with a total of nine main campuses, as well as three independent research centers (Foundation of Research and Technology Hellas, Hellenic Centre for Marine Research, Orthodox Academy of Crete). Hence, there is a total of twelve campuses on the island. Thus, the results of this study could motivate all academic institutions on the island and play a vital role towards the energy transition, while also providing a useful case study for power systems with similar characteristics around the world. In particular, apart from the high density of university/research campuses, Crete has a high maximum load of almost 700 MW, especially due to the tourism industry that is active during the summer months. Crete also hosts a large capacity of renewable energy generation sources, especially through wind and sun. As it is a weakly interconnected insular power system, the maximum level of RES units has been reached and the installation of new ones for electricity generation is not allowed, until the new HVDC interconnection with the Hellenic Transmission System is established. Hence, novel concepts on the demand side, such as campus microgrids, will significantly support the grid operation reducing the load demand locally, especially in periods of high system stress.

In this paper, the smart campus microgrid concept is considered and its applicability towards a sustainable energy transition is emphasized through the performed case study. The case study investigates such an approach for the campus of the Hellenic Mediterranean University (HMU) in Heraklion, Crete, Greece. More specifically, the performance of the microgrid is examined for two operational cases: (1) HMU to act as an autoproducer and (2) the installation of an independent MV station between HMU load and the grid, which offers an increased power transfer capability. According to the microgrid optimal sizing results, based on financial or environmental indices, the cost of batteries seems to be a crucial factor; thus, a sensitivity analysis of the battery cost is also implemented. Additionally, through the sensitivity analysis, the impact of solar and wind potential is studied using weather data from university campuses located in northern Europe. Overall, it is shown that this kind of investment can be feasible for the second case of operation (MV station) and especially in areas that have a high RES potential, while it further supports the main power system with a significant load reduction and its behavior as a RES producer for many hours of the year. Moreover, due to the large number of university and research campuses on the island, the concept of campus microgrids can lead to significant effects in the plans for energy transition on the island of Crete, a fact that highlights the scope of this paper. To support the replicability of the performed case study in other areas with or without similar characteristics, policy directions that could make such investments sustainable are also highlighted.

This paper is organized as follows: In Section 2, the concept of smart campus microgrids is reviewed and their benefits are categorized with regards to the possible directions through which their expansion could be facilitated. In Section 3, an introduction to microgrid planning is performed. In Section 4, the case study of this paper, which concerns the campus of HMU in Crete, is presented. Finally, Section 5 concludes the paper.

2. Smart Campus Microgrids as Part of the Energy Transition

2.1. Motivation

Sustainability has been identified as a major issue in recent decades. Currently, the United Nations has set 17 goals for sustainable development (SDGs) [18]. Some of these goals refer to affordable and clean energy and the sustainability of cities and communities. With regards to these directions, recent national and international policies have focused on RES investments and utilization of the concept of smart cities. Moreover, a variety of studies have also considered the transformation of university campuses into RES-based microgrids, paving the way towards sustainable electricity. In fact, many universities are already transforming (or planning) their campus into smart microgrids. An important aspect regarding university and research campuses is that the novel technologies that facilitate the green transition can at the same time be employed for educational and research activities, hence, increasing the value of the investment [19]. In the sequel, a brief review of existing solutions regarding smart campus microgrids will be performed.

2.2. Brief Review of Existing Solutions

Aiming to perform a brief review of existing solutions of the smart campus microgrid concept, both existing and under-planning campus microgrids around the world will be discussed, with emphasis given on the employed components and the operation philosophy.

With regards to existing smart campus microgrids, the University of Texas in Austin is usually referred to as the largest and most elaborate microgrid in the United States of America [20]. The Austin microgrid follows a net zero policy and its energy production is based on natural gas, covering also the heat vector, apart from electricity. The University of California, San Diego, microgrid hosts cogeneration plants, solar resources, storage units and fuel cells among others, while it also follows a demand response plan [21]. The Wesleyan University microgrid uses PV systems and cogeneration systems and also employs an energy management system to govern energy consumption [22]. The Princeton University microgrid is based on a CHP plant and PV systems and has the ability to transit between grid-connected and islanded modes [23]. At the Lappeenranta University of Technology in Finland, an islanded operation is also possible during power system contingencies [6,17]. The microgrid of the Universiti Teknologi Petronas consists of two gas turbine generators [24]. Moreover, bidirectional electric vehicle chargers are becoming an important part of campus microgrids, such as the one at Coimbra University in Portugal [25].

With regards to campus microgrids under planning or updating, the Illinois Institute of Technology is expanding its campus by considering a looped architecture that will increase microgrid reliability [26]. Deakin University in Victoria, Australia, is planning a microgrid comprised of PV and BES systems that will also provide grid support services and develop a digital twin model for testing [27]. Additionally, the University of California, Berkeley, currently plans a RES-based microgrid with the capability of islanded operation during power outages [28]. Finally, the 3DMicroGrid project collaboration between academic and industrial partners is working on the development of directions for the design and development of university campus microgrids [29].

Most of the existing approaches are the outcome of a university's administration and research projects' actions. Nevertheless, wider national or transnational actions would significantly impact the energy transition of those areas. This can especially be the case in areas with a high density of academic institutions, such as Crete, which will serve as the case study of this paper. In the sequel, possible directions for promoting the smart campus microgrid concept towards energy transition are given.

2.3. Systematic Campus Microgrids Development towards a Sustainable Energy Transition

Despite there being many cases around the world where universities are moving towards smart campus microgrids, wider and systematic approaches towards this direction could lead to even more significant results and support the sustainability of such investments. In this subsection, the benefits of campus microgrids are categorized through possible policy directions for such systematic approaches, which are also aligned with relevant SDGs. Based on the campus microgrids' characteristics, the directions for their establishment towards energy transition actions can be categorized as follows:

- Environmental impact: Most of the smart campus microgrids utilize RES for their electricity needs and thus reduce the environmental burden. Hence, campus microgrids could be boosted through actions regarding their environmental footprint. This direction is clearly aligned with SDGs 7 and 13.
- Friendly social and academic environment: Universities and research centers correspond to a friendly environment for green investments, while through the concept of real-life living laboratories, the campus microgrid can be also employed for research and teaching purposes. Thus, campus microgrids can fit very well into the universities' development master plans. This direction is also aligned with SDG 4.
- Potential use of the "energy community" concept: Adjacent universities and research centers can easily collaborate through energy communities to optimize their environmental footprint and make their microgrid investments more sustainable in financial terms. This direction is aligned with SDG 17.
- Financial ease in areas with high RES potential: As will also be showcased through this paper's case study, campus microgrids investments are more sustainable financially in areas with high RES potential, and therefore important pilots could be established in such areas through research and development projects. This direction is aligned with SDG 7.
- Peak shaving in touristic areas: The summer tourism industry coincides with the low loading period of universities, i.e., May to October. In that period, the campus microgrids can act as significant grid-supporting units by utilizing the RES and storage units.

Indeed, university and research campuses have characteristics that can facilitate energy transition, while they are usually a relatively friendly environment for such investments.

Moreover, a special case for the application of smart campus microgrids arises in areas with a high density of academic institutions, where even more benefits arise. In this case, campuses correspond to a significant amount of the power system's load and their contribution towards energy transition can be critical.

The above-described directions are supported by the case study in the sequel of this paper, which highlights the financial ease of an investment towards a smart campus microgrid in an area with high RES potential, which also happens to have a high density of academic institutions, i.e., the island of Crete.

3. Optimal Sizing during the Planning Procedure of Microgrids

With the aim of performing a case study that will justify the benefits of campus microgrids, a representative smart campus microgrid will be financially and environmentally sized in the next section. Note that a crucial component of microgrids, that usually take care of the power balance in the islanded mode or of the optimal operation in the grid-connected mode, is the BESS. Hence, apart from the microgrid planning, optimization becomes a vital tool for the real-time microgrid operation too, since the operation of the BESS is subject to strict constraints that need to be considered during the microgrid operation. These constraints mainly relate to the acceptable state of charge (SOC) ratios of the BESS unit and their capacity. With regards to the BESS control hierarchy, the inner control algorithm of a BESS unit is described in [24]. Then, the bidirectional converter responsible for interfacing the battery to the grid controls the power flow, usually based on an outer voltage control scheme, while also respecting the SOC limits. On top of this local control action, an EMS is used to optimize the overall microgrid operation, as previously explained [6,30].

It is clear that optimization in microgrids has different applications at different stages, i.e., the planning and the operation. The different microgrid planning problems are reviewed in [15], with the most important being the microgrid sizing, sitting and scheduling issues, which can all be addressed through optimization techniques. In this paper, emphasis is given on the sizing part of the planning procedure instead of the operational EMS development [13,31]. Conventionally, the sizing of microgrids has been performed through software tools such as HOMER [32]. Optimization in HOMER is implemented through complete enumeration, which ensures that the solution is the optimal one, based on the measured or estimated load, but it can be very time-consuming in complex problems that contain many alternative combinations. Complete enumeration will be also used in the simple in-house optimizer of this paper, as the RES-based examined systems consist of three components (photovoltaics, wind turbines and BESS units), and the optimal solution is obtained after only a few minutes of simulation runs.

4. Planning of a Smart Campus Microgrid in Crete

In this section, the planning in terms of sizing of a smart campus microgrid in Crete will be performed, with the aim of showcasing the benefits that it can bring to the environment, the power system and the institution in financial terms. The methodology that will be followed for the case study of this paper is highlighted in the flow chart of Figure 1, while a sensitivity analysis will be further performed, considering the weather data of the microgrid and the cost of the storage system.

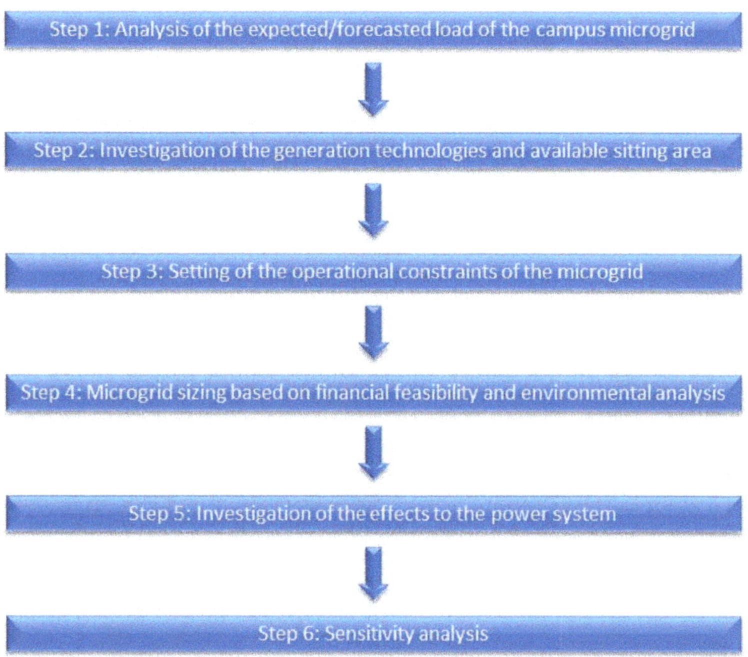

Figure 1. Hourly time series for the load demand of a smart campus microgrid.

The Hellenic Mediterranean University is a young university on the island of Crete, Greece. HMU consists of five campuses in the major cities of Crete (Heraklion, Chania, Rethymno, Agios Nikolaos and Siteia). Among the five campuses, the Heraklion campus hosts the largest number of academic departments and students. Moreover, the Heraklion campus is home to the Laboratory of Energy and Photovoltaic Systems (LEPS) and to the Institute of Environment, Energy and Climate Change (IEECC) of the university's research center, i.e., institutions that perform teaching and research in the smart grids area.

4.1. Data Elaboration

An analysis of the load profile of the HMU campus in Heraklion, in its current state, is presented in this section. The analysis is based on hourly data of the campus's load demand during the year 2020–2021 (which is the first 12-month period that such data were stored), referring to its (main) grid power supply at medium voltage level (20 kV). At the moment, the overall electricity demand of the campus is exclusively serviced from the main electrical grid. Therefore, the aim is to significantly reduce the demand supplied from the electrical grid, taking as input the data of the following study.

Figure 2 shows the hourly time series of load demand for the HMU campus, with the average value of load demand being equal to 167.49 kW. The peak load demand is equal to 394.65 kW and was observed during a cold winter day in January. However, the corresponding value for the summer period is only 9.85 kW less than the season peak (384.8 kW), which shows that the load demand depends, at a significant level, on the needs for heating and cooling. It is worth mentioning that the peak load demand for each month is observed during 10 am–2 pm, which are working and teaching hours. From the duration curve of Figure 3, it can be concluded that for over half of the year's hours, the load demand exceeds 150 kW.

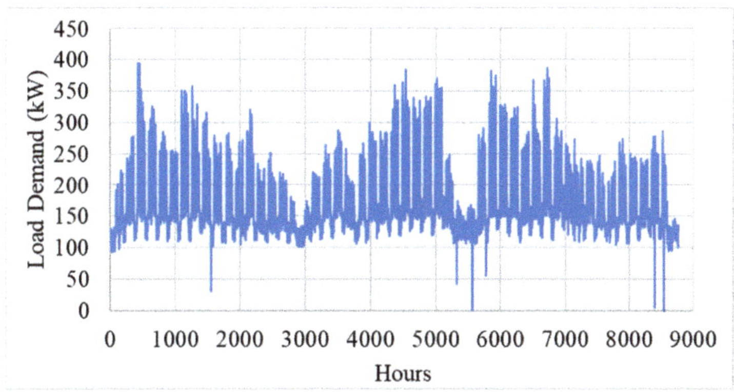

Figure 2. Hourly time series for load demand for the HMU campus.

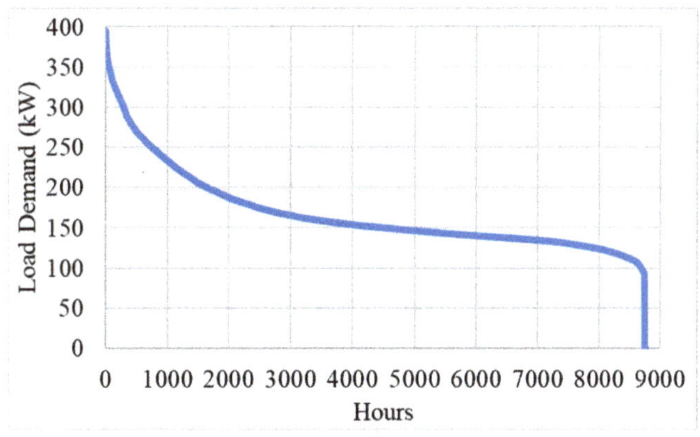

Figure 3. Duration curve of load demand for the HMU campus.

As previously mentioned, a factor affecting the campus's energy consumption is the weather conditions. July, with 144,652 kWh, is the month with the highest total energy consumption, as shown in Figure 4. The neutral climate conditions, in combination with the public holidays due to the Orthodox Easter, make April the month with the lowest energy consumption (107,987.05 kWh). Similar results are also obtained for August due to the summer holidays that are generally applied to all levels of education, including universities and research centers.

Figure 5 shows that the highest energy consumption appears on weekdays instead of weekends mainly due to the usage of academic premises (teaching rooms, offices). However, there are no significant variations in the rate of consumption between weekdays. Making a step forward, the hours of the day with the highest energy consumption are the three hours of 12–3 pm which account for 16.15% of the annual energy consumption, as shown in Figure 6. Thus, the utilization of PV technology seems to be an ideal means of electrical energy saving for the examined load profile.

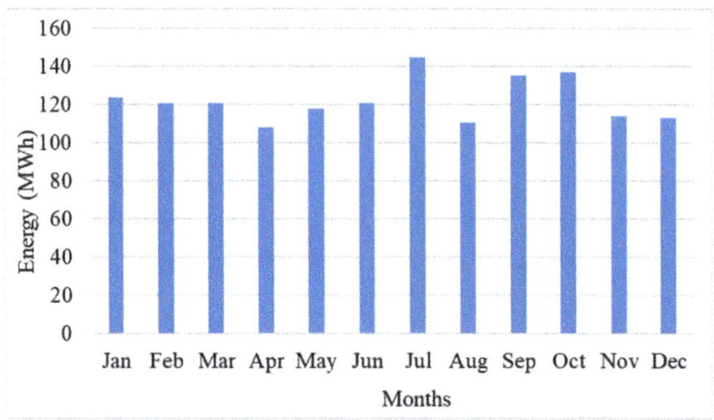

Figure 4. Monthly energy consumption for the HMU campus.

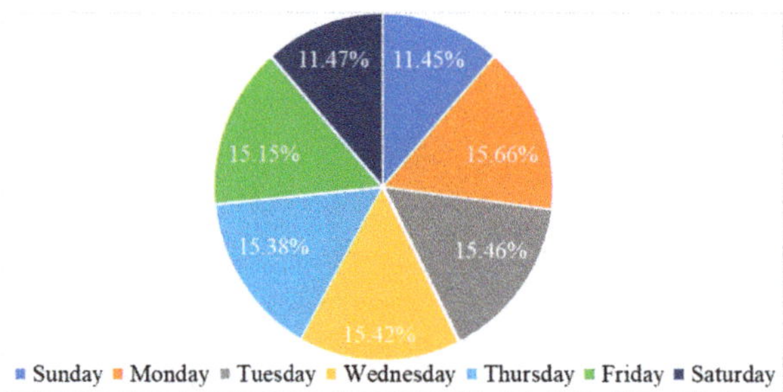

Figure 5. Energy consumption classification per day of the week.

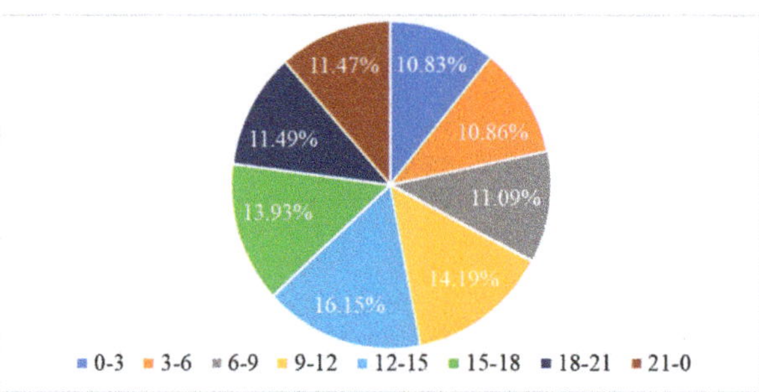

Figure 6. Energy consumption classification per hour of the day.

4.2. DERs Considerations

Indeed, based on the estates' department sitting areas investigations, the HMU campus microgrid could include photovoltaic (PV) systems, small wind turbines (WT) and batteries. Thereafter, all implemented calculations are based on components' technical data and HMU meteorological data, provided on an hourly basis.

Regarding the PV array, its output power P_{PV} (kW) is calculated from [33]:

$$P_{PV} = f_{PV} P_{STC} \frac{G_A}{G_{STC}} (1 + (T_C - T_{STC}) C_T), \qquad (1)$$

where is the f_{PV} is the PV derating factor which accounts for losses related to dust cover, aging and the unreliability of the PV array; P_{STC} (kW) is the peak power of the PV array; G_A is the global solar radiation incident on the PV array (kW/m^2); G_{STC} = 1 kW/m^2 is the solar radiation under PV Standard Test Conditions (STC); T_C is the temperature of the PV cells (°C); T_{STC} is the PV temperature under STC (25 °C) and C_T is the PV temperature coefficient (−0.36%/°C). The value of f_{PV} is considered to be 0.90. T_C can be estimated from the ambient temperature T_α (°C) and the global horizontal solar irradiance G (kW/m^2) [34]:

$$T_c = T_a + \frac{(NOCT - 20)}{0.8}G,\qquad(2)$$

where $NOCT$ is the normal operating cell temperature, which is considered equal to 42 °C. From the description above, the needed weather data for P_{PV} estimation are global horizontal solar irradiance G and ambient temperature T_α. Global solar radiation incident on the PV array (G_A) is calculated from the HDKR model [35], which for each hour takes into account the current value of G, the orientation of the PV array, the location on the Earth's surface, the time of year and the time of day.

Regarding the small WT, the Aeolos-H 60 kW model is considered. Its rated power is on the upper bound of small WTs according to the Greek Legislative Framework. Moreover, this model has a large diameter (22.3 m) considering its rated power. This characteristic provides improved performance of the WT for small and medium wind speeds, which are typical in most of the university campuses. Figure 7 shows the power curve of this WT. The required weather data for WT estimation include wind speed, which has been measured at a 10 m height. In order to estimate wind speed at WT hub height (30 m), the power law has been used with a power law exponent equal to $\alpha = 1/7$, which is a typical value for flat terrain [36].

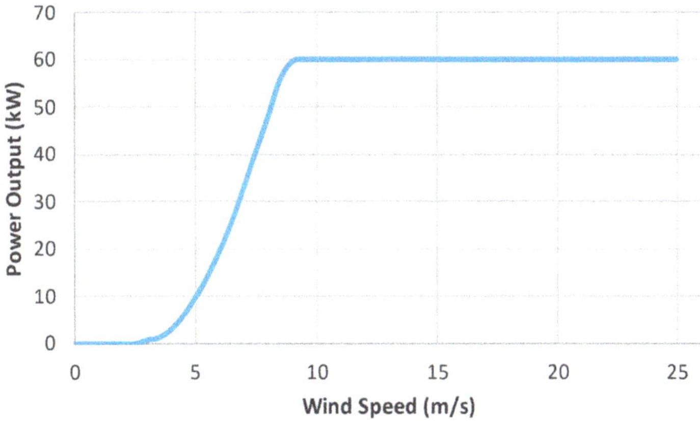

Figure 7. Aeolos-H 60 kW WT power curve [37].

For the BESS components, two alternative container models provided by Narada were examined and compared. Both models have an installed capacity of 250 kWh, installed power of 500 kW and efficiency of 98.5%. The first model is a lead–carbon battery with an expected life of 5000 daily cycles (around 15 years) operating at 40% depth of discharge (DOD). The second model is a lithium battery with an expected life of 6000 daily cycles (around 18 years) operating at 60% DOD.

The basic financial and technical characteristics of the above-mentioned components are summarized in Table 1. In all cases, the lifetime of the project is considered equal to the batteries' lifetimes.

Table 1. Characteristics of the HMU campus microgrid components.

Component	Capital Cost	O&M Cost	Lifetime	Salvage Value
PV array	400 €/kWp	10 €/kWp	25 years	20% of capital cost
Small WT (60 kW)	120,000€	3000 €/(WT·yr)	20 years	0
Lead–carbon battery	300 €/kWh	2500 €/(unit·yr)	15 years at 40% DOD	20% of capital cost
Lithium battery	500 €/kWh	5000 €/(unit·yr)	18 years at 60% DOD	20% of capital cost

4.3. Load Considering Electric Vehicle Charging

The evaluation of the HMU campus microgrid is based on its current electric load consumption, with the addition of 20 electric vehicle (EV) chargers of 7 kW AC. The estimation of the additional load due to EV charging in a university campus took into account information of arrival time at the Campus (85% of arrivals are between 7–9 am) [38], EV charging time [14] and reduced EV charging in certain periods of the year (weekends, Christmas and Easter vacations, July and August). The comparison of current and expected load duration curves is shown in Figure 8. Most of the additional load consumption happens in the morning and noon hours. The study of the load curves shows that EV chargers increase the annual load consumption by 6% and the peak load demand by more than 17%.

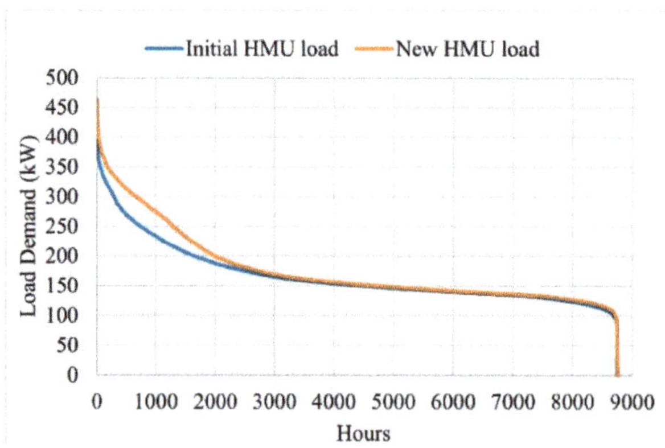

Figure 8. Comparison of load duration curves in the HMU campus considering the addition of EV chargers.

4.4. Scenarios of HMU Microgrid Operation

Regarding the operation of the HMU campus microgrid, two cases were examined and compared: (1) HMU to act as an autoproducer and (2) installation of an independent MV station that contains PV/WTs and batteries as well as the university load. In the first case, the maximum power that can be exchanged with the grid is equal to the agreed power between the HMU and the power system operator (1000 kVA). Moreover, HMU purchases power at retail cost (80 EUR/MWh) and sells electricity to the grid at the system's marginal price (SMP). In the second case, the maximum power that can be purchased or sold to the grid can be increased (in our case, it is considered equal to 2000 kVA). Moreover, grid purchases and grid sales are implemented in the system's SMP.

Figure 9 shows the annual variation of the SMP for Crete Island. It must be noted that these values correspond to the autonomous operation of the Cretan power system. From late 2021, the first phase of Cretan interconnection (weak interconnection) with the Hellenic Transmission System is under operation. This project consists of the construction of 150 kV AC 2 × 200 MVA interconnection between Crete and the Peloponnese Peninsula. The second phase (strong interconnection) through 400 kV HVDC 2 × 500 MW interconnection is expected to start its operation in 2024.

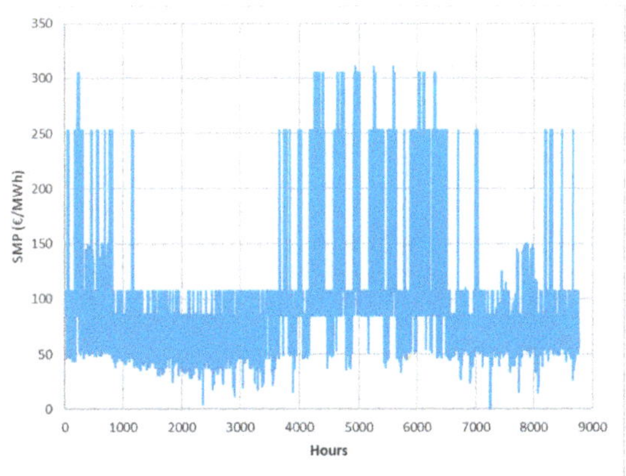

Figure 9. Annual variation of the SMP on Crete Island (current situation).

4.5. Sizing and Financial Feasibility of the Campus Microgrid

The financial evaluation of alternative configurations of the HMU microgrid is implemented using the criterion of Net Present Value (NPV). The annual discount rate is equated to 6%, whereas the evaluation takes into account the capital costs, O&M costs and salvage values of the components, the new load demand, and the electricity purchases and sales with the main grid. The maximum WT number is set at 5 due to spatial constraints provided by the estates' department. The step size for the PV array size is equated to 10 kWp. Table 2 shows the optimal results for the autoproducer case. It can be seen that although there is no configuration with NPV > 0, optimal solutions tend to have large PV array sizes (1000 kWp is the maximum considered value), no WTs (although their installation does not significantly increase overall costs) and no batteries due to their high costs. Regarding energy storage installation, the most preferable option includes one battery container, 1000 kWp of PV and no WT, with NPV equal to EUR −148,055 for lead–carbon batteries and EUR −195,744 for lithium batteries.

Table 2. Optimal results of the HMU campus microgrid for the autoproducer case.

WT	PV (kWp)	Lead–Carbon	Lithium	NPV (€)
0	1000	0	0	−24,919
1	1000	0	0	−33,713
0	990	0	0	−39,169
2	1000	0	0	−42,614
5	1000	0	0	−45,667
1	990	0	0	−47,960
4	1000	0	0	−48,263
3	1000	0	0	−48,826

The operation of the HMU microgrid considering the existence of a MV station requires battery installation, because its operation is based on purchasing electricity at a low SMP and selling electricity at a high SMP. Table 3 shows the optimal results for this case. It can be seen that the best configurations have positive NPVs and tend to have large PV array sizes (2000 kWp is the maximum considered value due to spatial constraints), no WTs (although their installation does not significantly increase overall costs) and one container of lead–carbon batteries, which is their lowest possible size. The configuration with lithium batteries with the highest NPV contains 2000 kWp of PV, no WT and one battery container, with NPV equal to EUR 572,098.

Table 3. Optimal results of the HMU campus microgrid considering the installation of a MV station.

WT	PV (kWp)	Lead–Carbon	Lithium	NPV (€)
0	2000	1	0	684,148
0	1990	1	0	672,369
1	2000	1	0	661,471
0	1980	1	0	660,589
1	1990	1	0	649,717
0	1970	1	0	648,807
1	1980	1	0	638,094
2	2000	1	0	637,747

Complementary to the aforementioned financial analysis, an environmental analysis is also implemented in relation to the annual greenhouse gas (GHG) emissions reduction in the entire insular power system of Crete, due to the installation of RES technologies (WTs and PVs) in the HMU microgrid. An important factor in this analysis is the definition of the fuel mix for electricity generation on the island of Crete. These data are shown in Table 4 and refer to the year 2020, having been extracted from [39]. The next step in the analysis is the calculation of the annual GHG emission reduction for each MWh that is generated from WTs and PVs in the HMU campus, which was provided by RETScreen 4.0 software [40]. The results indicate that each MWh generated from RES technologies reduces the GHG emissions of the Cretan power system by 0.307 tCO_2(eq.)/year. The term (eq.) stands for equivalent and takes into account the emissions of other GHGs apart from CO_2, such as CH_4 and N_2O. The final step of the analysis relates to the estimation of a non-dominated solutions set according to financial and environmental criteria, i.e., NPV and annual GHG emissions reduction. This set consists of solutions that are better in one criterion and worse in the other criterion, and is also known as the Pareto-optimal set [16]. Tables 5 and 6 present the Pareto-optimal sets for the two considered cases (autoproducer and MV station installation).

Table 4. Fuel mix for electricity generation on Crete Island (year 2020).

Fuel	Electricity Generation (MWh)	Percentage
Mazut (heavy oil)	1,503,131.8	54.09%
Diesel	622,574.2	22.40%
WTs	516,309.3	18.58%
PVs	136,554.3	4.91%
Small hydro	531.5	0.02%
Total	2,779,101.1	100.00%

Table 5. Pareto-optimal set of the HMU campus microgrid for the autoproducer case.

WT	PV (kWp)	Lead–Carbon	Lithium	NPV (€)	Annual CHG Emissions Reduction (tCO_2(eq.))
0	1000	0	0	−24,919	1127.8
1	1000	0	0	−33,713	1218.9
2	1000	0	0	−42,614	1310.0
5	1000	0	0	−45,667	1583.2

Table 6. Pareto-optimal set of the HMU campus microgrid considering the installation of a MV station.

WT	PV (kWp)	Lead–Carbon	Lithium	NPV (€)	Annual CHG Emissions Reduction (tCO$_2$(eq.))
0	2000	1	0	684,148	2255.7
1	2000	1	0	661,471	2346.7
2	2000	1	0	637,747	2437.8
3	2000	1	0	596,056	2528.9
4	2000	1	0	555,190	2619.9
5	2000	1	0	534,433	2711.0

The results provided in Tables 5 and 6 show that the optimal solutions of the Pareto-sets contain the maximum installed PV power for each case, whereas the addition of WTs in most cases reduces the NPV and increases the annual GHG emissions reduction. An exception to the latter is the installation of 3 and 4 WTs (for 1000 kWp installed PVs) in the autoproducer case, where the NPV is slightly smaller compared to this in the case of 5 WTs. The benefit of these Pareto-sets is that the optimal solution will be included in these independently of the criterion (or the combination of financial–environmental criteria) that will be used. For the analyses in the sequel of this article, the financially optimal solutions will be considered.

4.6. Effect to the Power System

Table 7 summarizes the annual simulation results for the financially optimal solutions of the autoproducer and MV station cases (second rows of Tables 2 and 3). For the autoproducer case (no battery is included), it can be seen the surplus of PV generation over load demand on an annual basis is equal to the surplus of excess electricity over purchases from the grid (energy balance). In the case of the MV station installation, the lead–carbon battery can store electricity either from PVs or from the grid, depending on the hour of the day and the load demand characteristics.

Table 7. Annual results of the HMU campus microgrid operation for optimal configurations.

Annual Simulation Results	Autoproducer	MV Station
Load demand (MWh)	1554	1554
PV generation (MWh)	1857	3715
WT generation (MWh)	–	–
Energy stored to battery (MWh)	–	42,584
Purchases from grid (MWh)	784	766
Purchases from grid (EUR)	62,702	68,081
Excess electricity (MWh)	1087	2549
Sales to grid (EUR)	104,755	243,596

Since all the results have considered an additional loading due to EV chargers, Figure 10 shows the impact of EV chargers' installation in the HMU campus compared to its current load demand (see Figure 2), as it depicts the annual variation of the additional load taking into account the considerations of Section 4.3. More specifically, the limited additional load around hour 3000 refers to Easter vacations and between hours 4000–6000 refers to the July and August reduced HMU operation, whereas the beginning and end of the year relates to Christmas vacations. On a typical weekday, the load addition can reach up to 120 kW for specific morning and midday hours.

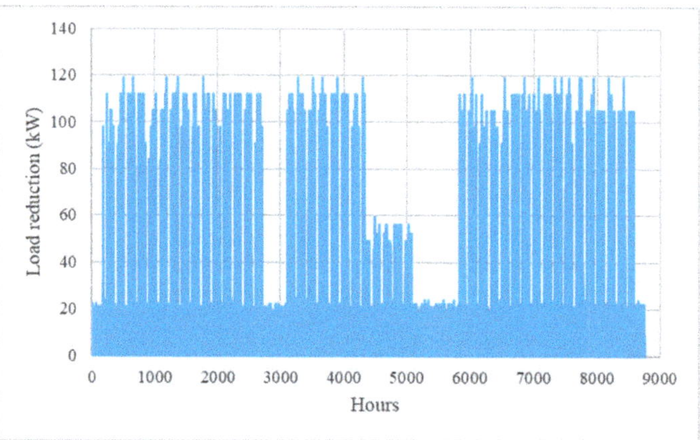

Figure 10. Annual variation of the additional load in the HMU campus after the installation of EV chargers.

Regarding the autoproducer case (considering the EV chargers' operation) and for its financially optimal solution from Table 2, an overall load reduction exists during daytime (when PVs generate electricity), whereas the maximum load reduction is observed during the hours of the year that the PVs are operating under conditions similar to STC (Standard Test Conditions: 1000 W/m^2 of solar irradiance and 25 °C of PV temperature). The maximum value of approximately 1000 kW is explained by the installed PV power (1000 kW$_p$). Figure 11 shows the load duration curve that refers to an external transformer in the HMU campus microgrid. It can be seen that for approximately 3000 h of the year, the HMU microgrid exports electricity to the grid.

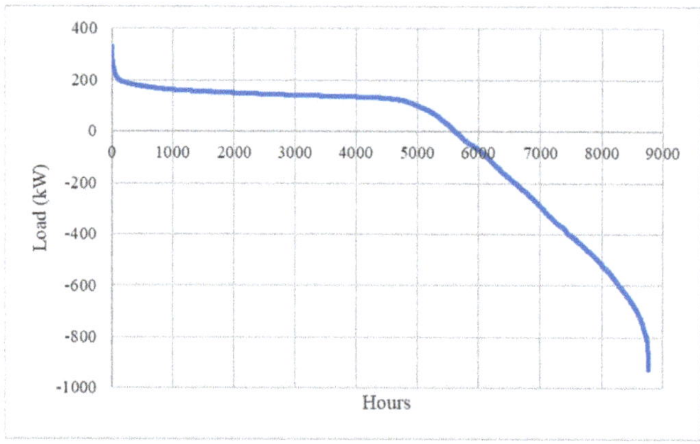

Figure 11. Load duration curve that refers to an external transformer in the HMU campus microgrid (optimal autoproducer case solution).

For the case of the MV bus installation and its financially optimal solution shown in Table 3, the maximum overall load reduction is approximately double the previous case due to the double PV installed capacity (2000 kW$_p$). Figure 12 shows the load duration curve that refers to an external transformer in the HMU campus microgrid. It can be concluded that the number of hours that the microgrid exports electricity to the grid is slightly increased, while the amount of maximum power that is exported is also approximately double.

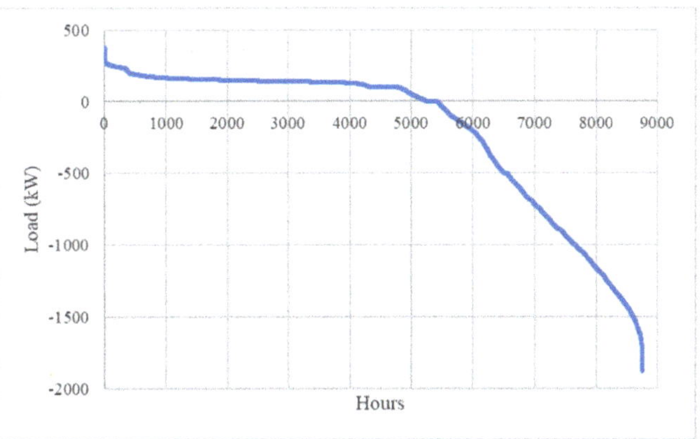

Figure 12. Load duration curve that refers to an external transformer in the HMU campus microgrid (optimal MV station case solution).

Based on the above results, the HMU campus microgrid acts as a RES producer for many hours during the year, hence ensuring a decrease in grid congestion and therefore supporting the overall power system operation, apart from driving the energy mix to a greener manner, which was highlighted in the previous section.

4.7. Sensitivity Analysis

From the results of Section 4.5, it can be concluded that in both cases, the financially optimal solution contains the smallest available number of batteries. As a result, the cost of batteries is an essential factor affecting the results. According to recent studies [41], the cost of utility-scale lithium batteries is expected to decrease by 50% by the end of this decade, so the effect of battery cost reduction by 25% and 50% is examined in Tables 8 and 9, respectively. In the autoproducer case, the reduction of battery costs does not have any effect on optimal results, so the solutions are identical to these provided in Table 2. Regarding MV station installation, the reduction of battery costs increases the NPV (as expected), though the configurations with the minimum available batteries number remain dominant.

Table 8. Optimal results of the HMU campus microgrid considering 25% battery cost reduction.

WT	PV (kWp)	Lead–Carbon	Lithium	NPV (EUR)
		Autoproducer		
0	1000	0	0	−24,919
1	1000	0	0	−33,713
0	990	0	0	−39,169
2	1000	0	0	−42,614
5	1000	0	0	−45,667
		Installation of MV station		
0	2000	1	0	707,403
0	1990	1	0	695,624
1	2000	1	0	684,726
0	1980	1	0	683,844
1	1990	1	0	672,973

Table 9. Optimal results of the HMU campus microgrid considering 50% battery cost reduction.

WT	PV (kWp)	Lead–Carbon	Lithium	NPV (EUR)
		Autoproducer		
0	1000	0	0	−24,919
1	1000	0	0	−33,713
0	990	0	0	−39,169
2	1000	0	0	−42,614
5	1000	0	0	−45,667
		Installation of MV station		
0	2000	1	0	730,658
0	1990	1	0	718,879
1	2000	1	0	707,981
0	1980	1	0	707,099
1	1990	1	0	696,228

An additional criterion that significantly affects the results is the campus microgrid location, since renewable energy production may vary significantly. Table 10 presents these variabilities for three different university locations: (1) HMU in Heraklion, Crete, Greece; (2) Imperial College London, UK; and (3) Stockholm University, Sweden, assuming the HMU loading case, available space for RES and electricity prices. Data for London and Stockholm have been extracted from Typical Model Years (TMYs) provided by the PV-GIS software tool [42]. Table 10 shows that HMU presents significantly better PV potential, whereas Stockholm University provides better WT electricity production.

Table 10. Comparison of PV and WT annual electricity production for three different locations of university campuses.

Location	Annual PV Production (kWh/kWp)	PV CF	Annual WT Production (kWh/Unit)	WT CF
HMU, Heraklion	1857	21.2%	149,975	28.53%
Imperial College London	1155	13.18%	108,547	20.65%
Stockholm University	1051	12%	188,244	35.82%

Tables 11 and 12 provide the optimal results for HMU load (considering EV charging) using weather data from Imperial College and Stockholm University. The analysis of the results and the comparison with Tables 11 and 12 show that the location of the microgrid plays a crucial role in the financial feasibility of its operation. Both locations have significantly lower solar potential compared to HMU and, as a result, all configurations for each case (autoproducer and MV station) have large negative NPVs. In all cases, the most dominant solutions contain large PV arrays and the lowest possible number of batteries. In the case of the MV station, lead–carbon batteries are preferable compared to lithium batteries.

Table 11. Optimal results for Imperial College London.

WT	PV (kWp)	Lead–Carbon	Lithium	NPV (EUR)
		Autoproducer		
0	1000	0	0	−663,316
0	990	0	0	−669,523
0	980	0	0	−675,727
0	970	0	0	−681,927
0	960	0	0	−688,122
		Installation of MV station		
0	2000	1	0	−453,815
0	1990	1	0	−459,193
0	1980	1	0	−464,707
0	1970	1	0	−471,065
0	1960	1	0	−476,398

Table 12. Optimal results for Stockholm University.

WT	PV (kWp)	Lead–Carbon	Lithium	NPV (EUR)
		Autoproducer		
5	1000	0	0	−739,719
5	990	0	0	−743,849
5	980	0	0	−749,980
4	1000	0	0	−753,880
5	970	0	0	−756,109
		Installation of MV station		
0	2000	1	0	−647,240
0	1990	1	0	−650,663
0	1980	1	0	−655,029
1	2000	1	0	−659,023
0	1970	1	0	−660,710

The comparison of the two alternative cases shows that the installation of an MV station provides lower costs than the autoproducer operation. However, the differences in NPV between these two cases are small compared to the HMU campus. The effect of increased wind potential in Stockholm University is obvious in autoproducer results, as the optimal solutions contain a large number of WTs. In the MV station operation, the battery that helps to sell electricity at higher prices plays a more significant role than WT installation.

In summary, from the financial analysis of all case studies, it can be concluded that the autoproducer case is not economically viable, as it presents negative NPV even in the HMU campus microgrid, which presents the best combination of weather conditions from all examined locations. Moreover, these results do not depend on the expected battery cost reduction by the end of this decade, as they do not contain any electricity storage technology. Regarding the case of MV station installation, the variation of NPV is even larger for different locations and can provide significant financial benefits for the HMU case. The optimal solutions present a small number of batteries and the reduction of storage costs does not significantly improve the results.

5. Conclusions

Campus microgrids are a promising solution as nations move towards their energy transition. The significant benefits they bring can support their utilization around the world, while they can have a vital impact in areas with high RES potential and a high density of university and research campuses.

Indeed, in the performed case study, it was shown that campus location plays a significant role, while the MV station case is preferable to the autoproducer case (which always presents a negative NPV). In particular, in the case of Crete, the MV station case can provide positive financial results. In all cases (even in northern locations), PVs are the preferred RES technology due to their lower installation and O&M costs. Batteries are an expensive component, and even their expected cost reduction does not significantly changes the results. Finally, it was shown that a significant load reduction (for the main grid) is obtained through the development of the HMU campus microgrid, which also behaves as a RES electricity producer supporting the power system for many days of the year. This feature is especially important during hours of stress and in reducing the environmental burden, which was also observed through an elaborate environmental analysis.

Future research will focus on evaluating the total impact of all campuses in Crete towards the energy transition of the island, as well as investigating the impact of the COVID-19 pandemic on the load of the HMU campus and the relevant results of microgrid sizing.

Author Contributions: Conceptualization, A.P. and E.K.; methodology, Y.K.; software, Y.K.; validation, Y.K., K.F. and A.P.; formal analysis, K.F.; investigation, A.P.; resources, E.K.; writing—original draft preparation, A.P.; writing—review and editing, A.P., K.F., Y.K. and E.K. All authors have read and agreed to the published version of the manuscript.

Funding: This research received no external funding.

Conflicts of Interest: The authors declare no conflict of interest.

References

1. Kourgiozou, V.; Commin, A.; Dowson, M.; Rovas, D.; Mumovic, D. Scalable Pathways to Net Zero Carbon in the UK Higher Education Sector: A Systematic Review of Smart Energy Systems in University Campuses. *Renew. Sustain. Energy Rev.* **2021**, *147*, 111234. [CrossRef]
2. Demand Response Integration in Microgrid Planning as a Strategy for Energy Transition in Power Systems-Mina-Casaran-2021-IET Renewable Power Generation-Wiley Online Library. Available online: https://ietresearch.onlinelibrary.wiley.com/doi/full/10.1049/rpg2.12080 (accessed on 29 July 2021).
3. Sun, L.; Li, J.; Chen, L.; Xi, J.; Li, B. Energy Storage Capacity Configuration of Building Integrated Photovoltaic-Phase Change Material System Considering Demand Response. *IET Energy Syst. Integr.* **2021**, *3*, 263–272. [CrossRef]
4. Hatziargyriou, N. *Microgrids: Architectures and Control*; Wiley-IEEE Press: Hoboken, NJ, USA, 2013; p. 317.
5. Hassan, M.A.; Worku, M.Y.; Eladl, A.A.; Abido, M.A. Dynamic Stability Performance of Autonomous Microgrid Involving High Penetration Level of Constant Power Loads. *Mathematics* **2021**, *9*, 922. [CrossRef]
6. Makkonen, H.; Lassila, J.; Tikka, V.; Partanen, J.; Kaipia, T.; Silventoinen, P. Implementation of Smart Grid Environment in Green Campus Project. In Proceedings of the CIRED 2012 Workshop: Integration of Renewables into the Distribution Grid, Lisbon, Portugal, 29–30 May 2012; pp. 1–4.
7. Husein, M.; Chung, I.-Y. Optimal Design and Financial Feasibility of a University Campus Microgrid Considering Renewable Energy Incentives. *Appl. Energy* **2018**, *225*, 273–289. [CrossRef]
8. Alrashed, S. Key Performance Indicators for Smart Campus and Microgrid. *Sustain. Cities Soc.* **2020**, *60*, 102264. [CrossRef]
9. Paspatis, A.G.; Konstantopoulos, G.C.; Dedeoglu, S. Control Design and Small-Signal Stability Analysis of Inverter-Based Microgrids with Inherent Current Limitation under Extreme Load Conditions. *Electr. Power Syst. Res.* **2021**, *193*, 106929. [CrossRef]
10. Papadopoulos, T.A.; Giannakopoulos, G.T.; Nikolaidis, V.C.; Safigianni, A.S.; Panapakidis, I.P. Study of Electricity Load Profiles in University Campuses: The Case Study of Democritus University of Thrace. In Proceedings of the Mediterranean Conference on Power Generation, Transmission, Distribution and Energy Conversion (MedPower 2016), Belgrade, Serbia, 6–9 November 2016; pp. 1–8. [CrossRef]
11. Elenkova, M.Z.; Papadopoulos, T.A.; Psarra, A.I.; Chatzimichail, A.A. A Simulation Platform for Smart Microgrids in University Campuses. In Proceedings of the 2017 52nd International Universities Power Engineering Conference (UPEC), Heraklion, Greece, 28–31 August 2017; pp. 1–6.
12. Ahmad, F.; Alam, M.S. Optimal Sizing and Analysis of Solar PV, Wind, and Energy Storage Hybrid System for Campus Microgrid. *Smart Sci.* **2018**, *6*, 150–157. [CrossRef]
13. Javed, H.; Muqeet, H.A.; Shehzad, M.; Jamil, M.; Khan, A.A.; Guerrero, J.M. Optimal Energy Management of a Campus Microgrid Considering Financial and Economic Analysis with Demand Response Strategies. *Energies* **2021**, *14*, 8501. [CrossRef]
14. Hovet, S.; Farley, B.; Perry, J.; Kirsche, K.; Jerue, M.; Tse, Z.T.H. Introduction of Electric Vehicle Charging Stations to University Campuses: A Case Study for the University of Georgia from 2014 to 2017. *Batteries* **2018**, *4*, 27. [CrossRef]

15. Gamarra, C.; Guerrero, J.M. Computational Optimization Techniques Applied to Microgrids Planning: A Review. *Renew. Sustain. Energy Rev.* **2015**, *48*, 413–424. [CrossRef]
16. Katsigiannis, Y.A.; Georgilakis, P.S.; Karapidakis, E.S. Multiobjective Genetic Algorithm Solution to the Optimum Economic and Environmental Performance Problem of Small Autonomous Hybrid Power Systems with Renewables. *IET Renew. Power Gener.* **2010**, *4*, 404–419. [CrossRef]
17. Makkonen, H.; Tikka, V.; Lassila, J.; Partanen, J.; Silventoinen, P. Green Campus-Energy Management System. In Proceedings of the 22nd International Conference and Exhibition on Electricity Distribution (CIRED 2013), Stockholm, Sweden, 28–31 August 2017; p. 1137. [CrossRef]
18. THE 17 GOALS | Sustainable Development. Available online: https://sdgs.un.org/goals (accessed on 26 May 2021).
19. Tsikalakis, A.; Routsis, D.; Pafilis, A.; Kalaitzakis, K.; Stavrakakis, G. A Methodology Exploiting Geographical Information Systems to Site a Photovoltaic Park inside a Sustainable Community. *Int. J. Sustain. Energy* **2016**, *35*, 132–147. [CrossRef]
20. *Of Mice and Microgrids: A Profile of the US' Largest Microgrid*; Microgrid Knowledge: Newbury, NH, USA, 2014.
21. Inside the World's Most Advanced Microgrid: The University of California San Diego. Available online: https://microgridnews.com/inside-the-worlds-most-advanced-microgrid-university-of-california-san-diego/ (accessed on 27 July 2021).
22. Energy Initiatives, Sustainability-Wesleyan University. Available online: https://www.wesleyan.edu/sustainability/initiatives/energy/index.html (accessed on 27 July 2021).
23. Case Study: Microgrid at Princeton University | Facilities. Available online: https://facilities.princeton.edu/news/case-study-microgrid-princeton-university (accessed on 27 July 2021).
24. Uddin, M.; Romlie, M.F.; Abdullah, M.F.; Hassan, K.N.M.; Tan, C.K.; Bakar, A.H.A. Modeling of Campus Microgrid for Off-Grid Application. In Proceedings of the 5th IET International Conference on Clean Energy and Technology (CEAT2018), Kuala Lumpur, Malaysia, 5–6 September 2018; pp. 1–5.
25. Moura, P.; Correia, A.; Delgado, J.; Fonseca, P.; de Almeida, A. University Campus Microgrid for Supporting Sustainable Energy Systems Operation. In Proceedings of the 2020 IEEE/IAS 56th Industrial and Commercial Power Systems Technical Conference, Las Vegas, NV, USA, 29 June–28 July 2020; pp. 1–7.
26. Coming Full Circle: Four New Loops Will Enhance Campus Microgrid, Complete Original Vision. Available online: https://www.iit.edu/news/coming-full-circle-four-new-loops-will-enhance-campus-microgrid-complete-original-vision (accessed on 14 February 2022).
27. Deakin University's Enterprising Plans for Its New Microgrid. Available online: https://www.pv-magazine-australia.com/2021/04/27/deakin-universitys-enterprising-plans-for-its-new-microgrid/ (accessed on 27 July 2021).
28. Howland, E. *UC Berkeley Seeks Help as It Pursues Microgrid Project*; Microgrid Knowledge: Newbury, NH, USA, 2021.
29. Hadjidemetriou, L.; Zacharia, L.; Kyriakides, E.; Azzopardi, B.; Azzopardi, S.; Mikalauskiene, R.; Al-Agtash, S.; Al-hashem, M.; Tsolakis, A.; Ioannidis, D.; et al. Design Factors for Developing a University Campus Microgrid. In Proceedings of the 2018 IEEE International Energy Conference (ENERGYCON), Limassol, Cyprus, 3–7 June 2018; pp. 1–6.
30. Jung, S.; Yoon, Y.T.; Huh, J.-H. An Efficient Micro Grid Optimization Theory. *Mathematics* **2020**, *8*, 560. [CrossRef]
31. Muqeet, H.A.; Munir, H.M.; Javed, H.; Shahzad, M.; Jamil, M.; Guerrero, J.M. An Energy Management System of Campus Microgrids: State-of-the-Art and Future Challenges. *Energies* **2021**, *14*, 6525. [CrossRef]
32. Tozzi, P.; Jo, J.H. A Comparative Analysis of Renewable Energy Simulation Tools: Performance Simulation Model vs. System Optimization. *Renew. Sustain. Energy Rev.* **2017**, *80*, 390–398. [CrossRef]
33. Thomson, M.; Infield, D.G. Impact of Widespread Photovoltaics Generation on Distribution Systems. *IET Renew. Power Gener.* **2007**, *1*, 33–40. [CrossRef]
34. McEvoy, A.; Markvart, T.; Castaner, L. *Practical Handbook of Photovoltaics Fundamentals and Applications*, 2nd ed.; Academic Press: Cambridge, MA, USA, 2012.
35. Duffie, J.A.; Beckman, W.A.; Blair, N. *Solar Engineering of Thermal Processes, Photovoltaics and Wind*, 5th ed.; Wiley: Hoboken, NJ, USA, 2020.
36. Manwell, J.F.; McGowan, J.G.; Rogers, A.L. *Wind Energy Explained: Theory, Design and Application*, 2nd ed.; Wiley: Chichester, UK, 2010.
37. Aeolos-H 60 kW Wind Turbine. Available online: http://www.heliosystems.it/wp-content/uploads/2014/10/Aeolos-H_60kW_Brochure.pdf (accessed on 27 July 2021).
38. Caruso, M.; Livreri, P.; Miceli, R.; Viola, F.; Martino, M. Ev Charging Station at University Campus. In Proceedings of the 2017 IEEE International Telecommunications Energy Conference (INTELEC), Broadbeach, QLD, Australia, 1 October 2017; p. 623.
39. *Cretan Power System Annual Report 2020*; Hellenic Electricity Distribution Network Operator (HEDNO): Athens, Greece, 2021.
40. Canada, N.R. RETScreen®Clean Energy Management Software. Available online: https://www.nrcan.gc.ca/maps-tools-and-publications/tools/modelling-tools/retscreen/7465 (accessed on 22 February 2022).
41. Cole, W.J.; Frazier, A. *Cost Projections for Utility-Scale Battery Storage*; Technical Report; NREL: Golden, CO, USA, 2019.
42. JRC Photovoltaic Geographical Information System (PVGIS)-European Commission. Available online: https://re.jrc.ec.europa.eu/pvg_tools/en/tools.html#PVP (accessed on 27 July 2021).

 *mathematics*

MDPI

Article

Differential Evolution Based Algorithm for Optimal Current Ripple Cancelation in an Unequal Interleaved Power Converter

Julio C. Rosas-Caro [1,*], Pedro M. García-Vite [2], Alma Rodríguez [1,3], Abraham Mendoza [1], Avelina Alejo-Reyes [1], Erik Cuevas [3] and Francisco Beltran-Carbajal [4]

[1] Facultad de Ingenieria, Universidad Panamericana, Alvaro del Portillo 49, Zapopan 45010, Mexico; anrodriguez@up.edu.mx or alma.rvazquez@academicos.udg.mx (A.R.); amendoza@up.edu.mx (A.M.); aalejo@up.edu.mx (A.A.-R.)

[2] Tecnologico Nacional de Mexico, Instituto Tecnologico de Ciudad Madero, Av. 1o. de Mayo s/n Col. Los Mangos, Ciudad Madero 89440, Mexico; pedro.gv@cdmadero.tecnm.mx

[3] Departamento de Electrónica, Universidad de Guadalajara, CUCEI, Av. Revolución 1500, Guadalajara 44430, Mexico; erik.cuevas@cucei.udg.mx

[4] Departamento de Energía, Universidad Autónoma Metropolitana, Unidad Azcapotzalco, Mexico City 02200, Mexico; fbeltran@azc.uam.mx

* Correspondence: crosas@up.edu.mx; Tel.: +52-33-1918-1065

Citation: Rosas-Caro, J.C.; García-Vite, P.M.; Rodríguez, A.; Mendoza, A.; Alejo-Reyes, A.; Cuevas, E.; Beltran-Carbajal, F. Differential Evolution Based Algorithm for Optimal Current Ripple Cancelation in an Unequal Interleaved Power Converter. *Mathematics* 2021, 9, 2755. https://doi.org/10.3390/math9212755

Academic Editor: Nicu Bizon

Received: 28 September 2021
Accepted: 21 October 2021
Published: 29 October 2021

Publisher's Note: MDPI stays neutral with regard to jurisdictional claims in published maps and institutional affiliations.

Abstract: This paper proposes an optimal methodology based on the Differential Evolution algorithm for obtaining the set of duty cycles of a recently proposed power electronics converter with input current ripple cancelation capability. The converter understudy was recently introduced to the state-of-the-art as the interleaved connection of two unequal converters to achieve low input current ripple. A latter contribution proposed a so-called proportional strategy. The strategy can be described as the equations to relate the duty cycles of the unequal power stages. This article proposes a third switching strategy that provides a lower input current ripple than the proportional strategy. This is made by considering duty cycles independently of each other instead of proportionally. The proposed method uses the Differential Evolution algorithm to determine the optimal switching pattern that allows high quality at the input current side, given the reactive components, the switching frequency, and power levels. The mathematical model of the converter is analyzed, and thus, the decision variables and the optimization problem are well set. The proposed methodology is validated through numerical experimentation, which shows that the proposed method achieves lower input current ripples than the proportional strategy.

Keywords: Differential Evolution; metaheuristic algorithms; optimization; DC–DC converter

1. Introduction

Power electronics converters are essential in many fields of applications. Particularly, DC–DC power converters range from a few watts in battery-powered portable devices [1–3] to several kilowatts when employed in power systems [4,5]. Portable electronics devices usually operate at high frequencies in order to reduce their physical sizes [2]. This research focuses on a DC–DC step-up power converter. This kind of converter can be employed in renewable energy sources with low D.C. power generation, such as photovoltaic and hydrogen-based fuel cell sources that provide a few tens of volts [6]. DC–DC converters are constituted by a combination of solid-state switches and reactive components that process the input power to feed a resistive load, usually at different voltage levels. The switches are driven by a binary commanding function $q(t)$ at a fixed frequency. The reactive components capacitors (C) and inductors (L) are employed to store the energy coming from the source and then deliver such energy to the load.

Among the features specified in a high voltage-gain DC–DC converter is the voltage gain; this can be defined as the output voltage ratio to the input voltage and is independent of the reactive components. Another important specification is the input current quality, which is analyzed in terms of the switching ripple. This is the article's main focus, and it is related to current drawn from the source; having a high input current quality means that a pure (or almost pure) D.C. is extracted, without A.C. ripple or variation. This is an ideal condition since the ripple naturally appears due to the commutation process of the switches. Several topologies that provide high input current quality have recently appeared in the literature [7–10]. With the modification of a conventional boost converter presented in [11], it is possible to have a reduced input current ripple by adding extra components. More complex configurations with both capabilities have also been proposed.

This paper focuses on a recently proposed converter [12], made by the interleaved connection of two unequal power stages, the classical boost converter [13] and a modified (resonant version of the) three-switch high-voltage TSHV converter [14].

The interleaved interconnection of classical boost power stages is a proven technology [15]. It leads to a special situation in which the input current ripple is zero (for $D = 0.5$); in this zero-ripple situation, the input current ripple (high-frequency variation) of one power stage cancels perfectly with the other one (for example, in a two-phases converter), and the input current of the entire converter is a pure D.C. signal. However, this happens only in a particular operation point; for different duty cycles ($D \neq 0.5$), the ripple must be evaluated.

The contribution made in [12] was not only the interconnection of two unequal power stages. The interconnection, along with a particular PWM strategy (in which duty cycles of power stages were opposite or complementary), made it possible to choose the zero-ripple point or zero ripple duty cycle (it can be $D = 0.5$ or any other). Another contribution related to this converter was reported in [16], in which a different PWM strategy, and design procedure, were introduced. In ref. [16], the duty cycles of power stages are proportional instead of complementary to each other, allowing a better operation of the converter and reducing the input current ripple of the entire converter.

This article proposes a third PWM strategy for the discussed converter. In this case, duty cycles of the unequal power stages are independent of each other, and this makes it possible to further reduce the input current ripple of the entire operating range of the converter without changing the hardware (physical components) of the circuit. Having independent duty cycles brings a particular challenge; an infinite number of combinations of duty cycles provide a specific voltage gain. This challenge is solved by using the Differential Evolution DE algorithm. The DE algorithm is used to select the combination of duty cycles that comply with the required output voltage, and at the same time, minimizes the input current ripple of the converter. The proposed methodology is validated through numerical experimentation, which shows that the proposed method achieves a better result (lower input current ripple) than the proportional strategy for a predefined operation range.

Optimization algorithms are widely used in the electrical engineering field [17–22]. Other related approaches that use metaheuristic algorithms, particularly for DC–DC converters, have been recently proposed. These methods include the Harris Hawks Optimization (HHO) algorithm, which is applied for tuning the parameters of two PID controllers used in a buck converter by minimizing the maximum overshoot [18]. In addition, a hybrid method was proposed in [19] that uses the Firefly (F.A.) and Particle Swarm Optimization (PSO) algorithms to adjust the parameters of the PID controller used in a buck converter. Furthermore, the same problem of tuning parameters of the PID controller in a DC–DC buck converter was managed by a hybrid strategy that integrated the Whale Optimization Algorithm and Simulated Annealing to enhance the transient response of the converter [20]. Similar approaches were proposed in [21,22] to face the parameter tuning problem in the PID controllers of converters of this type.

2. Converter Configuration

The interconnection of two or more power stages or cells is a common practice to build more complex converters, and the proposition of new topologies is a very active research field. There are many composed topologies; for example, the interleaved boost converter [15]. Despite the large number of different combinations, a large portion belongs to two main configurations. Considering two individual power cells (Cell 1 and Cell 2) with the voltage gain represented by the functions $g_1(d_1)$ and $g_2(d_2)$, respectively, two possible configurations can be obtained (see Figure 1). The first configuration shown in Figure 1a is the cascaded interconnection; the output of the first power stage is the input of the second one. The second configuration is the input-parallel output-series, in which power stages are connected in parallel at the input, and in series at the output.

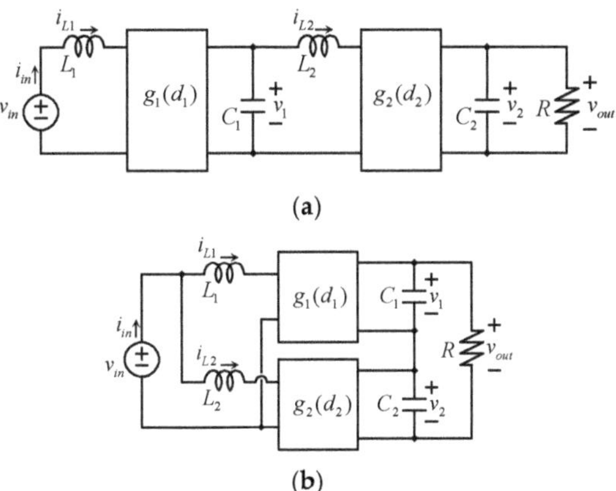

Figure 1. Two common configurations of two power cells to obtain high voltage gain: (**a**) cascaded configuration; (**b**) input-parallel output-series configuration.

The independent variables d_1 and d_2 are the duty cycles corresponding to each converter. Figure 1a shows a cascaded connection where the output voltage v_o is the input voltage times the product of the individual gains. In contrast, in Figure 1b, the value of v_o is the voltage times the summation of the individual gains of each power stage. The common feature of both configurations is that the source current is diverted into the two cells through its corresponding input inductor. Good examples of both configurations can be found in [16] for a photovoltaic application and in [23] adapted for DC-source renewable generation.

In the second case, the input-parallel output-series configuration, it is true that:

$$i_{in}(t) = i_{L1}(t) + i_{L2}(t) \tag{1}$$

The proposed methodology is applied to a power converter belonging to the input-parallel output-series configuration [12,22].

3. Converter Understudy

In order to apply the proposed methodology, the mathematical model of the selected power converter is first derived. Figure 2 shows the converter understudy, initially proposed in [12]. The converter is composed of two cells in an input-parallel output-series connection, following the configuration presented in Figure 1b.

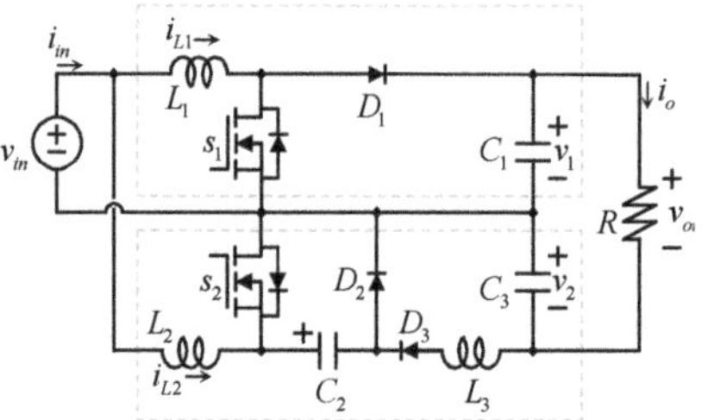

Figure 2. Configuration of the converter under study.

The cells that build the selected converter are: (i) A boost converter [13] composed of switch S_1, diode D_1, inductor L_1, and capacitor C_1. (ii) A modified (resonant version of the) three-switch high-voltage TSHV [14] converter constituted by the rest of the devices, that is, switch S_2, diode D_2, inductors L_2 and L_3, and capacitors C_2 and C_3; note that diode D_3 has been added to block negative current through L_3.

The load resistor R is fed by the summation of the individual output voltage of each cell, that is, v_{C1} for the boost converter and v_{C3} for the modified TSHV converter; therefore, the instantaneous output voltage $v_o(t)$ is determined by (2).

$$v_o(t) = v_{C1}(t) + v_{C3}(t) \tag{2}$$

It is worth mentioning that there are other hybrid topologies, such as [24], but the structure of Figure 2, driven by two independent transistors, makes it possible to perform the input current ripple cancelation, as will be explained.

3.1. Mathematical Model of the Power Converter

Transistors are the manipulated devices that allow power processing through the reactive components. Therefore, in the selected topology, two switching functions are independently operated, which are defined as (3).

$$q_j(t) = \begin{cases} 1 & \rightarrow & S_j & is & closed \\ 0 & \rightarrow & S_j & is & open \end{cases} \tag{3}$$

with $j = 1, 2$ corresponding to switches S_1 and S_2. Since there are two switching functions that are independently controlled, four possible combinations are obtained.

For practical realization, the switching functions of (3) can be implemented in digital or analog fashion in the well-known pulse-width modulation (PWM) technique, which consists of commutating the transistor at constant frequency $F_{S.}$, and hence, at constant period $T_S = 1/F_S$. Assuming that $q_j(0) = 1$, the transition occurs at $t = dT_S$, as illustrated in Figure 3.

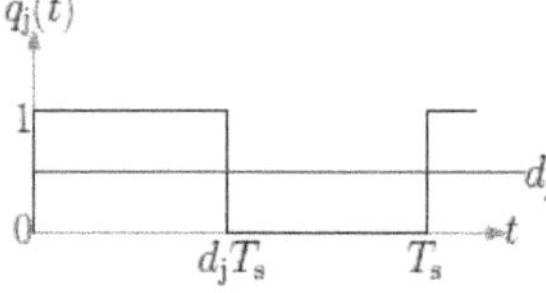

Figure 3. Graphical representation of the switching functions $q_j(t)$ as a PWM technique.

The ratio of the time during which the transistor is active (on-state) to the time when it is inactive (off-state) is called the duty cycle $d_j(t)$ and is equivalent to the average value of the switching function, as (4) indicates.

$$d_j(t) = \frac{1}{T_S} \int_{t}^{t+T_S} q_j(\tau)d\tau \tag{4}$$

Note that $d_j(t)$ is usually taken as a function of time. It represents the manipulated variable that can vary from one switching period to another during the transient response when a closed-loop is implemented [25].

3.2. Equivalent Circuits

In the converter understudy, the input current depends on the inductor's current, and the action of the switching functions drives the inductor's current.

The best case to minimize the input current (see Figure 4) is when transistors switch complementarily, which means when S_1 is open, S_2 is closed, and vice versa. This strategy was originally proposed in [12] and consisted of commutating between the equivalent circuits of Figure 4a,b. All switching states are summarized in Table 1.

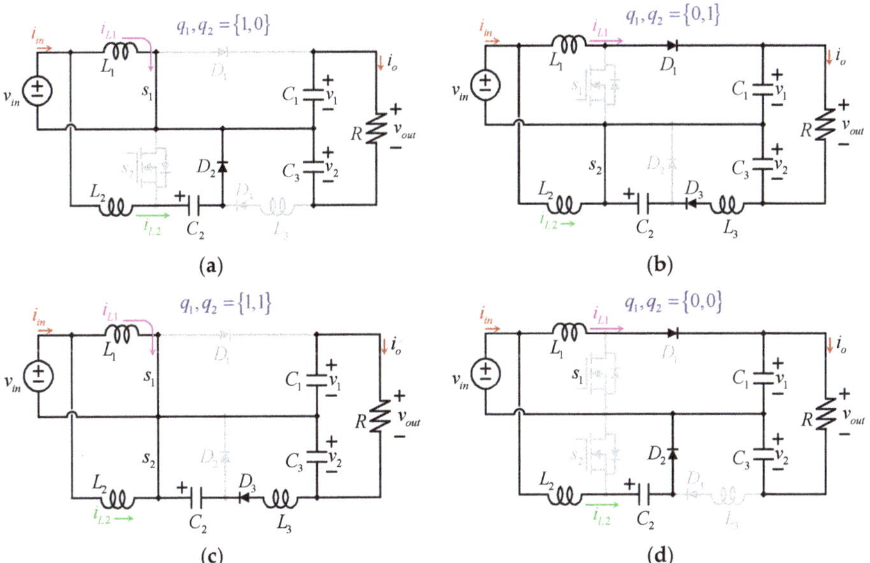

Figure 4. Equivalent circuits for the various commutation states: (**a**) state: $q_1, q_2 = \{1,0\}$, current through L_1 increases whereas that which runs through L_2 decreases; (**b**) state: $q_1, q_2 = \{0,1\}$, opposite to the state in (**a**); (**c**) state: $q_1, q_2 = \{1,1\}$, both inductors' currents increase; and (**d**) state: $q_1, q_2 = \{0,0\}$, both inductors' currents decrease.

Table 1. State.

S_1	S_1	Circuit
0	0	a
0	1	b
1	0	c
1	1	d

Average Model of the Converter

The instantaneous voltage $v_x(t)$ across an inductor L_x is defined as:

$$v_x(t) = L_x \frac{di_x(t)}{dt} \tag{5}$$

Since, in power electronics converters, the voltage across an inductor during a commutation state, as well as the inductance, is constant, it can be observed that the derivative is constant, and the current changes (rises or falls) linearly. For a positive voltage of $v_x(t)$, the current i_x increases with a constant slope, whereas the current decreases with a constant slope for negative values of $v_x(t)$. This consideration is usually called small ripple approximation or linear ripple approximation [13], and it is justified considering that passive components are correctly chosen for this purpose.

Therefore, due to the commutation action, the inductor currents possess a triangular-shaped waveform.

From the circuits of Figure 4a,b and following the average model theory, it is possible to write the inductor voltage along the entire switching period, as in (6) and (7).

$$\frac{di_{L1}}{dt} = d_1 \frac{v_{in}}{L_1} + (1 - d_1)\frac{v_{in} - v_{C1}}{L_1} \tag{6}$$

$$\frac{di_{L2}}{dt} = d_2 \frac{v_{in}}{L_2} + (1 - d_2)\frac{v_{in} - v_{C2}}{L_2} \tag{7}$$

The steady state can be calculated by considering that if the derivatives in (6) and (7) are zero, the capacitors' voltages are found to be:

$$V_{C1} = \frac{v_{in}}{(1 - D_1)} \tag{8}$$

$$V_{C2} = \frac{v_{in}}{(1 - D_2)} \tag{9}$$

It is worth noting that the variables are capitalized in order to denote the average values of the steady state or equilibrium.

4. Existing Methods

This article proposes a modulation strategy for a converter for which two modulation strategies have been used; we can call them (i) the complementary strategy, proposed in [12], and (ii) the proportional strategy, proposed in [16].

4.1. The Complementary Strategy

The complementary strategy, proposed in [12], consists of defining the duty cycles, as shown in (10) and (11).

$$D_1 = (1 - D) \tag{10}$$

$$D_2 = D \tag{11}$$

Important waveforms related to the complementary strategy can be appreciated in Figure 5.

Taking (2), (8), and (9) into account, the voltage gain ($G_{comp} = V_o / V_{in}$) for the complementary strategy can be calculated as (12).

$$G_{comp} = \frac{1}{D(1 - D)} \tag{12}$$

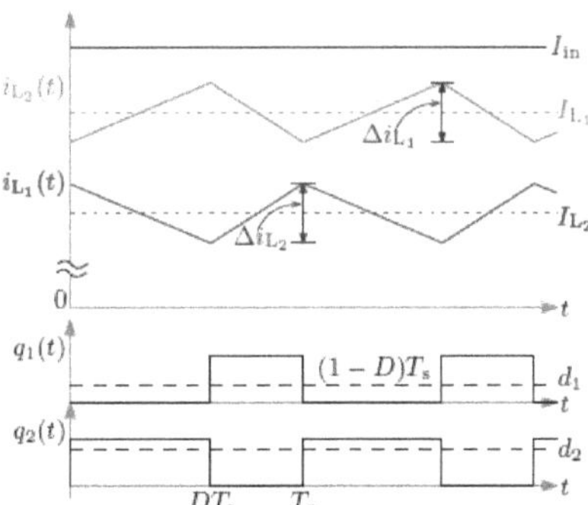

Figure 5. Graphical representation of the switching functions $q_j(t)$: complementary PWM technique.

For the proposed strategy presented in [12], it is easy to find the average current through inductors, as shown in (13) and (14):

$$I_{L1} = \frac{1}{D}I_o \tag{13}$$

$$I_{L2} = \frac{1}{(1-D)}I_o \tag{14}$$

where the output current $I_o = V_o/R$. Moreover, from the equivalent circuits for $q_1,q_2 = \{1,0\}$ and $q_1,q_2 = \{0,1\}$, the current ripples in each inductor can be readily quantified, as shown in (15) and (16).

$$\Delta i_{L1} = \frac{V_{in}}{L_1}\frac{(1-D)}{F_S} \tag{15}$$

$$\Delta i_{L2} = \frac{V_{in}}{L_2}\frac{(D)}{F_S} \tag{16}$$

Following (1), the input current ripple can be expressed as shown (17) and (18).

$$\Delta i_{in} = \Delta i_{L1} + \Delta i_{L2} \tag{17}$$

However, the inductor current ripples have a phase displacement of $180°$ (see Figure 5); in other words, they always have different signs, for which the input current ripple can be expressed as the difference among (15) and (16), and not their summation. Therefore, the input current ripple is defined as:

$$\Delta i_{in} = \left| \frac{V_{in}}{F_S}\left(\frac{D}{L_2} - \frac{1-D}{L_1} \right) \right| \tag{18}$$

From (18), it is easy to solve for D to obtain a zero-current ripple at the converter's input.

This strategy can fully cancel the current ripple at a preestablished operating point. However, as the converter moves away from such a point, the input current increases undesirably.

4.2. Proportional Modulation Strategy

The strategy described above is able to draw a D.C. pure current from the power source. The proportional strategy, proposed in [16], also makes it possible to perform the zero-ripple operation for a particular operation point, but converters usually operate not only on a specific operation point but into an operating range. It was shown in [16] that both strategies provide the same performance in the zero ripple operation points. Still, in different operation points, the proportional strategy provided a smaller input current ripple. That means, for a predefined operation range, the proportional strategy performs better.

Considering (2), (8), and (9) the voltage gain G can be expressed as (19) and (20) for duty cycles D_1 and D_2.

$$G = \frac{1}{(1 - D_1)} + \frac{1}{(1 - D_2)} \tag{19}$$

$$G = \frac{2 - D_1 - D_2}{1 - D_1 - D_2 + D_1 D_2} \tag{20}$$

Equations (19) and (20) apply for the converter in all different strategies.

It is evident that the desired voltage gain G_1 can be obtained with an infinite number of combinations of duty cycles employed in (19) and (20). The strategy proposed in [16] assigns a proportional relationship, as shown in (21) and (22).

$$D_1 = kD \tag{21}$$

$$D_2 = D \tag{22}$$

where the factor k is a constant value, determined during the design process as a function of the inductors in the following manner:

$$k = \frac{L_1}{L_2} \tag{23}$$

Considering (20)–(22), the voltage gain can be expressed as (24).

$$G_k = \frac{2 - D - kD}{1 - D + D + kD^2} \tag{24}$$

As it was presented in [16], the input current ripple can be analyzed in two different cases; let's consider d^* as the duty cycle in which the input current ripple is zero (as in [16]). The input current ripple is expressed differently when $d < d^*$ and when $d > d^*$. Figure 6 shows important waveforms of the operation: (i) in the case for $d < d^*$, and (ii) in the case for $d > d^*$. Figure 6 shows the graphical representation of both cases.

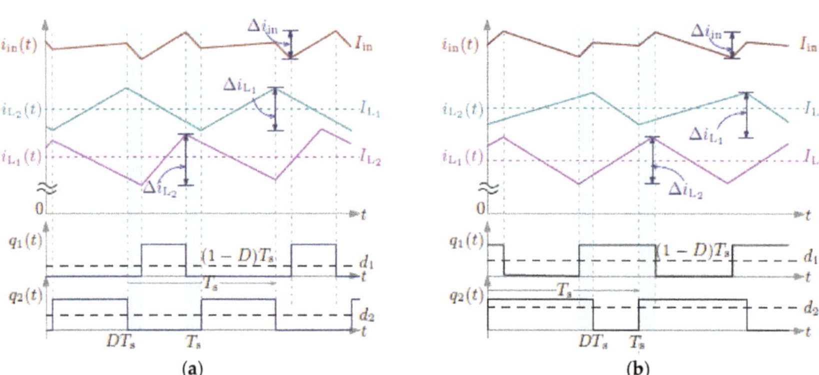

Figure 6. (a) $D_1 = 30\%$, $D_2 = 50\%$; (b) $D_1 = 30\%$, $D_2 = 70\%$.

Although the input current ripple behaves slightly differently in both cases, the input current ripple can be defined as the maximum ripple during the operation. Then, the largest vale is provided by the calculation in both cases [16].

It is worth mentioning that a PID controller still may control the voltage gain of the converter to achieve the desired output voltage. The discussed algorithms are designed to decide the proportions of duty cycles or of voltage gain in individual power stages of the composing converter. The design of the PID controller can still be based on an optimization algorithm [18–21].

5. Differential Evolution Algorithm

Differential Evolution (D.E.) [26–31] is a search method that is widely recognized in the evolutionary community and has proven its effectiveness and robustness in finding global solutions. The DE algorithm is a classic and popular method that is extensively used to solve complex optimization problems. Its low complexity and high performance have made it widely used for applications in different fields such as image processing, operation research, electronics engineering, mechanical engineering, manufacturing design, power engineering, and many others [30,31]. Although other sophisticated evolutionary methods have been recently proposed, the D.E. algorithm still maintains its prestige in the metaheuristic community, mainly because of its low computational cost and effectiveness even in constrained and high-dimensional optimization problems [29].

The DE technique considers different operators in its search strategy that allow the particles to improve as they evolve in each generation. The DE method process involves three basic operators, which are mutation, crossover, and selection. These operators are part of the algorithm search process and will be described in detail in this section.

5.1. Mutation

The mutation process generates a mutant vector that results from combining the information from three different vectors, x_{r1}, x_{r2}, and x_{r3}, taken randomly from the population. The combination of these vectors considers the scaled difference between two of the three vectors. Then, this difference is added to the third vector. The mathematical representation of the mutation operator is described as:

$$v = x_{r3} + F(x_{r1} - x_{r2}) \tag{25}$$

Each vector represents an individual from the population given in the presented equation, while every element of the vector represents one dimension of the optimization problem. The scale factor F from the same equation controls the difference between the vector x_{r2} and the vector x_{r3}. According to the authors of the D.E. algorithm, the scale factor can take values within the interval $[0, 2]$. The scale factor is also known as the differential weight since it regulates the difference $(x_{r1} - x_{r2})$.

5.2. Crossover

The crossover operator combines the information from one individual with the information from the mutant vector. This mechanism gives diversity to the population and, therefore, prevents the algorithm from stagnating in suboptimal solutions. The combination consists of randomly taking elements from the mutant vector and elements from one individual from the population with the aim of building a test vector u. The crossover operation includes a crossover probability P_{cross} that adjusts the contribution of the mutant vector in the generation of the test vector. The crossover probability strongly influences the effectiveness of the D.E. method in finding promising solutions. Therefore, it is considered a parameter to be adjusted by selecting values within the interval $[0, 1]$. The crossover operator definition is expressed as (26).

$$u_j = \begin{cases} v_{ij} & r(0,1) \leq P_{cross} \\ x_{ij} & otherwise \end{cases} \tag{26}$$

5.3. Selection

The selection operator is the last stage in the D.E. method's flow of operations. The objective of the selection method is to determine whether an individual in the population is replaced by the test vector generated with the mutation and crossover operators. The replacement of the individual with the test vector is carried out based on the evaluation of the quality of both solutions. The individual who is selected through this evaluation process becomes part of a new generation of individuals. In this way, this mechanism ensures that individuals evolve by selecting only the best particles to be part of a new population.

The selection mechanism contemplates the fitness value of the test vector u and the fitness value of the individual x_i. The selection follows a simple rule; if the fitness value of the individual is worse than the fitness value of the test vector, then the individual is replaced by the test vector. Otherwise, no replacement is made. Thus, the selection can be formulated as:

$$x_i = \begin{cases} u, & f(u) \leq f(x_i) \\ x_i, & otherwise \end{cases} \tag{27}$$

6. The Proposed Strategy

The proposed strategy considers two independent duty cycles, in contrast to the former two strategies, namely the complementary strategy [12] defined in (10) and (11), and the proportional strategy [16] defined by (21) and (22). Equations (2), (8), and (9) still define the output voltage, which can also be expressed as (19) and (20). However, in this case, there is not a particular equation that relates both duty cycles.

The strategy can be defined as (28) and (29).

$$D_1 = k_d D \tag{28}$$

$$D_2 = D \tag{29}$$

where the factor k_d is not a constant value, as in (21) and (22), but is instead a variable that has to be found at the same time as D_1 and D_2.

The problem can be defined as an optimization problem. The objective is to minimize the input current ripple while, at the same time, finding the duty cycles D_1 and D_2. Finding D_1 and D_2 is equivalent to finding D and k_d (see (28) and (29)). An important restriction of the problem is the need to comply with a required amount of voltage gain. The input current ripple can be defined as the largest of Equations (30) and (31).

$$\Delta i_{g1} = \frac{V_{in}}{k_L f s L_2}(k_L - k_d D - k_L k_d D) \tag{30}$$

$$\Delta i_{g2} = \frac{V_{in}}{k_L f s L_2}(1 - D - k_L D) \tag{31}$$

where k_L represents the relations among inductors (same as (23) for [16]), which must be provided to the optimization problem.

An optimizer is ideal for this problem since the input current ripple is variable, and their minimum value depends on the operating condition. The optimizer can minimize the input current ripple without the need for a particular desired value (or setpoint) for the ripple.

6.1. The Objective Function

The optimization problem is formulated according to the objective function defined in (32). This objective function minimizes the higher value of the input current ripple. This means that it evaluates Equations (30) and (31) and considers the highest value of the two current ripple calculations to change parameters D and k_d in order to decrease this value.

$$\min_{D, k_d \in \mathbb{R}} f(D, k_d) = \begin{cases} \Delta i_{g1}, & \Delta i_{g1} > \Delta i_{g2} \\ \Delta i_{g2}, & otherwise \end{cases} \tag{32}$$

The above equation is subject to:

$$\frac{V_{out}}{V_g} \leq \frac{2 - D - k_d D}{(1 - k_d D)(1 - D)} \leq \frac{V_{out}}{V_g} + \delta \tag{33}$$

$$0 \leq D \leq 1 \tag{34}$$

$$0 \leq k_d \leq 1 \tag{35}$$

The objective function is subject to some constraints given by Equations (33)–(35). Equations (34) and (35) define the range of possible values that parameters D and k_d can adopt. On the other hand, Equation (33) restricts the value of the output voltage. It ensures that the desired voltage gain is satisfied. Equation (33) also contemplates a tolerance in the output voltage, which is 1% of the estimated voltage gain. The solution obtained by the proposed method must accomplish three such restrictions in order to consider the solution as a feasible one.

6.2. Computational Procedure

The proposed strategy considers the implementation of the Differential Evolution algorithm to find the optimal values of the parameters D and k_d, which control the duty cycle of the hybrid switched-capacitor boost converter. Currently, the optimal operation of this type of converter is given by a specific duty cycle, which is obtained from a dependent relationship between parameters D and k_d. The dependency between these parameters limits a further reduction in the input current ripple. Therefore, this article proposes a different strategy to find the optimal switching pattern that allows high quality at the input current side, thereby providing the reactive components with the required switching frequency and power levels.

The optimization algorithm considers the set of values D and k_d as a possible solution. Thus, a population of possible solutions is created so that every individual i from the population is defined as (36).

$$x_i = \{D, k_d\} \tag{36}$$

The population size is determined by the parameter N, which must be set at the beginning of the search process. Additionally, every individual from the population is initialized, by assigned random values, to the set $\{D, k_d\}$. Then, each individual's fitness value is calculated employing the objective function, which determines the input current ripple. The assigned values to the set $\{D, k_d\}$ must satisfy the required gain both in the initialization phase and throughout the algorithm search process. Therefore, the objective function includes a constraint that identifies the particles that do not meet this requirement. The identified solutions that violate this constraint are penalized by increasing their fitness values. This mechanism avoids unfeasible solutions, and, at the same time, it guides the search towards the space of feasible solutions. In that sense, to include the penalization, the fitness function is rewritten as (37).

$$fitness = \min_{x_i \in \mathbb{R}^2} f(x_i) + h \tag{37}$$

From the above equation, the penalty function "h" is responsible for increasing the fitness value of the particles that have violated the restriction. The mathematical expression of this function is given by (38).

$$h = aC \left| \frac{V_{out}}{V_g} - \frac{2x_{i1} - x_{i2}x_{i1}}{(1 - x_{i1}x_{i2})(1 - x_{i1})} \right| \tag{38}$$

Here, a is a constant factor that controls the penalization degree of every unfeasible solution. Whenever there is an unfeasible solution, the activation function C sets its value to one to perform the penalty function. Otherwise, the penalty function is switched off. The definition of the activation function can be expressed as (39).

$$C = \begin{cases} 0, & \frac{2 - x_{i1} - x_{i2}x_{i1}}{(1 - x_{i2}x_{i1})(1 - x_{i1})} \geq \frac{V_{out}}{V_g} \\ 1, & otherwise \end{cases} . \tag{39}$$

The general structure of the D.E. method consists of initializing the population and evaluating its fitness value. Then, the mutation operation is performed for each particle to generate its mutant vector. After that, the crossover operation is also applied to every candidate solution to create its test vector. Finally, the test vectors are evaluated in terms of the fitness function. The best fitness between every particle and its respective test vector is considered as the selection criterion to determine the set of individuals that will remain for the next generation. This routine continues repeating for a specific number of generations, denoted as G_{max}. After all, the best solution found up to this point will represent the values of parameters D and k_d. The described procedure is summarized in Algorithm 1.

Algorithm 1 The general structure of the D.E. method

---

1. *Initialize parameters $\boldsymbol{P_{cross}}$, a, N, G_{max}*
2. *Initialization of the population*
3. *Evaluation of the population in the fitness function*
4. *For each particle*
5.   *Creation of the mutant vector $\boldsymbol{v}$*
6.   *Creation of the test solution $\boldsymbol{u}$*
7.   *Selection of the best-found solution between $\boldsymbol{u}$ and $\boldsymbol{x_i}$*
8. *End for*
9. *Update the global-best so far*
10. *If the maximum number of generations has not been achieved*
11.   *Go to step 4*
12. *Else*
13.   *End the search*
14. *End if-else*

---

7. Results

A simulation framework was implemented to validate the performance of the proposed method. The experiment contemplated a variable-voltage energy source that provided the required voltage to the converter. The converter input voltage varied from 28 V to 40 V, which caused the voltage gain to range from 5 to 7.15.

On the other hand, some D.E. algorithm parameters needed to be set, such as the factor a, the crossover probability P_{cross}, the number of generations G_{max}, and the population N (the number of individuals). This configuration was carried out by observing the performance of the D.E. method for different combinations of values. Since the D.E.'s performance is evaluated considering the achieved input current ripple Δ_{ig}, a sensitivity test was implemented to choose the best parameter configuration for reducing the input current ripple. Different combinations of parameter values were set to execute the algorithm for finding the input current ripple. For the crossover probability, we used the values of 0.1, 0.2, and 0.3. The constant factor a was considered with the values of 1, 10, and 100. The population size included in the test was 20, 30, and 50 individuals. Finally, the maximum number of generations was 100, 200, and 300. After several tests, the parameter combination that best decreased the input current ripple was used in the numerical experiments. These settings are listed in Table 2.

Table 2. Parameters of the D.E. method.

Crossover probability P_{cross}	0.2
Constant factor a	10
Population size N	50
Maximum number of generations G_{max}	300

Furthermore, the converter of interest involved several parameters that needed to be configured, namely the inductor constant factor, the input voltage, the inductance, and the switching frequency. Table 3 reports the stated parameters, whose values were assigned considering the standard specifications for this kind of converter.

Table 3. Parameters of the hybrid switched-capacitor boost converter.

Input voltage V_g	28–40 V
Switching frequency f_S	50 kHz
Inductor factor k_L	0.4
Inductor L	200 µH

Numerical experiments were carried out to compare the proposed method's performance against the proportional strategy. Only the proportional strategy was considered since it was previously proven that the proportional strategy achieves a lower input current ripple compared to the complementary strategy for a defined operating range; if the proposed strategy provides superior performance compared to the complementary strategy, it can be considered the best strategy.

This test considered several operating points to demonstrate the robustness of the proposed approach. Every operating point represented a different input voltage to the converter, which varied from 28 V to 40 V. Thus, the experiments considered 44 operating points in which the proposed method and the traditional strategy were simulated to obtain 44 input current ripples, one for every operating point.

Furthermore, the values of the parameters described in tables a and b were used in the experiments. Both methods were simulated under the same conditions to guarantee that the comparison was fair. The best-achieved outcomes from the proposed technique, considering 30 independent executions (due to its stochasticity nature), are reported in Table 4. Likewise, the former strategy results are listed in the same Table, where the best input current ripple Δi_g attained between both techniques is emphasized in boldface.

Table 4. Optimal parameter values and the input current ripple obtained from the numerical experiments considering the former strategy and the proposed method.

	Former Strategy			Proposed Method		
G	D	k_d	Δi_g	D	k_d	Δi_g
5.00	0.7218	0.4000	0.0526	0.7215	0.4054	**0.0502**
5.05	0.7254	0.4000	0.0770	0.7246	0.4079	**0.0715**
5.10	0.7289	0.4000	0.1003	0.7279	0.4111	**0.0937**
5.15	0.7323	0.4000	0.1224	0.7313	0.4134	**0.1154**
5.20	0.7357	0.4000	0.1441	0.7341	0.4162	**0.1336**
5.25	0.7389	0.4000	0.1641	0.7373	0.4186	**0.1533**
5.30	0.7421	0.4000	0.1837	0.7404	0.4194	**0.1722**
5.35	0.7452	0.4000	0.2022	0.7431	0.4230	**0.1882**
5.40	0.7483	0.4000	0.2205	0.7460	0.4252	**0.2056**
5.45	0.7513	0.4000	0.2377	0.7488	0.4268	**0.2219**
5.50	0.7542	0.4000	0.2540	0.7514	0.4298	**0.2369**
5.55	0.757	0.4000	0.2694	0.7544	0.4295	**0.2527**
5.60	0.7598	0.4000	0.2845	0.7569	0.4334	**0.2665**
5.65	0.7625	0.4000	0.2987	0.7597	0.4359	**0.2816**
5.70	0.7652	0.4000	0.3126	0.7622	0.4356	**0.2944**
5.75	0.7678	0.4000	0.3257	0.7646	0.4378	**0.3063**
5.80	0.7704	0.4000	0.3386	0.7669	0.4411	**0.3174**
5.85	0.7729	0.4000	0.3507	0.7694	0.4424	**0.3299**
5.90	0.7753	0.4000	0.3619	0.7720	0.4435	**0.3421**
5.95	0.7777	0.4000	0.3730	0.7742	0.4463	**0.3523**

Table 4. *Cont.*

	Former Strategy			Proposed Method		
G	D	k_d	Δi_g	D	k_d	Δi_g
6.00	0.7801	0.4000	0.3839	0.7764	0.4470	**0.3626**
6.05	0.7823	0.4000	0.3935	0.7785	0.4491	**0.3713**
6.10	0.7846	0.4000	0.4034	0.7807	0.4506	**0.3809**
6.15	0.7868	0.4000	0.4127	0.7828	0.4517	**0.3898**
6.20	0.789	0.4000	0.4218	0.7852	0.4537	**0.4001**
6.25	0.7911	0.4000	0.4302	0.7869	0.4554	**0.4067**
6.30	0.7932	0.4000	0.4384	0.7890	0.4555	**0.4152**
6.35	0.7952	0.4000	0.4460	0.7910	0.4571	**0.4229**
6.40	0.7972	0.4000	0.4534	0.7930	0.4579	**0.4303**
6.45	0.7992	0.4000	0.4608	0.7948	0.4608	**0.4372**
6.50	0.8011	0.4000	0.4675	0.7971	0.4589	**0.4459**
6.55	0.803	0.4000	0.4740	0.7986	0.4625	**0.4504**
6.60	0.8049	0.4000	0.4805	0.8004	0.4646	**0.4569**
6.65	0.8067	0.4000	0.4864	0.8022	0.4648	**0.4629**
6.70	0.8085	0.4000	0.4922	0.8043	0.4641	**0.4703**
6.75	0.8103	0.4000	0.4979	0.8057	0.4678	**0.4739**
6.80	0.812	0.4000	0.5029	0.8076	0.4667	**0.4803**
6.85	0.8137	0.4000	0.5080	0.8094	0.4689	**0.4859**
6.90	0.8154	0.4000	0.5129	0.8111	0.4678	**0.4911**
6.95	0.817	0.4000	0.5173	0.8124	0.4722	**0.4943**
7.00	0.8186	0.4000	0.5216	0.8140	0.4726	**0.4988**
7.05	0.8202	0.4000	0.5258	0.8157	0.4719	**0.5034**
7.10	0.8218	0.4000	0.5300	0.8172	0.4751	**0.5073**
7.15	0.8233	0.4000	0.5336	0.8189	0.4758	**0.5122**

In Table 4, the best parameter combination values (D, k_d) obtained from both strategies are reported. Additionally, it shows the obtained input current ripple Δi_g using those values for every voltage gain G. From the Table, a close inspection demonstrates that the lowest input current ripple was reached by the proposed approach for all the experiments. These results are a consequence of the proposed optimization method's search process, which was able to find the optimal values for the duty cycles in such a way that these values led to a decrease in the input current ripple without a change in the required voltage gain.

The comparison between the proposed method and the former strategy, in terms of the input current ripple obtained from the numerical experiments, is reported in Figure 7.

Figure 7a–d can be visually analyzed to observe the extent to which the proposed method outperformed the former strategy. For every voltage gain from 5 V to 7.15 V, the statistics show the obtained input current ripple considering both approaches, revealing that the proposed technique found the best parameter combination to further reduce the input current ripple.

In summary, the proposed method proved its efficiency in finding the best parameter combination to decrease the input current ripple and maintain the required voltage gain. The experimental results support the proposed strategy, in which a different mathematical model that considers independent duty cycles, and the implementation of the D.E. method to optimize the parameter values, achieved better results than the traditional strategy without modifying the converter at a hardware level, instead only making software modifications.

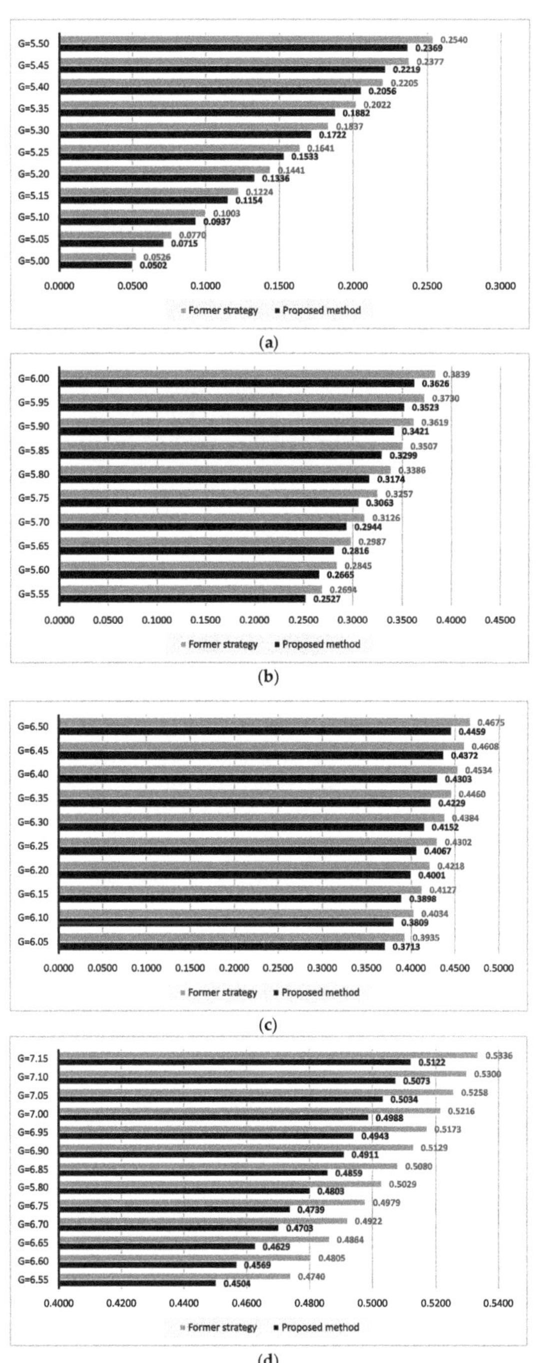

Figure 7. Input current ripple obtained from the former strategy and the proposed method: (**a**) for voltage gain from 5.00 V to 5.50 V; (**b**) for voltage gain from 5.55 V to 6.00 V; (**c**) for voltage gain from 6.05 V to 6.50 V; and (**d**) for voltage gain from 6.65 V to 7.15 V.

8. Conclusions

A novel strategy for reducing the input current ripple of the hybrid switched-capacitor boost converter is proposed in this article. This work aims to optimize the duty cycles of the converter by generating a new model that considers independent duty cycles. However, finding the optimal values for duty cycles when they are independent is not an easy task. Therefore, the proposed strategy includes the implementation of the Differential Evolution algorithm to find the best value combination for the switching pattern to reach high quality at the input current side while satisfying the required voltage gain.

The main objective of the proposed method is to improve the traditional strategy used to reduce the input current ripple. This former strategy considers dependent duty cycles, limiting the possibility of further decreasing the input current ripple and increasing the input current's quality. Furthermore, the proposed approach does not make any changes to the converter at a hardware level. Instead, it only makes software changes.

The proposed method's performance and the former strategy were compared under several experiments, which considered different input voltage values to the converter. The experiments demonstrate that the proposed technique outperforms the former strategy by obtaining a lower input current ripple for all operating points without affecting the required voltage gain.

In future work, an enhanced version of the D.E. algorithm or a different, more recent algorithm can be included. In addition, several swarm-based algorithms can be implemented in order to compare their performances and analyze which could be the best option for finding the optimal duty cycle values. The analysis can be from the perspective of the swarm-based methods and their effectiveness when applied to this kind of constrained optimization problem. Furthermore, if new PWM strategies are reported for the converter understudy, a comparison against this strategy can be performed.

Author Contributions: J.C.R.-C. and P.M.G.-V. contributed to the conceptualization of the article, from the power electronics point of view; F.B.-C., E.C. and A.M. contributed to the methodology; E.C., A.R. and A.A.-R. contributed to the software, validation, and formal analysis; P.M.G.-V., A.R. and J.C.R.-C. wrote the draft and prepared the manuscript. All authors have read and agreed to the published version of the manuscript.

Funding: J.C.R.-C. would like to thank Universidad Panamericana, for their support through the program "Fomento a la Investigación UP 2021" and the project "Estrategias de reducción de variaciones por conmutación y energía almacenada, y optimización en el diseño de convertidores de energía eléctrica", grant number: UP-CI-2021-GDL-01-ING.

Institutional Review Board Statement: Not applicable.

Informed Consent Statement: Not applicable.

Data Availability Statement: Not applicable.

Acknowledgments: The authors would like to thank Universidad Panamericana, Tecnologico Nacional de Mexico (Instituto Tecnologico de Ciudad Madero), Universidad de Guadalajara, and Universidad Autónoma Metropolitana.

Conflicts of Interest: The authors declare no conflict of interest.

References

1. Arias, M.B.; Bae, S. Design Models for Power Flow Management of a Grid-Connected Solar Photovoltaic System with Energy Storage System. *Energies* **2020**, *13*, 2137. [CrossRef]
2. Hou, D.; Lee, F.C.; Li, Q. Very High Frequency IVR for Small Portable Electronics with High-Current Multiphase 3-D Integrated Magnetics. *IEEE Trans. Power Electron.* **2017**, *32*, 8705–8717. [CrossRef]
3. Ahsanuzzaman, S.M.; Prodi´c, A.; Johns, D.A. An Integrated High-Density Power Management Solution for Portable Applications Based on a Multioutput Switched-Capacitor Circuit. *IEEE Trans. Power Electron.* **2016**, *31*, 4305–4323. [CrossRef]
4. Kurdkandi, N.V.; Nouri, T. Analysis of an efficient interleaved ultra-large gain DC–DC converter for DC microgrid applications. *IET Power Electron.* **2020**, *13*, 2008–2018. [CrossRef]

5. Liu, T.; Yang, X.; Chen, W.; Xuan, Y.; Li, Y.; Huang, L.; Hao, X. High-Efficiency Control Strategy for 10-kV/1-MW Solid-State Transformer in PV Application. *IEEE Trans. Power Electron.* **2020**, *35*, 11770–11782. [CrossRef]
6. Youn, H.S.; Yun, D.H.; Lee, W.S.; Lee, I.O. Study on Boost Converters with High Power-Density for Hydrogen-Fuel-Cell Hybrid Railway System. *Electronics* **2020**, *9*, 771. [CrossRef]
7. Balog, R.S.; Krein, P.T. Coupled-Inductor Filter: A Basic Filter Building Block. *IEEE Trans. Power Electron.* **2013**, *28*, 537–546. [CrossRef]
8. Nag, S.S.; Mishra, S.; Joshi, A. A Passive Filter Building Block for Input or Output Current Ripple Cancellation in a Power Converter. *IEEE J. Emerg. Sel. Top. Power Electron.* **2016**, *4*, 564–575. [CrossRef]
9. Cheng, M.; Pan, C.; Teng, J.; Luan, S. An Input Current Ripple-Free Flyback-Type Converter with Passive Pulsating Ripple Canceling Circuit. *IEEE Trans. Ind. Appl.* **2017**, *53*, 1210–1218. [CrossRef]
10. Kim, S.; Do, H. Soft-Switching Step-Up Converter with Ripple-Free Output Current. *IEEE Trans. Power Electron.* **2016**, *31*, 5618–5624. [CrossRef]
11. Al-Saffar, M.A.; Ismail, E.H. A high voltage ratio and low stress DC-DC converter with reduced input current ripple for fuel cell source. *Renew. Energy* **2015**, *82*, 35–43. [CrossRef]
12. Rosas-Caro, J.C.; Mancilla-David, F.; Mayo-Maldonado, J.C.; Gonzalez-Lopez, J.M.; Torres-Espinosa, H.L.; Valdez-Resendiz, J.E. A Transformer-less High-Gain Boost Converter with Input Current Ripple Cancelation at a Selectable Duty Cycle. *IEEE Trans. Ind. Electron.* **2013**, *60*, 4492–4499. [CrossRef]
13. Erickson, R.; Maksimovic, D. *Fundamentals of Power Electronics*, 2nd ed.; Springer: New York, NY, USA; Philadelphia, PA, USA, 2001.
14. Dongyan, Z.; Pietkiewicz, A.; Cuk, S. A three-switch high-voltage converter. *IEEE Trans. Power Electron.* **1999**, *14*, 177–183. [CrossRef]
15. Wang, C. Investigation on Interleaved Boost Converters and Applications. Ph.D. Thesis, Virginia Polytechnic Institute and State University, Blacksburg, VA, USA, 2009.
16. Soriano-Rangel, C.A.; Rosas-Caro, J.C.; Mancilla-David, F. An Optimized Switching Strategy for a Ripple-Canceling Boost Converter. *IEEE Trans. Ind. Electron.* **2015**, *62*, 4226–4230. [CrossRef]
17. Wan, W.; Bragin, M.A.; Yan, B.; Qin, Y.; Philhower, J.; Zhang, P.; Luh, P.B. Distributed and Asynchronous Active Fault Management for Networked Microgrids. *IEEE Trans. Power Syst.* **2020**, *35*, 3857–3868. [CrossRef]
18. Ekinci, S.; Hekimoğlu, B.; Demirören, A.; Kaya, S. Harris Hawks Optimization Approach for Tuning of FOPID Controller in DC-DC Buck Converter. In Proceedings of the 2019 International Artificial Intelligence and Data Processing Symposium (IDAP), Malatya, Turkey, 21–22 September 2019; pp. 1–9.
19. Ekinci, S.; Hekimoğlu, B.; Eker, E.; Sevim, D. Hybrid Firefly and Particle Swarm Optimization Algorithm for PID Controller Design of Buck Converter. In Proceedings of the 2019 3rd International Symposium on Multidisciplinary Studies and Innovative Technologies (ISMSIT), Ankara, Turkey, 11–13 October 2019; pp. 1–6.
20. Hekimoğlu, B.; Ekinci, S. Optimally Designed PID Controller for a DC-DC Buck Converter via a Hybrid Whale Optimization Algorithm with Simulated Annealing. *Electrica* **2020**, *8*, 19–27. [CrossRef]
21. Izci, D.; Ekinci, S.; Hekimoğlu, B. A novel modified Lévy flight distribution algorithm to tune proportional, integral, derivative and acceleration controller on buck converter system. *Trans. Inst. Meas. Control* **2021**, *8*, 014233122110365. [CrossRef]
22. Izci, D.; Hekimoğlu, B.; Ekinci, S. A new artificial ecosystem-based optimization integrated with Nelder-Mead method for PID controller design of buck converter. *Alex. Eng. J.* **2021**, *8*. [CrossRef]
23. Walker, G.R.; Sernia, P.C. Cascaded DC-DC converter connection of photovoltaic modules. *IEEE Trans. Power Electron.* **2004**, *19*, 1130–1139. [CrossRef]
24. García-Vite, P.M.; del Rosario Rivera-Espinosa, M.; Alejandre-López, A.; Castillo-Gutiérrez, R.; González-Rodríguez, A.; Hernández-Angel, F. Analysis and implementation of a step-up power converter with input current ripple cancelation. *Int. J. Circuit Theory Appl.* **2018**, *47*, 1338–1357. [CrossRef]
25. Karthikeyan, M.; Elavarasu, R.; Ramesh, P.; Bharatiraja, C.; Sanjeevikumar, P.; Mihet-Popa, L.; Mitolo, M. A Hybridization of Cuk and Boost Converter Using Single Switch with Higher Voltage Gain Compatibility. *Energies* **2020**, *13*, 2312. [CrossRef]
26. Ogata, K. *Moder Control Engineering*; Prentice Hall: Hoboken, NJ, USA, 2010.
27. Storn, R.; Price, K. A Simple and Efficient Heuristic for global Optimization over Continuous Spaces. *J. Glob. Optim.* **1997**, *11*, 341–359. [CrossRef]
28. Simon, D. *Evolutionary Optimization Algorithms*; Wiley: Hoboken, NJ, USA, 2013.
29. Erik, C.; Rodriguez, A. *Metaheuristic Computation with MATLAB*, 1st ed.; Taylor & Francis: Boca Raton, FL, USA, 2020.
30. Das, S.; Suganthan, P.N. Differential Evolution: A Survey of the State-of-the-Art. *IEEE Trans. Evol. Comput.* **2011**, *15*, 4–31. [CrossRef]
31. Pant, M.; Zaheer, H.; Garcia-Hernandez, L.; Abraham, A. Differential Evolution: A review of more than two decades of research. *Eng. Appl. Artif. Intell.* **2020**, *90*, 103479.

 *mathematics*

MDPI

Article

A Fast-Tracking Hybrid MPPT Based on Surface-Based Polynomial Fitting and P&O Methods for Solar PV under Partial Shaded Conditions

Catalina González-Castaño [1,†] , Carlos Restrepo [2,*,†] , Javier Revelo-Fuelagán [3,†] , Leandro L. Lorente-Leyva [4] and Diego H. Peluffo-Ordóñez [5,6,†]

[1] Department of Engineering Sciences, Universidad Andres Bello, Santiago 7500971, Chile; inv.cet@unab.cl
[2] Department of Electromechanics and Energy Conversion, Universidad de Talca, Curicó 3340000, Chile
[3] Department of Electronics Engineering, Faculty of Engineering, Universidad de Nariño, Pasto 520001, Colombia; javierrevelof@udenar.edu.co
[4] Postgraduate Center, Universidad Politécnica Estatal del Carchi, Tulcán 040101, Ecuador; leandro.lorente@upec.edu.ec
[5] Modeling, Simulation and Data Analysis (MSDA) Research Program, Mohammed VI Polytechnic University, Ben Guerir 47963, Morocco; diego.peluffo@sdas-group.com
[6] Corporación Universitaria Autónoma de Nariño, Pasto 520001, Colombia
*   Correspondence: crestrepo@utalca.cl
†   These authors contributed equally to this work.

Citation: González-Castaño, C.; Restrepo, C.; Revelo-Fuelagán, J.; Lorente-Leyva, L.L.; Peluffo-Ordóñez, D.H. A Fast-Tracking Hybrid MPPT Based on Surface-Based Polynomial Fitting and P&O Methods for Solar PV under Partial Shaded Conditions. *Mathematics* **2021**, *9*, 2732. https://doi.org/10.3390/math9212732

Academic Editor: Nicu Bizon

Received: 29 September 2021
Accepted: 26 October 2021
Published: 28 October 2021

Publisher's Note: MDPI stays neutral with regard to jurisdictional claims in published maps and institutional affiliations.

Abstract: The efficiency of photovoltaic (PV) systems depends directly on solar irradiation, so drastic variations in solar exposure will undoubtedly move its maximum power point (MPP). Furthermore, the presence of partial shading conditions (PSCs) generates local maximum power points (LMPPs) and one global maximum power point (GMPP) in the P-V characteristic curve. Therefore, a proper maximum power point tracking (MPPT) technique is crucial to increase PV system efficiency. There are classical, intelligent, optimal, and hybrid MPPT techniques; this paper presents a novel hybrid MPPT technique that combines Surface-Based Polynomial Fitting (SPF) and Perturbation and Observation (P&O) for solar PV generation under PSCs. The development of the experimental PV system has two stages: (i) Modeling the PV array with the DC-DC boost converter using a real-time and high-speed simulator (PLECS RT Box), (ii) and implementing the proposed GMPPT algorithm with the double-loop controller of the DC-DC boost converter in a commercial low-priced digital signal controller (DSC). According to the simulation and the experimental results, the suggested hybrid algorithm is effective at tracking the GMPP under both uniform and nonuniform irradiance conditions in six scenarios: (i) system start-up, (ii) uniform irradiance variations, (iii) sharp change of the (PSCs), (iv) multiple peaks in the P-V characteristic, (v) dark cloud passing, and (vi) light cloud passing. Finally, the experimental results—through the standard errors and the mean power tracked and tracking factor scores—proved that the proposed hybrid SPF-P&O MPPT technique reaches the convergence to GMPP faster than benchmark approaches when dealing with PSCs.

Keywords: maximum power point tracking; photovoltaic system; partial shading conditions; surface-based polynomial fitting

1. Introduction

Global energy demand has increased substantially in recent decades, and along with a greater need for electric power comes the expansion of alternative energy solutions that promote environmental protection and sustainability. For example, the solar energy industry has rapidly grown resulting in decreased manufacturing costs and the proliferation of affordable photovoltaic (PV) systems [1]. However, PV systems face significant challenges, mainly the irregular solar irradiation due to partial shading that affects their efficiency.

PV systems are prone to fluctuations in their efficiency related to the operating environment, so they need to work at their maximum power point (MPP) all the time regardless of atmospheric conditions. PV-generated power varies significantly due to rapid changes in irradiance caused by the shadows of passing clouds. The PV curve presents multiple peaks, several local maximum power points (LMPPs) and a global maximum power point (GMPP), under these partial shading conditions (PSCs).

Several works use the maximum power point tracking (MPPT) technique in their practical implementation to optimize PV output under PSCs [2–7]. In this vein, an experimental study about MPP characteristics of partially shaded strings is presented in [2]. Based on over 26,000 measured current-voltage curves, this work tests six and seventeen series-connected PV modules. The results proved that following the MPP closest to the nominal MPP voltage can significantly reduce the wide operating voltage range at the expense of small energy losses. To circumvent these shortcomings, a novel MPPT control for PV systems based on the search and rescue (SRA) optimization algorithm is presented in [3]. The improvements exhibited by the proposed technique are enhanced PV system performance, very low oscillations at global maximum (GM) and quick and efficient tracking of GM). Additionally, robustness, up to 99.93% power tracking efficiency in steady-state and implementation simplicity are other outstanding features of the proposed SRA control strategy.

In [8], a new framework based on a sliding-mode controller (SMC) is developed for the MPPT algorithm. This technique delivers precise tracking under changing weather conditions and performs better than conventional methods. Similarly, a new MPPT approach based on the adaptive fuzzy logic controller (AFLC) is introduced in [9]. In the AFLC method, with the goal to produce the optimal duty cycle for MPPT, the membership functions (MFs) are optimized through Grey Wolf Optimization (GWO). Testing under four shading patterns proves that the AFLC approach can track the global MPP for all conditions with enhanced speed, efficiency, and reduced oscillations.

As seen in [10,11], bio-inspired optimization has also been a source for developers to draw inspiration for designing MPPT algorithms. Improving the squirrel search algorithm (ISSA), Ref. [10] introduces a novel MPPT technique that reduces tracking time by half compared to the conventional SSA algorithm. Moreover, the results showed faster convergence and fewer power oscillations when tracking the GMPP.

An alternative method to the classical Marine Predator Algorithm (MPA) is proposed in [11] to cope with its implicit weaknesses. The MPAOBL-GWO method integrates the Opposition Based Learning (OBL) strategy with the Gray Wolf Optimizer (GWO), hence the name. This combination enhances the efficiency of the MPA and prevents it from descending into local points, as can be observed by the results.

Research similar to that of this study can also be found in the literature. For example, an MPPT optimal design based on a surface-based polynomial fit (MPPT-SPF) for a PV system is presented in [12]. The hardware-in-the-loop system is implemented using a high-speed, real-time simulator (PLECS RT Box 1) and a digital signal controller (DSC). In addition, this work applies an optimized version of the SPF technique for partial shading conditions. Likewise, a novel two-stage MPPT method is presented in [13]. In the first stage, the presence of (PSCs) is detected, and then, in the second stage, the global maximum power point (MPP) is reached using a new algorithm based on ramp change of the duty cycle and continuous sampling from the P-V characteristic of the array. Finally, the "Perturb and Observe" algorithm traces small changes of the new MPP.

The comprehensive review of online, offline, and hybrid optimization MPPT algorithms conducted in [14,15] indicates that most conventional MPPT algorithms track the GMPP correctly under conditions of uniform solar irradiance. Otherwise, under conditions of partial shading, rapidly switching, and conditions of nonuniform irradiance, it fails to obtain an accurate GMPP. However, under conditions of rapid solar irradiance change and PSCs, hybrid optimization algorithms are quick and precise in GMPP tracking. Unfortunately, these models are complex and, therefore, difficult to implement using integrated technologies. Therefore, this paper proposes a fast-tracking hybrid MPPT based on Surface-Based Polynomial Fitting and P&O for solar PV under PSCs.

This paper introduces the SPF-P&O GMPPT hybrid approach that exploit simultaneously the benefit of the classical P&O approach and a more data-driven curve-fitting method, as its name implies. This synergistic and complementary strategy improves the P&O algorithm efficiency using a polynomial approximation for global data fitting; the algorithm outputs the polynomial coefficients that best fit the PV panel characteristic curves. Specifically, the proposed approach is a curve-based type, which is considered to be data-driven since its parameters are optimized over the structure of the data. Following from this premise, curve-based fitting approaches have been successfully applied in different application, such as: variation of cosmic ray intensity with atmospheric pressure using a straight line fitting, calibration of a prism spectrometer using a polynomial curve, variation of viscosity of water with temperature using polynomial with equally spaced observations, variation of vapour pressure of ethyl alcohol with temperature using a generalized linear function, and the counting rate of a type I counter using non-linear functions, among others [16]. Finally, as for the MPPT stage, the resulting coefficients are used as the basis aimed at achieving a more accurate estimation and simplifying digital implementations into low-cost digital signal controllers.

The experimental results that validate the GMPPT algorithm come from a simulation in six testing scenarios and different transients from the 133 different cases available in the P-V characteristic data (shown in Figure 1). The analyzed case studies are: (i) system start-up, (ii) uniform irradiance variations, (iii) sharp change of the (PSCs), (iv) multiple peaks in the P-V characteristic, (v) dark cloud passing, and (vi) light cloud passing. The results for the six scenarios are evaluated using the standard errors and the mean power tracked and tracking factor scores.

The main contributions of this paper are as follows:

1. A hybrid SPF-P&O GMPPT algorithm is proposed to determine the GMPP of the PV system. This method can operate under uniform or nonuniform irradiance conditions.
2. The proposed SPF-P&O MPPT method is compared with the GMPPT P&O [13], obtaining results that prove a fast convergence and minimum steady-state oscillations for the PV system under 133 different cases of shading patterns.
3. For the validation of the proposed GMPPT algorithm, six scenarios with different transient of shading patterns are presented: (i) system start-up, (ii) uniform irradiance variations, (iii) sharp change of the (PSCs), (iv) multiple peaks in the P-V characteristic, (v) dark cloud passing, and vi) light cloud passing.
4. A two-stage strategy for the experimental PV system is proposed: (i) Modeling the PV array with the DC-DC boost converter using a real-time and high-speed simulator (PLECS RT Box), (ii) Implementing the proposed GMPPT algorithm and the double-loop controller of the DC-DC boost converter in a commercial low-priced digital signal controller (DSC).
5. The simulated and hardware-in-the-loop results of the six scenarios are evaluated using the standard errors, and the mean power tracked and tracking factor scores.
6. A nested control loop design is proposed to regulate—along with the SPF-P&O MPPT algorithm—the output voltage of the PV system under challenging environmental conditions. The double-loop control scheme consists of a current (inner-loop) controller and a voltage (outer loop) controller. Low steady-state error under demanding tests including irradiance variations, dynamic partial shading changes and system start-up, and the fast-tracking of control set points, are ensured by each proposed controller. In addition, the implementation of these loops guarantees an independent and fast dynamic response from the system.

The rest of this paper is structured as follows. First, Section 2 shows the PV system and its controllers. Then, the proposed SPF-P&O GMPPT algorithm is explained in Section 3. Next, the numerical simulations and Real-Time HIL Results are detailed in Section 4. Lastly, conclusions and future work are presented in Section 5.

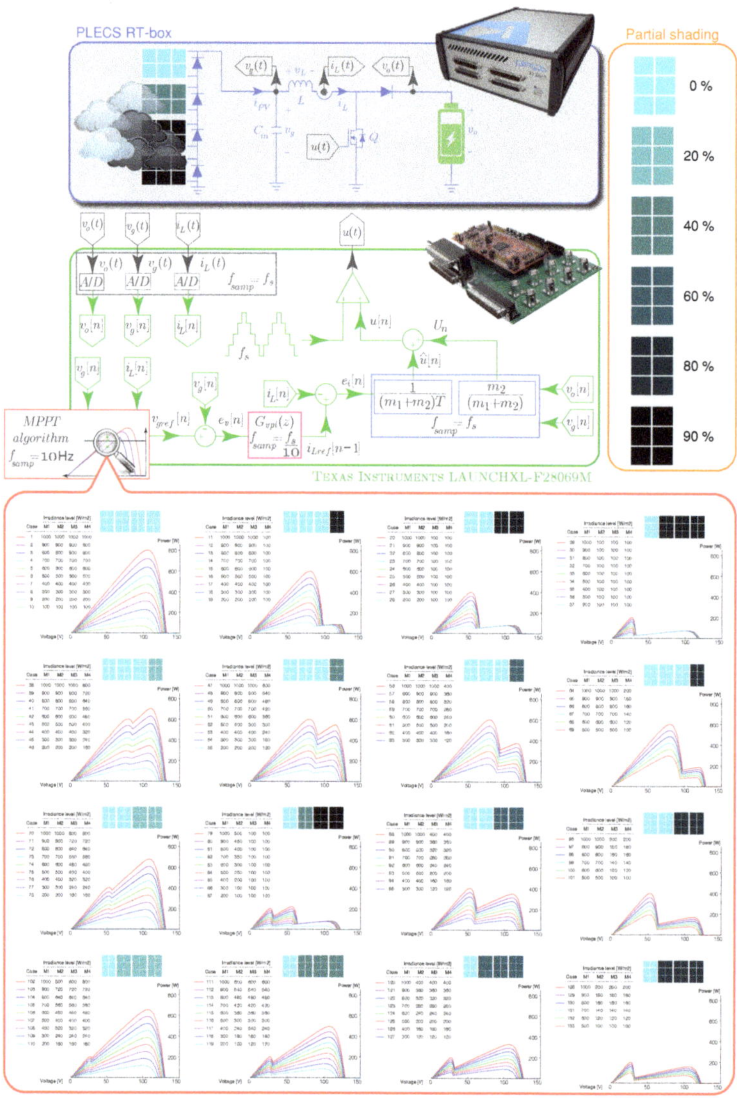

Figure 1. P-V characteristics of four PV module KC200GT under uniform and nonuniform irradiance conditions and connected in series.

2. PV Modeling

Through a setup of four PV modules, with its respective parallel bypass diodes and all connected in series, the essence of PV string characteristics is studied as shown in Figure 1. This PV string employs a DC-DC boost switching converter to charge a battery while a nested control loop tracks the MPP of the PV system.

Figure 1 depicts the MPPT algorithm and the double-loop controls. Using PLECS, the PV units are modeled as an array of four series-connected KC200GT solar modules whose electrical characteristics are presented in Table 1. Likewise, Figure 1 details the non-linear P-V characteristic for 133 different cases of shading patterns. There are 24 possible permutations for each of the 133 cases, totaling 3192 possible PSCs. Thus, the PV system can be studied under uniform irradiance levels (cases 1 to 10) with a unique peak in the P-V characteristic corresponding to the GMPP. On the other hand, different combinations of PSCs generate multiple peaks in the P-V characteristic, from two to three. Figure 1 shows the nonuniform irradiance levels on the PV modules that produce these PSCs. The GMPP moves with respect to the LMPPs at the multiple P-V characteristic peaks to test different situations with the MPPT algorithms.

The following are the differential equations for the boost converter [17–19]:

$$\frac{di_L(t)}{dt} = \frac{v_g - (1-u)\,v_o}{L} \tag{1}$$

$$\frac{dv_o(t)}{dt} = \frac{-v_o}{R_L C} + \frac{(1-u)i_L}{C}, \tag{2}$$

where v_o is the output voltage, u represents the control variable $\in \{0,1\}$ and i_L is the inductor current. For the boost converter, the duty cycle is [17–19]:

$$\bar{u} = 1 - \frac{\bar{v}_g}{\bar{v}_o}. \tag{3}$$

Table 1. Electrical characteristics of PV module KC200GT.

Electrical Parameters	Value
Maximum power P_{max}	200.0 W
Voltage at maximum power V_{mp}	26.3 V
Current at maximum power I_{mp}	7.61 A
Short-circuit current I_{sc}	8.21 A
Open-circuit voltage V_{oc}	32.9 V
Temperature coefficient of short-circuit current	3.18×10^{-1} A/°C
Temperature coefficient	-1.23×10^{-1} V/°C

2.1. Discrete-Time Sliding-Mode Current Control

Designing the inner-loop control for the DC-DC boost converter is complex due to the inherent non-linearity of the converter. Thus, a robust sliding-mode controller is used in this study. This section outlines the discrete-time sliding-mode current control (DSMCC) strategy with a fixed frequency. In order to ensure that the control surface (4) is reached in the next sampling period ($f_{samp} = f_s$), this approach computes the variable control $u[n]$ in the n-th time sample period. This control has been implemented for switching systems in [20–22].

$$s[n] = i_{Lref}[n-1] - i_L[n]. \tag{4}$$

Equation (2) is used to calculate the inductor current slopes of the boost converter, presented in Table 2. Assuming the averaged model of the inductor current slope of the converter $\frac{di_L}{dt} \approx \frac{i_L[n+1]-i_L[n]}{T}$, the Euler approximation heads to the next discrete-time inductor current expression:

$$i_L[n+1] = i_L[n] + T(m_1 + m_2)u[n] - m_2 T \tag{5}$$

being T the sampling or switching period. Consequently, the resulting duty cycle expression is:

$$u[n] = \frac{1}{(m_1 + m_2)T}e[n] + \frac{m_2}{m_1 + m_2}, \tag{6}$$

where $e[n] = i_{Lref}[n] - i_L[n]$, being $i_{Lref}[n] = i_L[n+1]$ (see Figure 1). Using the expressions for m_1 and $-m_2$ for the output current slopes from Table 2 in (6), the boost converter's control law is given by:

$$u[n] = \frac{L}{v_o[n]T}e[n] + 1 - \frac{v_g[n]}{v_o[n]}. \tag{7}$$

Table 2. Slope of the inductor current waveform.

Converter	m_1	$-m_2$
Boost	$\dfrac{v_g}{L}$	$\dfrac{v_g - v_o}{L}$

2.2. Discrete-Time PI Voltage Control

The input voltage of the boost converter v_g is regulated for the external loop using a proportional-integrator controller. The forward Euler method is used to express the controller transfer function in the z domain as follows:

$$G_{vpi}(z) = K_{pv} + \frac{K_{iv}T_{samp}}{z - 1}, \tag{8}$$

where $T_{samp} = 1/f_{samp}$,

$$K_{pv} = 2\pi\, C_{in}\, f_c, \tag{9}$$

and

$$K_{iv} = \frac{K_{pv}}{T_i}, \tag{10}$$

with C_{in} representing the input capacitor. For the voltage loop (f_c), the crossover frequency (CF) value should be lower than the current loop CF. Furthermore, the PI zero should be below f_c $(1/(2\pi T_i) < f_c)$.

3. SPF-P&O MPPT

3.1. Maximum Power Point Tracking (MPPT) Algorithm

The Maximum Power Point Tracking (MPPT) control keeps the power transfer at the highest efficiency, optimizing the PV system performance in any radiation and temperature conditions that the solar panels undergo. The MPPT approach allows the PV module to function at its maximum power point by controlling the switching converter [23]. This section briefly explains the "Perturb and Observe" benchmark method before introducing the offered MPPT method.

3.2. Conventional "Perturb and Observe" Method

The so-called "Perturb and Observe" (P&O) method is widely used due to its simplicity and low cost [24–26]. This algorithm provokes perturbations by either decreasing or increasing the reference voltage according to the output power PV module. If the current measured power $P[n]$ is higher than its previous sampled value $P[n-1]$, the voltage change continues in the same direction. Otherwise, it is reversed. Next, the PV module voltage is compared to the maximum voltage, to predict the MPP. Finally, a power step of the PV module [24] is produced through a small step of reference voltage. The P&O-based MPPT is abbreviated as MPPT-P&O.

3.3. Proposed MPPT Method

This work proposes a polynomial curve-fitting approach in favor of a more accurate MPP estimation. This approach, known as Surface-based Polynomial Fitting (MPPT-SPF), works as explained below:

3.3.1. Curve-Based Fitting

The fundamentals of a polynomial model ($y = f(x)$) for any curve can be expressed in mathematical terms as follows:

$$y\{\rho_N\} = \sum_{i=1}^{N+1} C_i x^{N+1-i}, \tag{11}$$

where x is the input times series, $y\{\rho_N\}$ is the output time series, n is the degree of the polynomial, such that $1 \leq N \leq 9$, and $N + 1$ is the order of the polynomial. In this case, the order is the number of coefficients to be adjusted, while the degree represents the highest power of the predictor variable.

Below, and based on their degree, polynomials are presented. For example, a four-degree polynomial is given by:

$$y\{\rho_4\} = f(x) = C_1 x^4 + C_2 x^3 + C_3 x^2 + C_4 x + C_5. \tag{12}$$

Polynomials are helpful when a simple empirical model is needed. Hence, a polynomial model is suitable for interpolation and extrapolation processes or characterizing data using a global fit.

Polynomial fitting is reasonably flexible when dealing with simple data structures. However, fitting can become unstable for high-degree polynomials. Moreover, although all polynomials fit correctly within a predefined data range, they diverge significantly outside of it.

Curve-fitting with high-degree polynomials can result in scale affectations because it uses predictor values as the basis for a matrix with high values. This problem can be addressed by preprocessing input data using z-score normalization, i.e., centering to zero mean and scaling to unit standard deviation [16].

3.3.2. Surface-Based Fitting

The output time series is represented as $z = f(x, y)$ when the fitting $f(\cdot)$ involves two input time series. For MPPT purposes, the variables are defined as follows:

- z: maximum power estimation for PV module current and voltage measurements (P_{max}),
- x: current $i_L[n]$,
- y: power measurement from the PV module $P_{pv}[n]$.

The following notation is used for polynomial surfaces: ρ_{ij} is the fitting type, where j represents the degree of y and, on the other hand, i the degree of x. Additionally, the maximum value for i and for j [27] is 5. The maximum between i and j is the overall degree of the polynomial. The degree of x is going to be less than or equal to i in each term. Likewise, in each term, the degree of y is going to be less than or equal to j. Accordingly, a surface with i and j degrees is denoted as $z\{\rho_{ij}\} = f(x, y)$. Some examples are mentioned in Table 3.

Table 3. Examples of polynomial models for surfaces.

Polynomial Models	Equations
ρ_{21}	$z[\rho_{21}] = C_{00} + C_{10}x + C_{01}y + C_{20}x^2 + C_{11}xy$
ρ_{13}	$z[\rho_{13}] = C_{00} + C_{10}x + C_{01}y + C_{11}xy + C_{02}y^2 + C_{12}xy^2 + C_{03}y^3$
ρ_{55}	$z[\rho_{55}] = C_{00} + C_{10}x + C_{01}y + \ldots + C_{14}xy^4 + C_{05}y^5$

Table 4 shows the degrees that make up the model terms. For example, for an x degree of 1 and a y degree of 3, the name of the model will be ρ_{13}. The mathematical foundation of the numerical curve-fitting methods is further detailed in [16].

Table 4. Polynomial model terms.

Degree of Term	0	1	2	3	4
0	1	y	y^2	y^3	y^4
1	x	xy	xy^2	xy^3	xy^4
2	x^2	x^2y	x^2y^2	x^2y^3	-
3	x^3	x^3y	x^3y^2	-	-
4	x^4	x^4y	-	-	-
5	x^5	-	-	-	-

A function is obtained from the ρ_{ij} approach that precisely fits the behavior and general trend of the analyzed data. Next, the most accurate fitting is found, considering both the criteria to quantify the adjusting procedure suitability and the polynomial to be tuned to represent the input data. The relationship between the obtained polynomial degree, the curve adjustment, and the values to interpolate is obtained from this adjustment. Finally, a five-degree polynomial is generated from a given data sequence in the form $(x[n], y[n])$, as shown below:

$$
\begin{aligned}
f(x,y) = \quad & C_{00} + C_{10}x + C_{01}y + C_{20}x^2 + C_{11}xy \\
& + C_{02}y^2 + C_{30}x^3 + C_{21}x^2y + C_{12}xy^2 + C_{03}y^3 \\
& + C_{40}x^4 + C_{31}x^3y + C_{22}x^2y^2 + C_{13}xy^3 + C_{04}y^4 \\
& + C_{50}x^5 + C_{41}x^4y + C_{32}x^3y^2 + C_{23}x^2y^3 + C_{14}xy^4.
\end{aligned}
\tag{13}
$$

The afore-developed analysis allows assessing how well the curve fits the data. Goodness-of-fit is measured by determining the accuracy of the coefficients based on the 95% confidence limits. The polynomial coefficients determined from the robust fitting of the data are:

$$
\begin{aligned}
C_{00} &= 186.8(184.7, 188.9), \\
C_{10} &= -14.14(-14.55, -13.73), \\
C_{01} &= 6.969(6.883, 7.056), \\
C_{20} &= 0.3872(0.3721, 0.4022), \\
C_{11} &= -0.3113(-0.316, -0.3065), \\
C_{02} &= 0.009194(0.008359, 0.01003), \\
C_{30} &= -0.004332(-0.004533, -0.00413), \\
C_{21} &= 0.006992(0.006861, 0.007124), \\
C_{12} &= -0.0004896(-0.0005158, -0.0004635), \\
C_{03} &= 7.545e - 06(4.966e - 06, 1.012e - 05), \\
C_{40} &= 2.297e - 05(2.184e - 05, 2.409e - 05), \\
C_{31} &= -7.728e - 05(-7.875e - 05, -7.581e - 05), \\
C_{22} &= 1.026e - 05(9.913e - 06, 1.061e - 05), \\
C_{13} &= -6.762e - 07(-7.171e - 07, -6.353e - 07), \\
C_{04} &= 2.86e - 08(2.565e - 08, 3.155e - 08), \\
C_{50} &= -4.633e - 08(-4.858e - 08, -4.407e - 08), \\
C_{41} &= 3.104e - 07(3.049e - 07, 3.158e - 07), \\
C_{32} &= -5.768e - 08(-5.93e - 08, -5.607e - 08), \\
C_{23} &= 5.3e - 09(5.014e - 09, 5.586e - 09), \text{ and} \\
C_{14} &= -2.523e - 10(-2.799e - 10, -2.248e - 10).
\end{aligned}
$$

generating the following polynomial:

$$
\begin{aligned}
f(x,y) = \ & 186.8 - 14.14x + 6.969y + 0.3872x^2 - 0.3113xy \\
& + 0.009194y^2 - 0.004332x^3 + 0.006992x^2y - 0.0004896xy^2 \\
& + 7.545e - 06y^3 + 2.297e - 05x^4 - 7.728e - 05x^3y \\
& + 1.026e - 05x^2y^2 - 6.762e - 07xy^3 + 2.86e - 08y^4 \\
& - 4.633e - 08x^5 + 3.104e - 07x^4y - 5.768e - 08x^3y^2 \\
& + 5.3e - 09x^2y^3 - 2.523e - 10xy^4.
\end{aligned}
\tag{14}
$$

As a result, the obtained fitting reaches a remarkable 0.9507 R-square, indicating a significant data trend and a good model fitting. Moreover, a 38.74 root mean square error (RMSE) and a $9.477e + 07$ sum of square error (SSE) estimation are reached. Figure 2 depicts the previous adjustment as a MPP surface in terms of P_{pv} and v_g.

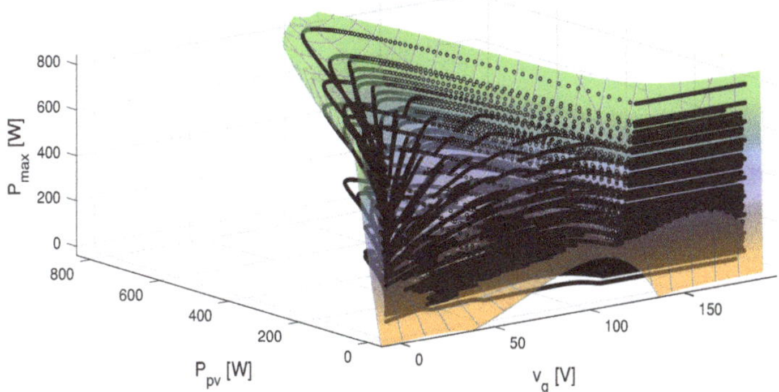

Figure 2. Plotting of the MPP surface in terms of P_{pv} and v_g.

Validation and implementation of the MPPT-SPF aim to maximize the available energy of the connected solar modules at any given moment during operation. Hence, MPPT continually samples the output of the PV cells and adjusts the voltage and current so that the PV system generates maximum power regardless of environmental conditions.

Additionally, by following the approach described in [28], uncertainty values for all the coefficients can be calculated. To do so, consider the vector $\mathbf{C}$ holding the estimates of the coefficients given by:

$$
\tilde{\mathbf{C}} = \left(\mathbf{\Phi}^\top \mathbf{\Phi} \right)^{-1} \mathbf{\Phi}^\top \mathbf{E},
\tag{15}
$$

where

$$
\mathbf{\Phi} = [\mathbf{1}_m, \mathbf{x}, \mathbf{y}, \mathbf{x}^2, \mathbf{xy}, \mathbf{y}^2, \mathbf{x}^3, \mathbf{x}^2\mathbf{y}, \mathbf{xy}^2, \mathbf{y}^3, \mathbf{x}^4, \mathbf{x}^3\mathbf{y}, \mathbf{x}^2\mathbf{y}^2, \mathbf{xy}^3, \mathbf{y}^4, \mathbf{x}^5\mathbf{x}^4\mathbf{y}, \mathbf{x}^3\mathbf{y}^2, \mathbf{x}^2\mathbf{y}^3, \mathbf{xy}^4],
\tag{16}
$$

and

$$
\mathbf{E} = [\varepsilon_1, \ldots, \varepsilon_m],
\tag{17}
$$

with m denoting the number of samples, $varepsilon_i$ being the error of approximation and $i \in \{1, \ldots, m\}$. Then, the uncertainty $u(C_j)$ for the coefficient j can be estimated as:

$$
u(C_j) = \sqrt{ \frac{ \left(\mathbf{\Phi}\tilde{\mathbf{C}} - \mathbf{E} \right)^\top \left(\mathbf{\Phi}\tilde{\mathbf{C}} - \mathbf{E} \right) }{ m - 3 } } \sqrt{\theta_{jj}},
\tag{18}
$$

where

$$\Theta = (\Phi^\top \Phi)^{-1}, \tag{19}$$

θ_{jj} is the entry jj of matrix Θ and $j \in \{1, \ldots, 20\}$. In this case, all the uncertainty values are admissible as can be noted in the following:

$$u(C_1) = u(C_{00}) = -8.92 \times 10^{11},$$
$$u(C_2) = u(C_{10}) = -4.09 \times 10^{9},$$
$$u(C_3) = u(C_{01}) = 2.49 \times 10^{5},$$
$$u(C_4) = u(C_{20}) = -2.79 \times 10^{6},$$
$$u(C_5) = u(C_{11}) = 7.36 \times 10^{3},$$
$$u(C_6) = u(C_{02}) = 3.61 \times 10^{1},$$
$$u(C_7) = u(C_{30}) = -4.53 \times 10^{2},$$
$$u(C_8) = u(C_{21}) = 4.33,$$
$$u(C_9) = u(C_{12}) = 2.5 \times 10^{-2},$$
$$u(C_{10}) = u(C_{03}) = 1.09 \times 10^{-4},$$
$$u(C_{11}) = u(C_{40}) = -1.75 \times 10^{-2},$$
$$u(C_{12}) = u(C_{31}) = 3.77 \times 10^{-4},$$
$$u(C_{13}) = u(C_{22}) = 2.13 \times 10^{-06},$$
$$u(C_{14}) = u(C_{13}) = 5.14 \times 10^{-08},$$
$$u(C_{15}) = u(C_{04}) = 4.55 \times 10^{-12},$$
$$u(C_{16}) = u(C_{50}) = -1.05 \times 10^{-07},$$
$$u(C_{17}) = u(C_{41}) = 4.06 \times 10^{-09},$$
$$u(C_{18}) = u(C_{32}) = 2.55 \times 10^{-11},$$
$$u(C_{19}) = u(C_{23}) = 1.52 \times 10^{-12}, \text{ and}$$
$$u(C_{20}) = u(C_{14}) = 5.12 \times 10^{-16}.$$

3.4. SPF-P&O Algorithm

The SPF-P&O GMPPT algorithm is detailed in Algorithm 1, which works as follows: Its objective is to obtain the input voltage reference (v_{gref}) for the boost converter from the measurement of the output voltage and output current of the PV module array. In this way, the maximum power characteristic (P_{max}) associated with the measured of current and voltage is obtaining from the expression (14). Once P_{max} is estimated, the voltage reference for the P&O is selected by searching from a register of possible solutions for power point maximum power (P_{mpp}).

When power changes, greater than or equal to reference ΔP_{pv} [%] occur, the MPPT-SPF algorithm is executed, so:

$$\frac{\left|P_{pvnew} - P_{pvlast}\right|}{P_{pvlast}} \geq \Delta P_{pv} \; [\%], \tag{20}$$

where P_{pvnew} is the actual power measurement and P_{pvlast} is the previous power measurement.

The GMPP search will again execute if the condition (20) is satisfied. This condition ensures that the expression (14) is evaluated to detect the optimal voltage solution for the GMPP even if there is any change in solar irradiance.

Algorithm 1: SPF-P&O GMPPT algorithm running at the microcontroller (see Figure 1).

Input: $v_g[n], i_L[n]$
Output: $v_{gref}[n]$
Main Function Main:

    Calculate $P_{pvnew}[n] = v_g[n] \cdot i_L[n]$ **if** *Inequation (20) = true* **then**
        Calculate P_{max} for the solution v_{gref} by Equation (14)

        $g_{op} = \infty$
        while $s <$ *number of solutions* **do**
            $g(s) = abs(P_{mpp}(s) - P_{max})$;

            if $(g(s) < g_{op})$ **then**
                $g_{op} = g(s)$;
                $s_{op} = s$;

            end

        end
        $v_{gref} = v_{mpp}(s_{op})$;

    else
        Calculate $P_{error} = P_{pvnew}[n] - P_{pvlast}[n]$ Calculate $v_{error} = v_g[n] - v_{glast}[n]$ $v_{gref}[n]=$ P&O $(P_{error}, v_{error}, v_{gref}[n])$
    end
    Update $P_{pvlast}[n+1] = P_{pvnew}[n]$ Update $v_{glast}[n+1] = v_g[n]$
return
Subfunction P&O $(P_{error}, v_{error}, v_{ref})$

    // Dynamic step size
    if *abs* $(P_{error}) > \Delta P$ **then**
        $\Delta v = \Delta v_{big}$
    else
        $\Delta v = \Delta v_{small}$
    end
    if $(P_{error} > 0)$ **then**
        if $(v_{error} > 0)$ **then**
            $v_{ref} = v_{ref} + \Delta v$
        else
            $v_{ref} = v_{ref} - \Delta v$
        end
    else
        if $(v_{error} > 0)$ **then**
            $v_{ref} = v_{ref} - \Delta v$
        else
            $v_{ref} = v_{ref} + \Delta v$
        end
    end
    return v_{ref}

---

4. Simulation and Real-Time HIL Results

In this section, the SPF-P&O MPPT algorithm is validated using a DC-DC boost converter through hardware in the loop (HIL). PLECS RT Box 1 implements the stage power converter and the PV array, and 6.6 μs is the sampled time to model the converter. And the values of the boost converter components are: $C_{in} = 200$ μF, $L = 1$ mH, $V_0 = 160$ V and $f_s = 25$ kHz. Through the TI 28069M LaunchPad, which is a low-cost Texas Instrument microcontroller, different controls that integrate the PV global system control scheme are implemented as presented Figure 1. Figure 3 shows the setup for HIL experiments.

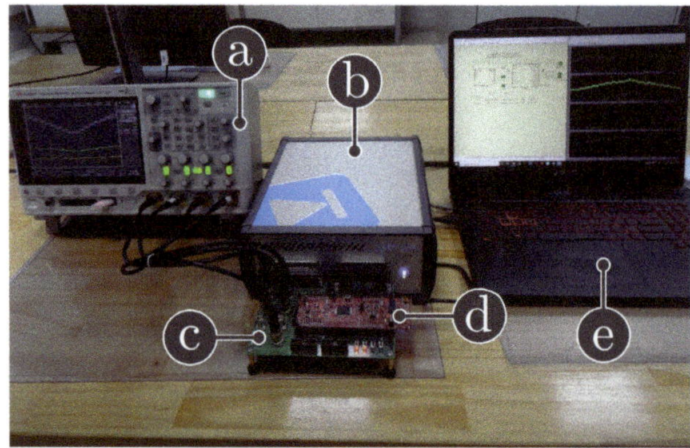

Figure 3. Hardware in-the-loop experimental setup: (a) oscilloscope, (b) PLECS RT box, (c) RT box launchPad interface, (d) Texas Instruments LAUNCHXL-F28069M, (e) Laptop.

The proposed SPF-P&O MPPT method is compared with a GMPPT P&O algorithm studied in [13], which realizes a voltage swept in partial shading conditions across the entire voltage range (from 0 to the open voltage value) for the PV system. The voltage reference that produces the maximum power is used to initialize after the voltage swept the P&O algorithm.

4.1. Inner-Loop Current Control Results

Figure 4 illustrates the transient responses for the inductor current of the boost converter in reaction to variations of the current reference. Where v_g, v_o and i_L are the signals sampled for the control, and 500 µs is the sampling time. The current reference is changed from 4 A to 8 A and back to 4 A, as can be observed in Figure 4a,b. to ensure a boost operation, the output voltage is $V_o = 160$ V and the input voltage is set in 100 V. Smooth transitions during reference changes are found, settling times near 250 µs and no overshoot. As shown, the inductor current is well regulated. The inductor currents followed the change in the current reference perfectly. During the current step reference change, the current control performance is validated.

4.2. Double-Loop Results

To demonstrate the effectiveness of the external loop control, voltage reference variations from 100 V to 110 V with a step between variations of 5 V for external loop validation, are presented in Figure 4c. The crossover frequency (CF) corresponds to $f_c = 500$ Hz, allowing the proportional gain calculation according to (9). Since the location of the PI zero in Equation (10) is lower than f_c $(1/(2\pi T_i) < f_c)$, $T_i = 3.18\mathrm{e}{-3}$ s was chosen.

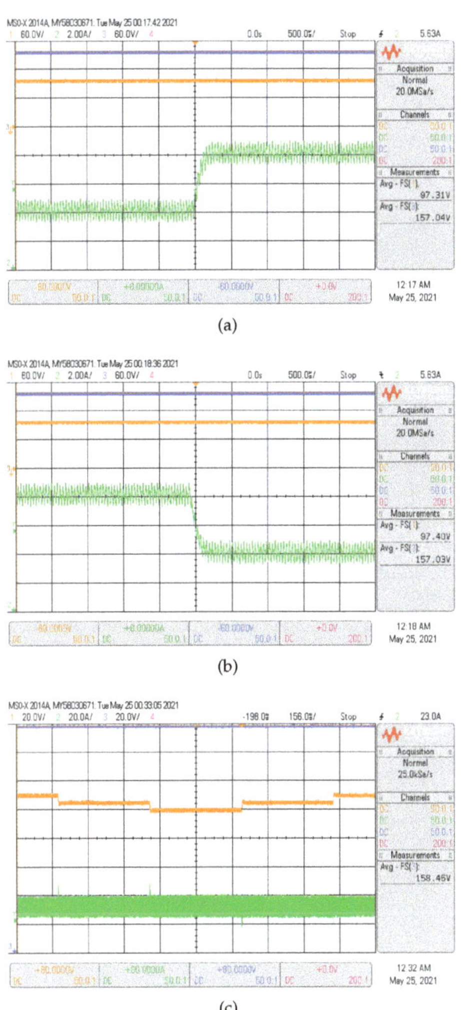

(a)

(b)

(c)

Figure 4. Experimental (**a**,**b**), responses of the sliding digital current input control when the reference i_{ref}: (**a**) changes from 4 A to 8 A, and (**b**) from 8 A to 4 A. The converter operates with $V_g = 100$ V and $V_o = 160$ V of input and output voltage, respectively. CH1: V_g (60 V/div), CH2: V_o (60 V/div), CH3: i_L (2 A/div) and a time base of 500 μs. (**c**) responses of the double-loop using sliding digital current control when the reference v_{ref} changes with steps of 5 V between 100 V to 110 V while the output voltage ($V_o = 160$ V) ensures a boost operation. CH1: v_g (20 V/div), CH2: V_o 20 V/div), CH3: i_L (20 A/div) and a time base of 156 ms.

Every 400 μs, the voltage regulator ($G_{vpi}(z)$) calculates the inductor current reference as shown in Figure 1. And current transitions, caused by the voltage changes, are smooth and the voltage reference is concisely tracked as can be appreciated from Figure 4c.

4.3. GMPPT P&O and Proposed SPF-P&O Method Comparison

The proposed method is compared with the GMPPT P&O studied in [13], every 100 ms the GMPPT algorithms are executed to provide a new voltage reference for the voltage loop control, as shown in Figure 1 for the GMPPT algorithm block. For the SPF-P&O, ΔP_{pv} [%] from expression (20) is set to 8%. The inner current loop is updating at 25 kHz, the outer voltage loop is calculating at 2.5 kHz, and the MPPT strategy is computing at 10 Hz, these strategies are implemented in the DSC.

These scale time differences between the systems are challenging but at the same time advantageous to implement an MPPT based on a finite-state machine.

For the validation of the proposed algorithm, six scenarios with different transient of the cases presented in Figure 1 are described below.

4.3.1. Scenario 1: System Start-Up

This scenario presents the start-up with the panels' irradiance levels corresponding to case 56 of Figure 1.

Figure 5 shows, corresponding to the maximum power, the transient behavior from zero current until an equilibrium point, where three modules have an irradiance of 1000 W/m^2 and a module of 400 W/m^2.

In Figure 5, both GMPPT algorithms reach the steady-state close to 1.3 s, with the proposed MPPT a uniform step voltage reference to tracking the MPP (599.9 W at 79.1 V) while the system starts up.

The proposed SPF-P&O algorithm works at the optimum point, without oscillation, as is shown in the input voltage signal after it has been tracked. Due to the GMPPT P&O realized a swept of voltage, the reference voltage is increased until 120 V to find the GMPPT at 79.1 V.

In Figure 6, a quantitative analysis can be observed of the proposed GMPPT method and GMPPT P&O method, and the results can be seen in Figure 5. These results confirm a similar performance for both methods through the start-up, where the GMPPT P&O algorithm has a higher tracking factor because the mean power tracked value is closer to the global power maximum.

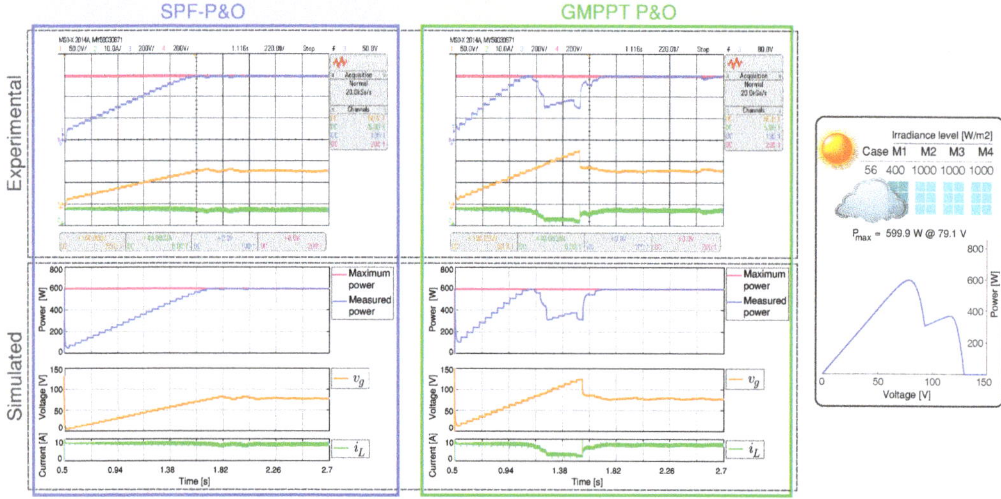

Figure 5. Simulated and experimental dynamic behavior of the MPPT algorithms for Scenario 1 and an output voltage $V_o = 160$ V. The proposed MPPT algorithm (**left**) is compared with GMPPT P&O (**right**). CH1: v_g (50 V/div), CH2: i_L (10 A/div), CH3: Maximum power (200 W/div), CH4: Measured power (200 W/div) and a time base of 220 ms.

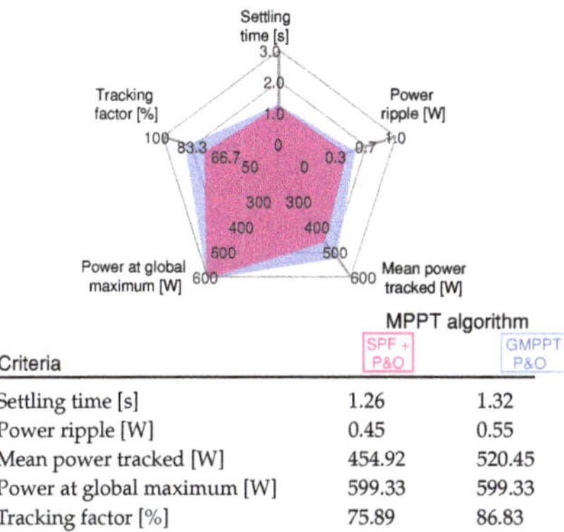

Criteria	MPPT algorithm	
	SPF + P&O	GMPPT P&O
Settling time [s]	1.26	1.32
Power ripple [W]	0.45	0.55
Mean power tracked [W]	454.92	520.45
Power at global maximum [W]	599.33	599.33
Tracking factor [%]	75.89	86.83

Figure 6. Comparative analysis of the MPPT methods for Scenario 1 shown in Figure 5.

4.3.2. Scenario 2: Uniform Irradiance Variations

Scenario 2 studies the results for the MPPT techniques under uniform irradiance variations as shown in Figure 7. The irradiance sequence corresponds to cases 1, 3, 5, 7, and 9 shown in Figure 1.

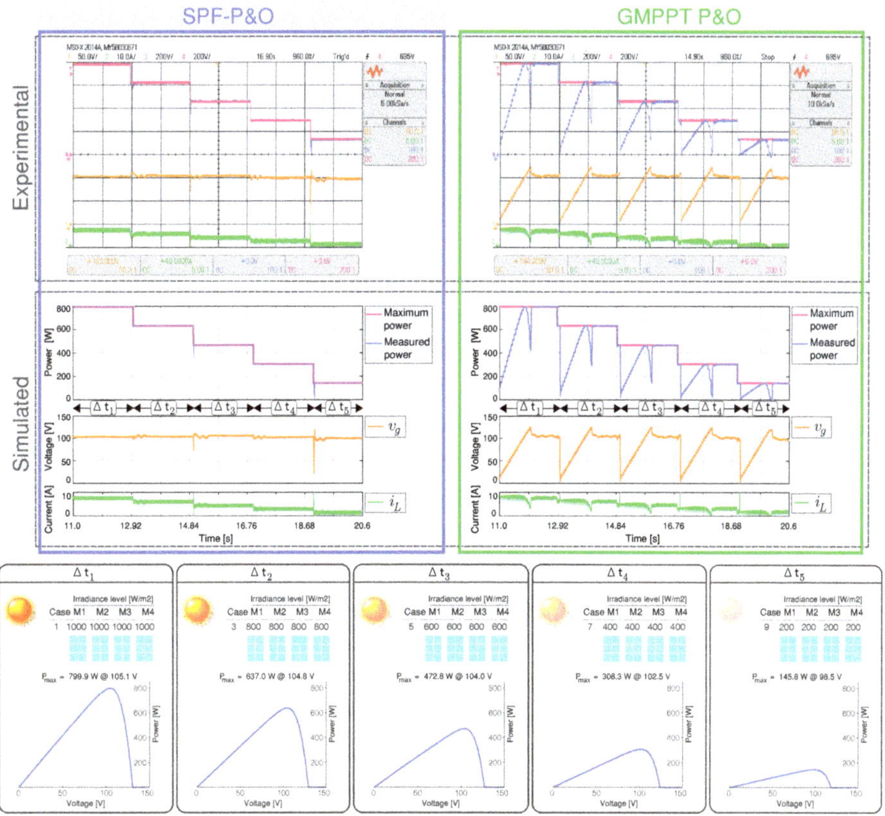

Figure 7. Simulated and experimental dynamic behavior of the MPPT algorithms for Scenario 2 with an output voltage $V_o = 160$ V. The proposed MPPT algorithm (right) is compared with the GMPPT P&O algorithm (right). CH1: v_g (50 V/div), CH2: i_L (10 A/div), CH3: Maximum power (200 W/div), CH4: Measured power (200 W/div) and a time base of 960 ms.

The proposed algorithm outputs the optimal voltage reference for the case and tracks the MPP faster while the GMPPT P&O has a slow convergence.

The proposed model facilitates the calculation and estimation of the variable of interest with high levels of accuracy; however, its proper evaluation through the results obtained to ensure its accuracy is crucial. Certainly, this evaluation can be performed by analyzing the errors, such as the mean absolute error (MAE), the relative error (RE) and the root mean square error (RMSE) and evaluate the performance of the results obtained [29,30], whose equations can be written as follows:

$$\text{RE} = \frac{\sum_{i=1}^{m}(P_{pvi} - P_{mpp})}{P_{mpp}} 100 \ [\%], \tag{21}$$

$$\text{MAE} = \frac{\sum_{i=1}^{m}|P_{pvi} - P_{mpp}|}{m}, \text{ and} \tag{22}$$

$$\text{RMSE} = \sqrt{\frac{\sum_{i=1}^{m}(P_{pvi} - P_{mpp})^2}{m}}, \tag{23}$$

where P_{pvi} represents the measured power of the PV module, P_{mpp} is the available MPP power of the solar module, and m the total number of sampling data. Figure 8 shows the sensitivity of the MPPT algorithms through MAE, RE, and RMSE for the results shown in Figure 7.

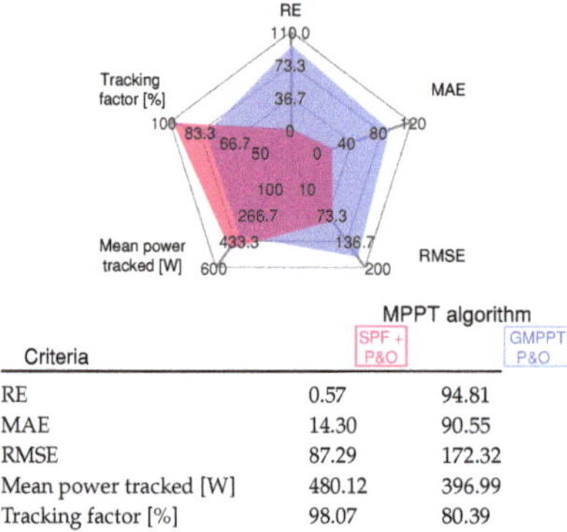

Criteria	MPPT algorithm	
	SPF + P&O	GMPPT P&O
RE	0.57	94.81
MAE	14.30	90.55
RMSE	87.29	172.32
Mean power tracked [W]	480.12	396.99
Tracking factor [%]	98.07	80.39

Figure 8. Comparative analysis of the MPPT methods under different uniform irradiance conditions for Scenario 2 shown in Figure 7.

Standard error values indicate that the performance of the proposed MPPT algorithm has higher effectiveness in tracking the maximum power point. This statistical analysis shows that the SPF-P&O method reaches the lowest value for the error compared to the GMPPT P&O algorithm. The proposed SPF-P&O algorithm has a tracking factor of 98.07%, while for the GMPPT P&O method, the tracking factor is 80.39%.

4.3.3. Scenario 3: Sharp Change of the PSCs

Scenario 3 presents a sharp change between case 103 and case 68, producing high PV power variations from 588 W to 355 W. Figure 9 shows simulated and HIL results of the GMPP tracking performance. The MPPT tracking efficiency for the GMPPT P&O method is 75.65%, while the 90.48% is achieved by the SPF-P&O proposed method. Figure 10 show the measure of the standard error for each GMPPT method, where the SPF-P&O GMPPT method has minimum error. As can be seen in the input voltage of the converter in Figure 9 and by the inductor current, the PV array always operates in an oscillating mode for the GMPPT P&O method.

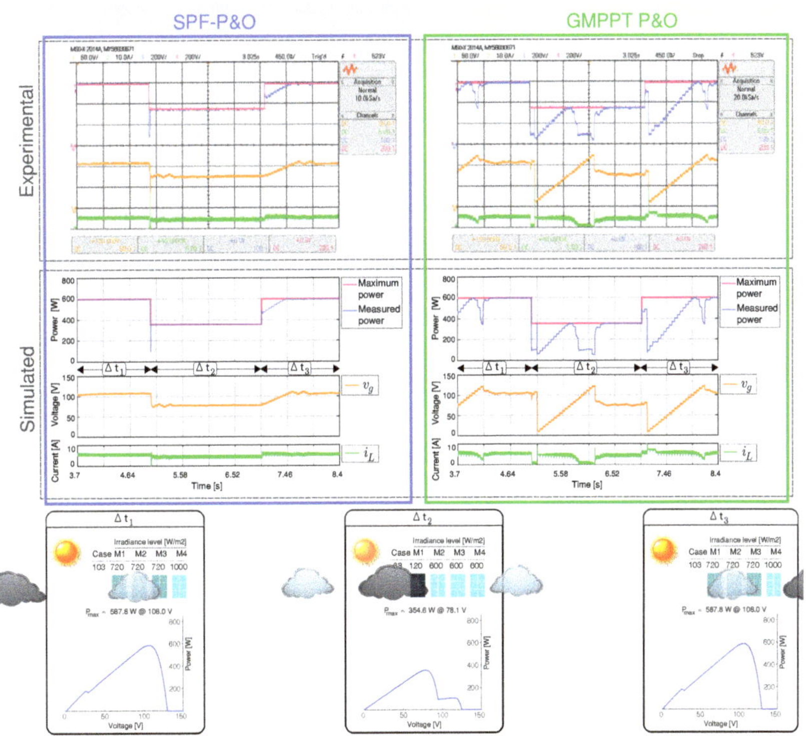

Figure 9. Simulated and experimental dynamic behavior of the MPPT algorithms for Scenario 3 with an output voltage $V_o = 160$ V. The proposed MPPT algorithm (right) is compared with the GMPPT P&O algorithm (right). CH1: v_g (50 V/div), CH2: i_L (10 A/div), CH3: Maximum power (200 W/div), CH4: Measured power (200 W/div) and a time base of 450 ms.

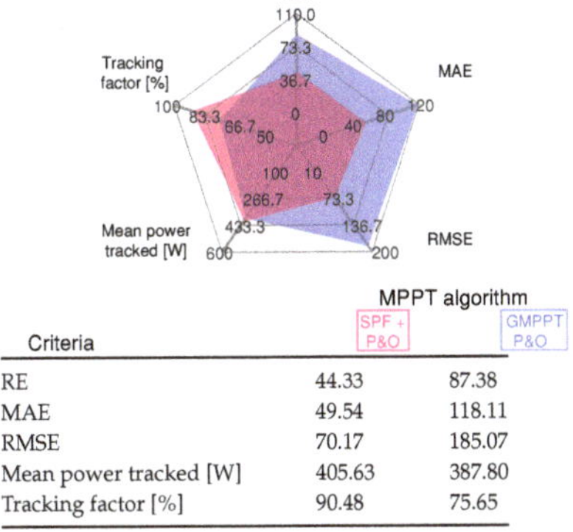

	MPPT algorithm	
Criteria	SPF + P&O	GMPPT P&O
RE	44.33	87.38
MAE	49.54	118.11
RMSE	70.17	185.07
Mean power tracked [W]	405.63	387.80
Tracking factor [%]	90.48	75.65

Figure 10. Comparative analysis of the MPPT methods for Scenario 3 shown in Figure 9.

Therefore, the proposed GMPPT method achieves a superior performance during abrupt irradiation variations than the GMPPT P&O method. Please note that during the change from case 68 to 103, the experimental result of the SPF-P&O algorithm did not estimate an optimal reference voltage. Nonetheless, when P&O operates, GMPPT is achieved, demonstrating the robustness of the SPF-P&O method.

4.3.4. Scenario 4: Multiple Peaks in the P-V Characteristic

Scenario 4 has the sequence of case 3 with only one peak (GMPP), case 52 with two peaks (One GMPP and one LMPP), and case 85 with three peaks (One GMPP and two LMPP) in the P-V characteristic. The results for each MPPT algorithms are shown in Figure 11.

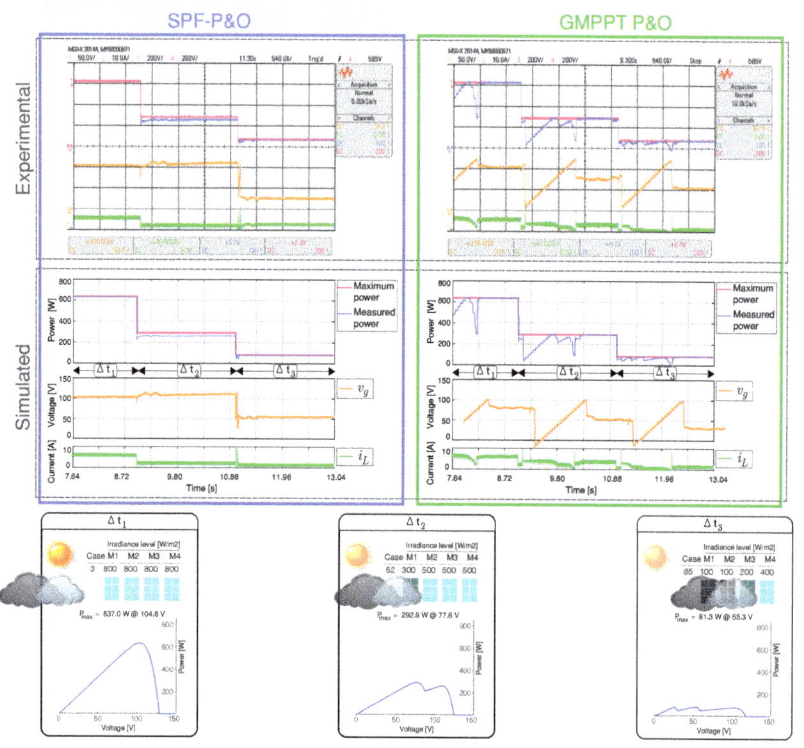

Figure 11. Simulated and experimental dynamic behavior of the MPPT algorithms for Scenario 4 with an output voltage $V_o = 160$ V. The proposed MPPT algorithm (right) is compared with the GMPPT P&O algorithm (right). CH1: v_g (50 V/div), CH2: i_L (10 A/div), CH3: Maximum power (200 W/div), CH4: Measured power (200 W/div) and a time base of 540 ms.

For case 3, the irradiance is uniform with a unique MPP at 104.28 V, which is tracked faster by the proposed algorithm. When the irradiance changes in case 52, there are two peaks, the proposed algorithm misidentifies the optimum voltage but archived the maximum power.

Finally, when the irradiation is changed in case 85, there are three peaks, where the algorithm has quickly identified and tracked the second peak (55.3 V, 81.3 W) as the GMPP. The SPF-P&O presents an overall GMPPT tracking efficiency of 90.86%, while the GMPPT P&O method presents a tracking efficiency of 79.85%.

The results for the cases' sequence are evaluated through standard errors (21)–(23), and the scores of mean power tracked and tracking factor as shown in Figure 12.

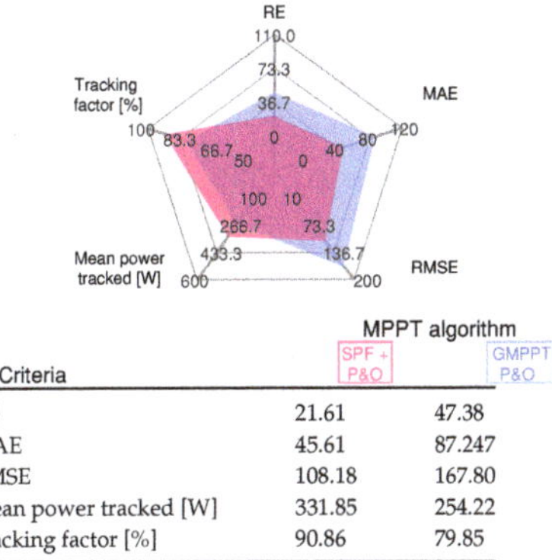

Criteria	MPPT algorithm	
	SPF + P&O	GMPPT P&O
RE	21.61	47.38
MAE	45.61	87.247
RMSE	108.18	167.80
Mean power tracked [W]	331.85	254.22
Tracking factor [%]	90.86	79.85

Figure 12. Comparative analysis of the MPPT methods for Scenario 4 shown in Figure 11.

From this figure it can inferred that the proposed GMPPT method has high MPP tracking capability regarding to GMPPT P&O method under multiple peaks in the P-V characteristic due to different PSCs.

4.3.5. Scenario 5: Dark Cloud Passing

A dark cloud passes in this scenario, obscuring each one of the PV modules. The sequence of the cases included in this scenario correspond to the cases 7, 17, 26, and 35 as presented in Figure 13.

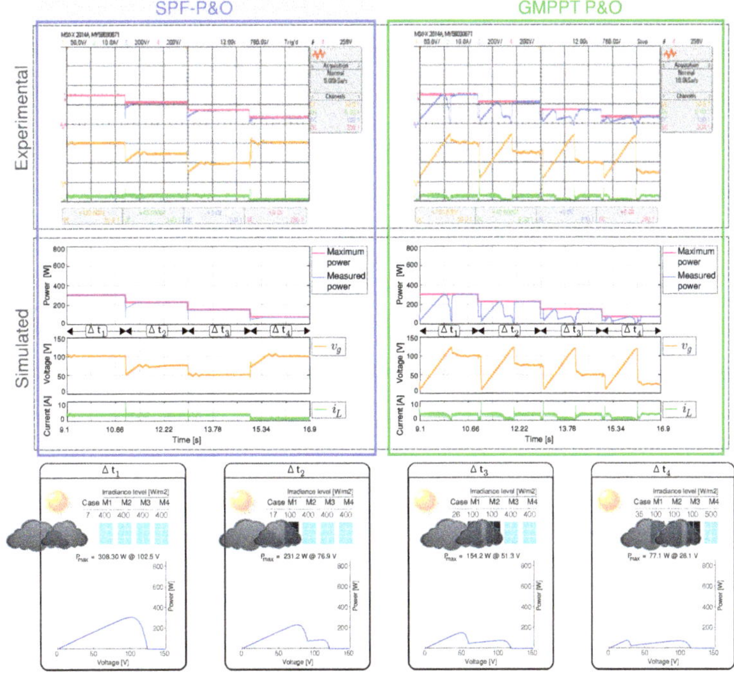

Figure 13. Simulated and experimental dynamic behavior of the MPPT algorithms for Scenario 5 with an output voltage $V_o = 160$ V. The proposed MPPT algorithm (right) is compared with the GMPPT P&O algorithm (right). CH1: v_g (50 V/div), CH2: i_L (10 A/div), CH3: Maximum power (200 W/div), CH4: Measured power (200 W/div) and a time base of 780 ms.

Cases 17, 26 and 35 correspond to (PSCs), which has two peaks, and the proposed GMPPT correctly identifies the optimal voltage for the maximum power point for the transition between cases 7, 17, and 26. Nevertheless, the reference voltage value for case 35 is incorrectly identified by the SPF-P&O algorithm.

This error is because the power between the GMPP (77.1 W at 26.1 V) and the LMPP located in the second peak (72.26 W at 100 V) are similar. The proposed MPPT tracked the GMPP much faster than the GMPPT P&O algorithm. Figure 14 shows the standard error and comparison indicators for the results presented in Figure 13.

The SPF-P&O algorithm exhibits smaller error values in comparison to the GMMP P&O method. The SPF-P&O algorithm presents a tracking factor of 93.53% while the GMPPT P&O method a tracking factor of 72.29%.

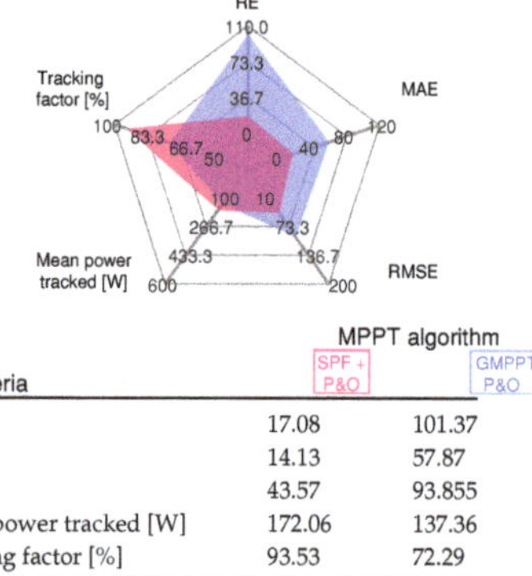

Criteria	MPPT algorithm	
	SPF + P&O	GMPPT P&O
RE	17.08	101.37
MAE	14.13	57.87
RMSE	43.57	93.855
Mean power tracked [W]	172.06	137.36
Tracking factor [%]	93.53	72.29

Figure 14. Comparative analysis of the MPPT methods under different nonuniform irradiance conditions for Scenario 5 shown in Figure 13.

4.3.6. Scenario 6: Light Cloud Passing

In this scenario a light cloud passes, partially obscuring one by one the PV modules.

The results of scenario 6 has the sequence of cases: 8, 63, 95, and 127, and the results are presented in in Figure 15.

Figure 16 illustrates that the proposed method provides the lowest value error for the results seen in Figure 15.

When the irradiance level is reduced by the transition of the cloud, the SPF-P&O algorithm identifies the new GMPPT power for the different cases without oscillations while the GMPPT P&O method has a big oscillation around the GMPPT, and this is reflected in the current and voltage waveforms.

The SPF-P&O algorithm presents a tracking factor of 97.1% while for the GMPPT P&O method a value of 76.27%.

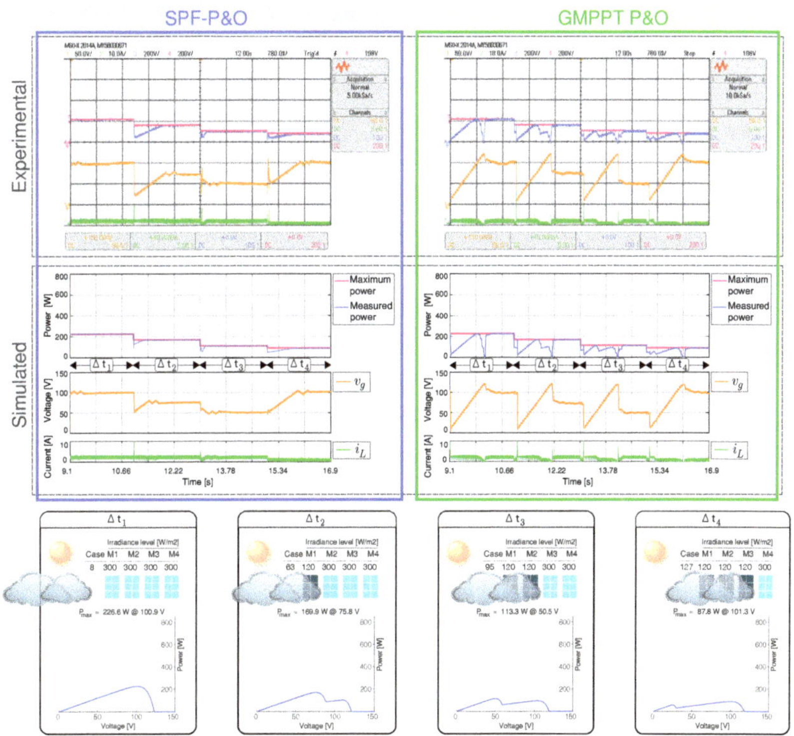

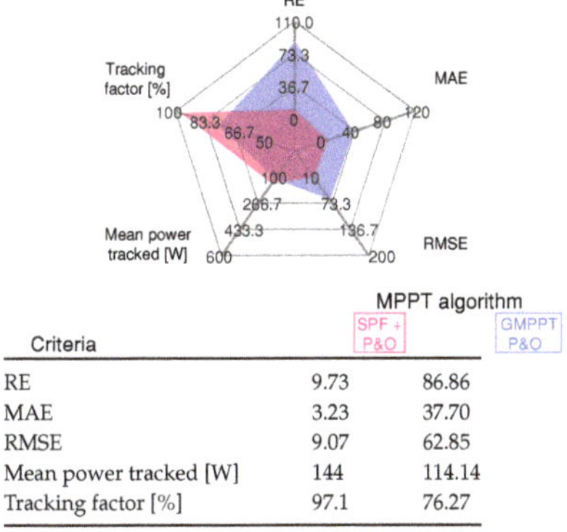

Figure 15. Simulated and experimental dynamic behavior of the MPPT algorithms for Scenario 6 with an output voltage $V_o = 160$ V. The proposed MPPT algorithm (right) is compared with the GMPPT P&O algorithm (right). CH1: v_g (50 V/div), CH2: i_L (10 A/div), CH3: Maximum power (200 W/div), CH4: Measured power (200 W/div) and a time base of 780 ms.

Criteria	MPPT algorithm	
	SPF + P&O	GMPPT P&O
RE	9.73	86.86
MAE	3.23	37.70
RMSE	9.07	62.85
Mean power tracked [W]	144	114.14
Tracking factor [%]	97.1	76.27

Figure 16. Comparative analysis of the MPPT methods for Scenario 6 shown in Figure 15.

5. Conclusions

A fast-tracking hybrid MPPT technique based on Surface-Based Polynomial Fitting and P&O has been presented for solar PV under PSCs. The SPF-P&O MPPT uses a polynomial model from the characterization of the PV module data, which is evaluated during irradiance variations. Meanwhile, the conventional P&O tracks the MPPT under uniform irradiance. The power circuit model was implemented using an RT BOX 1 tool.

A low-cost commercial DSC was used to implement the proposed MPPT algorithm and the double-loop strategies in a C programming software.

As experimentally demonstrated in this work, the introduced SPF-P&O GMPPT approach can tie together the estimation capability of the conventional P&O with a curve-fitting-based approach (namely surface-based polynomial fitting). Thus, it can be said that the hybrid approach represents a synergistic and complementary strategy to leverage the effectiveness of a P&O-type method with global data fitting using a polynomial approach. Different profile tests proved that the proposed hybrid method is robust and performs a faster and more effective MPP tracking with no steady-state oscillations for partial shading conditions.

In the future, a more extensive comparison between different MPPT techniques under PSCs using a low-cost commercial microcontroller will be conducted to accomplish deeper insights about the efficiency of PV systems, furthering the growth of photovoltaics as an affordable and sustainable energy source.

Author Contributions: Conceptualization, D.H.P.-O., C.R, L.L.L.-L., J.R.-F. and C.G.-C.; methodology, D.H.P.-O., C.R., L.L.L.-L., J.R.-F. and C.G.-C.; software, C.R. and C.G.-C.; validation, C.R. and C.G.-C.; formal analysis, D.H.P.-O., L.L.L.-L. and J.R.-F.; investigation, D.H.P.-O., C.R., L.L.L.-L. and C.G.-C.; resources, D.H.P.-O., C.R., L.L.L.-L., J.R.-F. and C.G.-C.; data curation, D.H.P.-O., C.R. and C.G.-C.; writing—original draft preparation, D.H.P.-O., C.R. and C.G.-C.; writing—review and editing, D.H.P.-O., C.R., L.L.L.-L., J.R.-F. and C.G.-C.; visualization, D.H.P.-O., C.R. and C.G.-C.; supervision, D.H.P.-O., C.R. and L.L.L.-L.; project administration, D.H.P.-O., L.L.L.-L. and C.R.; funding acquisition, D.H.P.-O., C.R. and J.R.-F. All authors have read and agreed to the published version of the manuscript.

Funding: This work was supported in part by the Chilean Government under projects ANID/ FONDECYT/ 1191680 and SDAS Research Group, www.sdas-group.com (accessed on 29 September 2021.)

Institutional Review Board Statement: Not applicable.

Informed Consent Statement: Not applicable.

Data Availability Statement: Not applicable.

Conflicts of Interest: The authors declare no conflict of interest. The funders had no role in the design of the study; in the collection, analyses, or interpretation of data; in the writing of the manuscript, or in the decision to publish the results.

References

1. Murdock, H.E.; Gibb, D.; Andre, T.; Sawin, J.L.; Brown, A.; Ranalder, L.; Collier, U.; Dent, C.; Epp, B.; Hareesh Kumar, C.; et al. Renewables 2021-Global Status Report. 2021. Available online: https://www.ren21.net/wp-content/uploads/2019/05/GSR2021_Full_Report.pdf (accessed on 29 September 2021).
2. Lappalainen, K.; Valkealahti, S. Experimental study of the maximum power point characteristics of partially shaded photovoltaic strings. *Appl. Energy* **2021**, *301*, 117436. [CrossRef]
3. Hamza Zafar, M.; Mujeeb Khan, N.; Feroz Mirza, A.; Mansoor, M.; Akhtar, N.; Usman Qadir, M.; Ali Khan, N.; Raza Moosavi, S.K. A novel meta-heuristic optimization algorithm based MPPT control technique for PV systems under complex partial shading condition. *Sustain. Energy Technol. Assess.* **2021**, *47*, 101367. [CrossRef]
4. Chen, X.; Du, Y.; Lim, E.; Wen, H.; Yan, K.; Kirtley, J. Power ramp-rates of utility-scale PV systems under passing clouds: Module-level emulation with cloud shadow modeling. *Appl. Energy* **2020**, *268*, 114980. [CrossRef]
5. Belhaouas, N.; Cheikh, M.S.A.; Agathoklis, P.; Oularbi, M.R.; Amrouche, B.; Sedraoui, K.; Djilali, N. PV array power output maximization under partial shading using new shifted PV array arrangements. *Appl. Energy* **2017**, *187*, 326–337. [CrossRef]
6. Bradai, R.; Boukenoui, R.; Kheldoun, A.; Salhi, H.; Ghanes, M.; Barbot, J.P.; Mellit, A. Experimental assessment of new fast MPPT algorithm for PV systems under non-uniform irradiance conditions. *Appl. Energy* **2017**, *199*, 416–429. [CrossRef]
7. Chao, K.H.; Lin, Y.S.; Lai, U.D. Improved particle swarm optimization for maximum power point tracking in photovoltaic module arrays. *Appl. Energy* **2015**, *158*, 609–618. [CrossRef]
8. Mohammadinodoushan, M.; Abbassi, R.; Jerbi, H.; Waly Ahmed, F.; Abdalqadir kh ahmed, H.; Rezvani, A. A new MPPT design using variable step size perturb and observe method for PV system under partially shaded conditions by modified shuffled frog leaping algorithm- SMC controller. *Sustain. Energy Technol. Assess.* **2021**, *45*, 101056. [CrossRef]

9. Laxman, B.; Annamraju, A.; Srikanth, N.V. A grey wolf optimized fuzzy logic based MPPT for shaded solar photovoltaic systems in microgrids. *Int. J. Hydrogen Energy* **2021**, *46*, 10653–10665. [CrossRef]
10. Fares, D.; Fathi, M.; Shams, I.; Mekhilef, S. A novel global MPPT technique based on squirrel search algorithm for PV module under partial shading conditions. *Energy Convers. Manag.* **2021**, *230*, 113773. [CrossRef]
11. Houssein, E.H.; Mahdy, M.A.; Fathy, A.; Rezk, H. A modified Marine Predator Algorithm based on opposition based learning for tracking the global MPP of shaded PV system. *Expert Syst. Appl.* **2021**, *183*, 115253. [CrossRef]
12. González-Castaño, C.; Lorente-Leyva, L.L.; Muñoz, J.; Restrepo, C.; Peluffo-Ordóñez, D.H. An MPPT Strategy Based on a Surface-Based Polynomial Fitting for Solar Photovoltaic Systems Using Real-Time Hardware. *Electronics* **2021**, *10*, 206. [CrossRef]
13. Ghasemi, M.A.; Foroushani, H.M.; Parniani, M. Partial shading detection and smooth maximum power point tracking of PV arrays under PSC. *IEEE Trans. Power Electron.* **2015**, *31*, 6281–6292. [CrossRef]
14. Ali, A.; Almutairi, K.; Padmanaban, S.; Tirth, V.; Algarni, S.; Irshad, K.; Islam, S.; Zahir, M.H.; Shafiullah, M.; Malik, M.Z. Investigation of MPPT Techniques Under Uniform and Non-Uniform Solar Irradiation Condition—A Retrospection. *IEEE Access* **2020**, *8*, 127368–127392. [CrossRef]
15. Bollipo, R.B.; Mikkili, S.; Bonthagorla, P.K. Hybrid, optimal, intelligent and classical PV MPPT techniques: A review. *CSEE J. Power Energy Syst.* **2021**, *7*, 9–33. [CrossRef]
16. Guest, P.G.; Guest, P.G. *Numerical Methods of Curve Fitting*; Cambridge University Press: New York, NY, USA, 2012.
17. Erickson, R.W.; Maksimovic, D. *Fundamentals of Power Electronics*, 2nd ed.; Kluwer Academic Publishers: Boston, MA, USA, 2001.
18. Hart, D.W. *Power Electronics*; Tata McGraw-Hill Education: New York, NY, USA, 2010.
19. Corradini, L.; Maksimovic, D.; Mattavelli, P.; Zane, R. *Digital Control of High-Frequency Switched-Mode Power Converters*; Wiley-IEEE Press: Piscataway, NJ, USA, 2015.
20. El Aroudi, A.; Martínez-Treviño, B.A.; Vidal-Idiarte, E.; Cid-Pastor, A. Fixed switching frequency digital sliding-mode control of DC-DC power supplies loaded by constant power loads with inrush current limitation capability. *Energies* **2019**, *12*, 1055. [CrossRef]
21. Vidal-Idiarte, E.; Marcos-Pastor, A.; Garcia, G.; Cid-Pastor, A.; Martinez-Salamero, L. Discrete-time sliding-mode-based digital pulse width modulation control of a boost converter. *IET Power Electron.* **2015**, *8*, 708–714. [CrossRef]
22. Vidal-Idiarte, E.; Marcos-Pastor, A.; Giral, R.; Calvente, J.; Martinez-Salamero, L. Direct digital design of a sliding mode-based control of a PWM synchronous buck converter. *IET Power Electron.* **2017**, *10*, 1714–1720. [CrossRef]
23. Farhat, M.; Barambones, O.; Sbita, L. A new maximum power point method based on a sliding mode approach for solar energy harvesting. *Appl. Energy* **2017**, *185*, 1185–1198. [CrossRef]
24. Kamran, M.; Mudassar, M.; Fazal, M.R.; Asghar, M.U.; Bilal, M.; Asghar, R. Implementation of improved Perturb & Observe MPPT technique with confined search space for standalone photovoltaic system. *J. King Saud Univ.-Eng. Sci.* **2018**, *32*, 432–441.
25. Kollimalla, S.K.; Mishra, M.K. A novel adaptive P&O MPPT algorithm considering sudden changes in the irradiance. *IEEE Trans. Energy Convers.* **2014**, *29*, 602–610.
26. Ahmed, J.; Salam, Z. A modified P&O maximum power point tracking method with reduced steady-state oscillation and improved tracking efficiency. *IEEE Trans. Sustain. Energy* **2016**, *7*, 1506–1515.
27. Chapra, S.C. *Applied Numerical Methods with MATLAB for Engineers and Scientists*; McGraw-Hill: New York, NY, USA, 2012.
28. Tomczyk, K.; Makowski, T.; Kowalczyk, M.; Ostrowska, K.; Beńko, P. Procedure for the Accurate Modelling of Ring Induction Motors. *Energies* **2021**, *14*, 5469. [CrossRef]
29. Stroock, D.W. *Probability Theory: An Analytic View*, 2nd ed.; Cambridge University Press: New York, NY, USA, 2010.
30. Zafar, M.H.; Al-shahrani, T.; Khan, N.M.; Feroz Mirza, A.; Mansoor, M.; Qadir, M.U.; Khan, M.I.; Naqvi, R.A. Group teaching optimization algorithm based MPPT control of PV systems under partial shading and complex partial shading. *Electronics* **2020**, *9*, 1962. [CrossRef]

Article

A Hybrid MCDM Approach in Third-Party Logistics (3PL) Provider Selection

Stefan Jovčić *[iD] and Petr Průša

Department of Transport Management, Marketing and Logistics, Faculty of Transport Engineering,
University of Pardubice, Studentská 95, 532 10 Pardubice, Czech Republic; petr.prusa@upce.cz
* Correspondence: stefan.jovcic@upce.cz; Tel.: +420-466-036-385

Abstract: Third-party logistics (3PL) is becoming more and more popular because of globalization, e-commerce development, and increasing customer demand. More and more companies are trying to move away from their own account transportation to third-party accounts. One reason for using 3PLs is that the company can focus more on its core activities, while the 3PL service provider can provide distribution activities in a more professional way, save costs and time, and increase the level of customer satisfaction. An emerging issue for companies in the logistics industry is how they can decide on the 3PL evaluation and selection process for outsourcing activities. For the first time, the entropy and the criteria importance through intercriteria correlation (CRITIC) methods were coupled in order to obtain hybrid criteria weights that are of huge importance to decide on the 3PL provider evaluation and selection process. The obtained criteria weights were further utilized within the additive ratio assessment (ARAS) method to rank the alternatives from the best to the worst. The introduced hybrid–ARAS approach can be highly beneficial, since combining two methods gives more robust solutions on one hand, while on the other hand eliminating subjectivity. Comparative and sensitivity analyses showed the high reliability of the proposed hybrid–ARAS method. A hypothetical case study is presented to illustrate the potentials and applicability of the hybrid–ARAS method. The results showed that 3PL-2 was the best possible solution for our case.

Keywords: 3PL logistics; decision making; ARAS; entropy; CRITIC

Citation: Jovčić, S.; Průša, P. A Hybrid MCDM Approach in Third-Party Logistics (3PL) Provider Selection. *Mathematics* **2021**, *9*, 2729. https://doi.org/10.3390/math9212729

Academic Editor: Nicu Bizon

Received: 16 September 2021
Accepted: 26 October 2021
Published: 27 October 2021

Publisher's Note: MDPI stays neutral with regard to jurisdictional claims in published maps and institutional affiliations.

1. Introduction

Third-party logistics (3PL) selection is becoming more and more popular because of globalization, e-commerce development, and increasing customer demand. Today's society, through the needs of different entities, represents a source of numerous new requirements and expectations for companies in the postal and logistics industries [1]. Systems for the distribution of goods, both at the national and international levels, are very important for appropriate business functioning and for the normal life of citizens [2]. Because of e-commerce development, there is increasing pressure on 3PL service providers all over the world. Wang et al. [3] emphasized that more trustworthy delivery, high inventory turnover, and inventory staged in forwarding locations near consumers are all effects of the e-commerce trend. According to Wang [4], e-commerce was sped up by the COVID-19 outbreak. According to Wang et al. [5], today's business globalization, customer satisfaction, and strong competition have forced many companies to work closely with external business partners. Third-party logistics service providers have had a significant impact on society on a global scale. They are not responsible not only for moving goods from the point of origin to the point of the destination, but in some cases for packaging, storage, etc. Third-party logistics services depend on the contracts they sign with collaborating companies.

Based on the aforementioned facts, the selection and evaluation process of 3PL service providers, is not an easy task for logisticians, since multiple factors affect the decision-making process. A poor choice of business partner can greatly negatively affect a company. The resulting losses can be financial, material, loss of reputation, loss of users, and many others. Furthermore, Soh [6] stated that if an appropriate 3PL provider was not selected, serious problems could occur, such as low-quality logistics services and contract nonfulfillment. He also emphasized that the decision-making problem for selecting the best 3PL provider had been receiving much attention recently among scholars as well as business practitioners. Nevertheless, Hsu et al. [7] stated that engaging 3PL could reduce fixed costs and increase flexibility, allowing organizations to focus on their core competencies and thereby enhancing their efficiency. Since 3PL evaluation and selection is a multidisciplinary field, there are many research questions that have been addressed by various authors in the field.

In this article, two research questions were dealt with: research question 1 (RQ1) refers to the combination of objective methods that allow the evaluation criteria for the 3PL provider selection process to obtain the best possible result; research question 2 (RQ2) is directed to the 3PL provider selection process.

To answer the aforementioned research questions, knowledge from various multidisciplinary fields was applied. Given the fact that no complete numerical data were available to create a real-life case study, given the time constraints of this research, some hypothetical data were used. The 3PL evaluation and selection process, in this paper, starts with a discussion with the experts in the field of logistics. The experts, according to their knowledge and experience, helped the authors define some of the criteria that should be of vital importance for companies that deal with the decision-making process. The number of alternatives (possible 3PL providers) depends on the company, but in this hypothetical case, there were five possible alternatives. Not all criteria were equally important, so it was necessary to assign them degrees of importance. For this part of the paper, the entropy and the CRITIC methods were coupled, and the new hybrid criteria importance is obtained. To obtain the final rank of the alternatives considered, the ARAS MCDM method is used.

The main contribution of this article lies in the proposed entropy–CRITIC (hybrid)–ARAS methodology, which, according to the authors' knowledge and the reviewed literature in the field, has not previously applied to this problem. The proposed methodology is general and can be applied to any other MCDM problem dealing with the selection of collaboration partners as well as any interrelated criteria. In addition, this paper contributes a combination of two objective methods to obtain hybrid criteria weights, since combining two methods gives more robust solutions on one hand while eliminating subjectivity on the other hand.

In this article, the methodology was applied only to a hypothetical example, which should be emphasized as a limitation. However, the authors intend to test the methodology for the real-life case study. Another weak point of the paper is that the considered criteria for 3PL selection were mostly part of the economic pillar. Nevertheless, future directions include addressing the other aspects such as environmental, social, and technical ones.

Since there is an increasing number of logistics companies that provide various logistics services as third parties, it is not easy to evaluate and select the best collaborative partner. The authors of this paper had the main motivation to propose methodology that on the one hand should be easy to implement and on the other hand should help decision makers select the best 3PL. The methodology upgraded the ARAS method by combining objective methods in order to eliminate subjectivity in the assessment of criterion weights and obtain more robust solutions using the ARAS method.

This paper is organized as follows: After the introductory section, Section 2 provides the current state of the methods used in the 3PL logistics field. The methodology proposed to solve the 3PL evaluation and selection problem is elaborated in Section 3. The application of the hybrid MCDM method to a hypothetical example is elaborated in Section 4. Section 5 gives some managerial insights into the 3PL logistics field. Section 6 is the conclusion and gives some future research directions.

2. Literature Review

As pointed out above, 3PL service provider evaluation and selection are not easy tasks for decision makers, given the fact that multiple criteria and many existing methods ought to be taken into consideration. Researchers have created many methods to solve the 3PL evaluation and selection problem. Most of these methods have belonged to the class of multicriteria decision-making methods. In addition to multicriteria analysis methods, many other methods, such as statistical or mathematical programming methods and integrated approaches, have been used. This section provides a review of the literature in the field of 3PL service providers based on the methods that other authors used to solve the 3PL evaluation and selection problem. Jovčić et al. [8] provided an extensive review of the literature regarding the most commonly used methods for 3PL provider evaluation and selection.

One of the most often used multicriteria analysis methods is the analytic hierarchy process (AHP). Saaty [9] originally developed this method. After its introduction, the AHP was widely used in many fields to solve multicriteria decision-making problems; one of those fields was logistics. Korpela and Touminen [10] applied the AHP to find the best possible solution of 3PL warehousing in the processing industry. Yahya and Kingsman [11] used the AHP to determine priorities in selecting suppliers. Akarte et al. [12] proposed a web-based AHP system to evaluate casting suppliers. Muralidharan et al. [13] developed a five-step AHP method to rank suppliers. Liu and Hai [14] applied the AHP to evaluate and select suppliers. So et al. [15] used the AHP to assess the quality of service of suppliers in Korea, while Göl and Çatay [16] applied the AHP to select the best 3PL service provider in a Turkish automotive company. Chan et al. [17] considered the supplier selection issue in the airline industry by using the AHP. Hou and Su [18] assessed and selected suppliers in the mass-customization environment by utilizing the AHP. Gomez et al. [19] proposed a model to evaluate the performance of suppliers by using the AHP. Hudymáčová [20] applied the AHP in supplier selection. Asamoah [21] applied the AHP in a pharmaceutical manufacturing company in Ghana. Hruška et al. [22] solved a 3PL selection problem by AHP in the production company in the Czech Republic. Jayant and Singh [23] applied an AHP–VIKOR hybrid MCDM approach for 3PL selection. Tuljak-Suban and Bajec [24] upgraded the AHP with the graph theory and matrix approach (GTMA). Aguezzoul and Pache [25] combined the AHP with the ELECTRE I methodology to solve the 3PL selection problem.

The analytic network process (ANP) is also a frequently used method for 3PL evaluation and selection problems. Meade and Sarkis [26] proposed a conceptual model to evaluate and select a third-party reverse logistics provider (3PRLP). Sarkis and Talluri [27] applied the ANP to evaluate and select the best supplier, considering seven evaluating criteria. Bayazit [28] applied the ANP to tackle the supplier selection problem. Jkharkharia and Shankar [29] applied the ANP to select the best logistics service provider. Further research regarding 3PRLP evaluation and selection was proposed by Zareinejad and Javanmard [30]. They applied ANP, intuitionistic fuzzy sets (IFS), and grey relation analysis (GRA). In their study, the ANP was used to identify the most important attributes in the selection and evaluation of 3PRLP. The technique for order of preference by similarity to ideal solution (TOPSIS) is one of the most frequently applied approaches in 3PL. Mostly, this method is coupled with fuzzy logic, ANP, AHP, etc. There have been many studies in the literature that may confirm it. Chen and Yang [31] proposed restricted fuzzy-AHP and fuzzy-TOPSIS to assess and choose the best supplier. Zeydan et al. [32] used a combination of fuzzy-AHP, fuzzy-TOPSIS, and DEA methods in the automotive industry to evaluate and select suppliers. Singh et al. [33] applied the TOPSIS method for supplier selection in the automotive industry as well. Jayant et al. [34] evaluated and selected reverse third-party logistics service providers (R3PLs) in the mobile phone industry by coupling the AHP (to evaluate the criteria for R3PLs) and TOPSIS (to select the best one). Laptate [35] used fuzzy modified TOPSIS for supplier selection problems in the supply chain. ELECTRE (ELimination Et Choice Translating REality) is a family of multicriteria decision analysis methods. Aguezzoul and Pires [36] used the ELECTRE method for 3PL performance evaluation and selection in a complex strategic decision process that involved various qualitative and quantitative criteria. Before that, Govindan et al. [37] used the fuzzy-ELECTRE method to rank 3PRL providers. The method was applied to a battery recycling case. When it comes to the combination of fuzzy logic with the multicriteria decision-making methods, there have been many studies in the scientific literature. According to Cheng [38] and Cheng et al. [39], a fuzzy-AHP approach handles issues that use the theory of fuzzy sets as well as hierarchical structure analysis. On the other hand, Ayhan [40] declared that the fuzzy-AHP approach was an extended AHP model into a fuzzy domain. This method has found application in various fields. For example, Kilincci and Onal [41] applied fuzzy-AHP to select suppliers for a washing machine company. When it comes to a low carbon supply chain, Shaw et al. [42] coupled the fuzzy-objective linear programming (LP) with the fuzzy-AHP in order to choose an optimal supplier solution. First, to determine the weights of the predetermined criteria, the Fuzzy-AHP was used. Second, the best supplier was determined by using fuzzy-objective LP. Zhang et al. [43], Zhang and Feng [44], Göl and Catay [16], and Soh [6] combined the AHP and fuzzy approaches to solve a 3PL service provider assessment issue. To compute the importance of individual parameters and subcriteria in fourth-party logistics (4PL), Cheng et al. [45] utilized the fuzzy-AHP approach. Arikan [46] dealt with the fuzzy-AHP method for multiple-objective supplier selection problems. Jagannath et al. [47] evaluated and selected 3PL providers from a sustainability perspective using the interval-valued fuzzy-rough approach. Rezaeisaray et al. [48] conducted a study on a pipe and fittings manufacturing company using a novel hybrid MCDM model for outsourcing supplier selection. They concluded that among the selective criteria for outsourcing, business development, focus on basic activities, and order delays were the three most important ones. They also ranked suppliers to facilitate decision making for selection. Sremac et al. [49] assessed logistics providers by combining the rough stepwise weight assessment ratio analysis and rough weighted aggregated sum product assessment approaches. Zarbakhshnia et al. [50] proposed a multiple attribute decision-making (MADM) model to rank and select 3PRLPs, using fuzzy stepwise weight assessment ratio analysis (SWARA) to weight the evaluation criteria. To rank and select sustainable 3PRLPs, the COPRAS (complex proportional assessment of alternatives) method was used. Özcan and Ahıskalı [51] solved the 3PL service provider selection problem by combining multicriteria decision-making methods with linear programming models. Some statistical methods that deal with the 3PL supplier selection problem can be found in the literature. For example, the correlation method was used by various authors [52–54]. Lai [55] conducted cluster analysis, which analyzed the service capability and performance of logistics service providers. Sinkovics and Roath [56] used descriptive statistics in 3PL relationships, considering six parameters: customer orientation, competitor orientation, operational flexibility, collaboration, logistics performance, and market performance. Knemeyer and Murphy [57] evaluated the performance of 3PL arrangements from a marketing perspective. Regarding mathematical programming methods (linear and nonlinear programming, dual and multiobjective programming, data envelopment analysis (DEA)), various research papers in the field of logistics service providers can be found. For example, Falsini et al. [58] carried out a study regarding logistic service provider evaluation and selection based on an integration of AHP, DEA, and linear programming methods. Zhou et al. [59] used the DEA method to evaluate the efficiency of Chinese 3PL. Hamdan and Rogers [60] evaluated the efficiency of 3PL operations using the DEA method. Kumar et al. [61] solved a multiobjective 3PL allocation problem for fish distribution. Tsai et al. [62] applied the new fuzzy DEA model to solving MCDM problems in supplier selection. Liu et al. [63] compared suppliers from a collaboration perspective for the new energy vehicle manufacturers in China. Hoseini et al. [3] used the combination of the fuzzy-best-worst method and the fuzzy inference system to solve the supplier selection issue in the construction industry. Kurniawan and Puspitasari [64] evaluated the criteria for supplier selection by applying fuzzy logic and the best-worst method. Whang et al. [65]

used the fuzzy-AHP and fuzzy-VIKOR Methods in sustainable supply chain third-party logistics. For better transparency, the aforementioned literature review on the methods for 3PL evaluation and selection is summarized in Table 1.

Table 1. Review based on the methods for 3PL evaluation and selection.

Author	Method
Korpela and Touminen [10]; Yahya and Kingsman [11]; Akarte et al. [12]; Liu and Hai [14]; So et al. [15]; Göl and Çatay [16]; Chan et al. [17]; Hou and Su [18]; Gomez et al. [19]; Hudymáčová [20]; Asamoah [21]; Hruška et al. [22]	AHP
Sarkis and Talluri [27]; Meade and Sarkis [26]; Bayazit [28]; Jkharkharia and Shankar [29]; Zareinejad and Javanmard [30]	ANP
Kumar et al. [61]; Zhou et al. [59]; Hamdan and Rogers [60]	DEA
Govindan et al. [37]	fuzzy-ELECTRE method
Chen and Yang [31]	fuzzy-AHP and fuzzy-TOPSIS (integrated approach)
Zeydan et al. [32]	fuzzy-AHP, fuzzy-TOPSIS, and DEA
Singh et al. [33]	TOPSIS
Falsini et al. [58]	AHP, DEA, and linear programming
Arikan [46]	fuzzy-AHP
Jayant et al. [34]	AHP–TOPSIS
Jayant and Singh [23]	AHP–VIKOR
Laptate [35]	fuzzy-modified TOPSIS
Rezaeisaray et al. [48]	DEMATEL, FANP, and DEA
Aguezzoul and Pires [36]	ELECTRE
Cheng [39]; Cheng et al. [42]; Zhang et al. [43]; Zhang and Feng [44]; Göl and Catay [16]; Cheng et al. [45]; Soh [6]; Kilincci and Onal [41]; Shaw et al. [42]; Ayhan [40]; Arikan [46]	fuzzy-AHP and fuzzy-objective linear programming (integrated approach)
Lai et al. [52]; Sinkovics and Roath [56]; Knemeyer and Murphy [57]; Sheen and Tai [53]; Yeung [54]; Lai [55]	Statistical methods
Sremac et al. [49]	rough SWARA, rough WASPAS, rough SAW, rough EDAS, rough MABAC, rough TOPSIS
Zarbakhshnia et al. [50]	SWARA, COPRAS
Jagannath et al. [47]	interval-valued fuzzy rough approach
Tuljak-Suban and Bajec [24]	AHP method with the Graph Theory and Matrix Approach (GTMA)
Aguezzoul and Pache [25]	AHP–ELECTRE I
Özcan and Ahıskalı [51]	MCDM–linear programming
Hoseini et al. [64]	fuzzy-best-Worst method and FIS
Whang et al. [3]	fuzzy-AHP and fuzzy-VIKOR
Kurniawan and Puspitasari [65]	fuzzy-best-worst method
Our study	Hybrid-ARAS method

Based on an extensive review of the literature in the field of third-party logistics, the most often used methods were multicriteria decision-making methods in combination with fuzzy logic. At present, in order to support the 3PL evaluation and selection process, there are many new multicriteria methods, such as SWARA, EDAS, MACBAC, and WASPAS, that are based on group decision making in a fuzzy environment. The main advantage of the methods mentioned above is the fact that when coupled with fuzzy logic, they have the power to help decision makers decide in an uncertain environment. In other words, the methods can help decision makers decide when the input data are not defined as crisp values but are given descriptively through linguistic statements. In this study, a hybrid MCDM approach is proposed to evaluate and rank the best 3PL service provider when crisp input data are given. Future research based on this paper should address the hybrid

MCDM approach in regard to the fuzzy domain as well. No previous research has been conducted that applies the hybrid–ARAS method in the way that is proposed here. One of the advantages of the method we propose is that it combines two objective methods to obtain more robust solutions and at the same time eliminate subjectivity.

3. Methodology

This paper combined three possible methods to solve a hypothetical example of the 3PL selection problem. The data for 3PL providers is usually, as a rule, privately owned. Moreover, some data are not freely available to the general public or the scientific community, probably because of corporate policy to protect proprietary information. Furthermore, the COVID-19 crisis additionally hampered obtaining data. However, the input data for 3PL service providers were formulated based on interviews with experts. The methodology proposed is presented in Figure 1. The first phase was problem formulation. In this case, the 3PL service provider selection problem was considered. From the extensive literature review as well as the Experts' opinions, the authors of this paper identified the criteria for 3PL provider selection. The criteria that were taken into consideration were mostly part of the economic pillar. The CRITIC (criteria importance through intercriteria correlation) and entropy methods were used to find the criteria importance for the 3PL service provider selection. By combining those two methods, the hybrid criteria weights were determined. The main reason for coupling the CRITIC and entropy methods was that both are used to obtain objective criteria weights, and when they are coupled together, a more robust solution is obtained (the subjectivity is eliminated). To find the final rank of the best possible 3PL provider, the criteria importance obtained by the hybrid method were further used within the ARAS method. The ARAS method is a relatively new MCDM method used to rank alternatives. and, according to the authors' knowledge and the literature review, it has not been previously applied and coupled with the hybrid criteria weights as it was here.

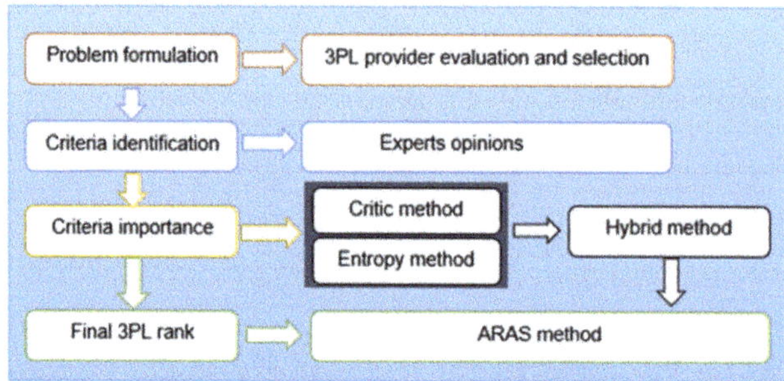

Figure 1. A flowchart of the methodology used for 3PL service provider evaluation and selection (Source: Authors).

In the second part of the paper, the hybrid criteria importance was integrated into the ARAS method to obtain the best 3PL solution.

3.1. CRITIC (Criteria Importance through Intercriteria Correlation) Method

In a decision-making process, the importance of criteria plays an important role, since not all criteria are equally important [66]. In this article, we applied the CRITIC method, since it is very useful in obtaining objective criteria importance. The CRITIC method was originally developed by Diakoulaki et al. [67]. The criteria importance calculated by the CRITIC method considers both conflicts among criteria and contrast intensity of each criterion [66]. Ghorabaee et al. [68] declared that in this method, the correlation coefficient is used to observe the conflict between criteria, while the standard deviation is used to consider the contrast intensity of each criterion. According to Diakoulaki et al. [67], the CRITIC method should be described as follows:

Step 1 computes the transformations of performance values (x_{ij}) and obtains criteria vectors. It is calculated by following Equation (1):

$$x_{ij}{}^T = \begin{cases} \dfrac{x_{ij}-x_j^-}{x_j^*-x_j^-} & if\ j\ \epsilon\ B; \\[2ex] \dfrac{x_j^- - x_{ij}}{x_j^- - x_j^*} & if\ j\ \epsilon\ N; \end{cases} \tag{1}$$

where $x_{ij}{}^T$ explains the transformed value, x_j represents the vector of jth criterion, and x_j^* and x_j^- represent the ideal and anti-ideal values with respect to jth criterion. If $j\ \epsilon\ B$ then $x_j^* = max_i x_{ij}$ and $x_j^- = min_i x_{ij}$. If $j\ \epsilon\ N$ then $x_j^* = min_i x_{ij}$ and $x_j^- = max_i x_{ij}$.

Step 2 calculates the standard deviation σ_j of each criterion utilizing the corresponding vector. Step 3 defines a square ($m \times m$) matrix R with r_{jk} elements, where $k = 1, 2, \ldots, m$:

$$R = \left[r_{jk}\right]_{m \times m} \tag{2}$$

The elements of this matrix are the linear correlation coefficients between the x_j and x_k vectors.

Step 4 computes the information measure of each criterion by using Equation (3):

$$H_j = \sigma_j \sum\nolimits_{k=1}^{m}(1 - r_{jk}) \tag{3}$$

Step 5 calculates the criteria importance by utilizing Equation (4):

$$W_j = \frac{H_j}{\sum_{k=1}^{m} H_k} \tag{4}$$

3.2. Entropy Method

According to Zhang [69], the entropy weight method was originally a concept of thermodynamics, which was first added into the information theory by C.E. Shannon and is now applied widely in the fields of engineering technology, social economy, etc. When it comes to multicriteria, Ranđelović et al. [70] emphasized that entropy was mainly used to determine the priority of an alternative. According to Ranđelović et al. [70] the method of entropy is described as follows: let us assume that $c_j = (a_{1j}, a_{2j}, \ldots a_{mj})$ describes a priority vector according to an exact criterion j, $j = 1, \ldots, n$. The entropy value for this vector can be calculated by applying Equation (5):

$$H_{Wj} = - \sum_{i=1}^{m} a_{ij} \ln(a_{ij}),\ j = 1, \ldots n \tag{5}$$

In addition, Ranđelović et al. [70] emphasized that in the theory of information, the entropy value H_{Wj} could be defined as a unit of discrete random variable X uncertainty, which could have a value from the fixed set $(x_1, x_2, \ldots x_n)$ in such a way that the feasibility that X is equal to x_j is given by w_j and may be presented by Equation (6):

$$P(X = x_j) = w_j \tag{6}$$

3.3. Hybrid Criteria Weights

To obtain the hybrid criteria weights, the authors of this paper combined the criteria weights obtained by the entropy and CRITIC methods. The combination of those two methods is demonstrated in the following equation:

$$Hybrid\ Weight\ (n^*) = 0.5 \cdot Entropy\ Weight\ (n^*) + 0.5 \cdot CRITIC\ Weight\ (n^*) \tag{7}$$

where (n^*) represents the hybrid weight of the nth criterion.

The hybrid criteria weights were chosen to test how those weights may affect the ranking of the final alternatives, as well as to notice whether there was any difference if the entropy and CRITIC methods were coupled with the ARAS method separately. In addition, the combination of two objective methods gave more robust and stable results.

3.4. The Additive Ratio Assessment (ARAS) Method

The additive ratio assessment (ARAS) method is an MCDM method originally developed by Zavadskas and Turskis [71]. Bošković et al. [66] declared that the ARAS method was very efficient and easy to implement in situations where multiple criteria are considered. There are several steps of the ARAS method described by Zavadskas and Turskis [71]:

Step 1 formulates an initial decision-making matrix which consists of m alternatives (rows) compared on n criteria (columns). The initial decision-making matrix is presented below:

$$X = \begin{bmatrix} x_{01} & \cdots & x_{0j} & \cdots & x_{0n} \\ \vdots & \ddots & \vdots & \ddots & \vdots \\ x_{i1} & \cdots & x_{ij} & \cdots & x_{in} \\ \vdots & \ddots & \vdots & \ddots & \vdots \\ x_{m1} & \cdots & x_{mj} & \cdots & x_{mn} \end{bmatrix}; i = \overline{0,\ m},\ j = \overline{1,n} \tag{8}$$

where m represents a number of alternatives, n represents a number of criteria describing each alternative, x_{ij} describes the performance value of the ith alternative in terms of the jth criterion, and x_{0j} shows an optimal value of jth criterion.

If the optimal value of jth criterion is unknown, then:

$$\begin{aligned} x_{0j} &= max_i x_{ij}, \text{ if } max_i x_{ij} \text{ is preferable;} \\ x_{0j} &= min_i x_{ij}^*, \text{ if } min_i x_{ij}^* \text{ is preferable.} \end{aligned} \tag{9}$$

Usually, the performance values x_{ij} and the criteria weights W_j are considered as the entries of a DMM. The system of criteria and the values and initial weights of criteria were determined by experts. The information can be corrected by the interested parties by considering their goals and opportunities.

Step 2 is the normalization of the input data of the initial decision-making matrix from the step 1. The normalization means that all the input data should be between an interval from 0 to 1. The normalized values $\overline{x_{ij}}$ in the normalized decision-making matrix $\overline{X}$ are obtained by applying Equations (11) and (12):

$$\overline{X} = \begin{bmatrix} \overline{x}_{01} & \cdots & \overline{x}_{0j} & \cdots & \overline{x}_{0n} \\ \vdots & \ddots & \vdots & \ddots & \vdots \\ \overline{x}_{i1} & \cdots & \overline{x}_{ij} & \cdots & \overline{x}_{in} \\ \vdots & \ddots & \vdots & \ddots & \vdots \\ \overline{x}_{m1} & \cdots & \overline{x}_{mj} & \cdots & \overline{x}_{mn} \end{bmatrix}; i = \overline{0,\ m},\ j = \overline{1,n}; \tag{10}$$

For the criteria with the highest preferable merits, the normalization is computed by utilizing Equation (11):

$$\overline{x_{ij}} = \frac{x_{ij}}{\sum_{i=0}^{m} x_{ij}}; \tag{11}$$

For the criteria with the lowest preferable merits, the normalization is obtained by Equation (12):

$$x_{ij} = \frac{1}{x_{ij}^*}; \overline{x_{ij}} = \frac{x_{ij}}{\sum_{i=0}^{m} x_{ij}}; \tag{12}$$

Step 3 involves defining a normalized-weighted matrix $\hat{X}$. It is possible to evaluate criteria with weights $0 < W_j < 1$. Only well-founded weights should be used, because weights are always subjective and influence the solution. The values of weight W_j are usually determined by the expert evaluation method. The sum of the weights W_j is limited as follows:

$$\sum_{j=1}^{n} w_j = 1; \tag{13}$$

$$\hat{X} = \begin{bmatrix} \hat{x}_{01} & \cdots & \hat{x}_{0j} & \cdots & \hat{x}_{0n} \\ \vdots & \ddots & \vdots & \ddots & \vdots \\ \hat{x}_{i1} & \cdots & \hat{x}_{ij} & \cdots & \hat{x}_{in} \\ \vdots & \ddots & \vdots & \ddots & \vdots \\ \hat{x}_{m1} & \cdots & \hat{x}_{mj} & \cdots & \hat{x}_{mn} \end{bmatrix}; i = \overline{0, \, m}, \, j = \overline{1, n} \tag{14}$$

Normalized-weighted values of all the criteria are calculated as follows:

$$\hat{x}_{ij} = \overline{x}_{ij} \cdot W_j; i = \overline{0, \, m}; \tag{15}$$

where W_j is the weight (importance) of the jth criterion and $\overline{x}_{ij}$ is the normalized rating of the jth criterion.

Step 4 determines the value of optimality function:

$$S_i = \sum_{j=1}^{n} \hat{x}_{ij}; \, i = \overline{0, \, m}; \tag{16}$$

where the optimality function of ith alternative is marked with S_i.

The maximum value of S_i reflects the best option, while on the contrary, the minimum value reflects the worst option. In other words, the greater the value of the optimality function S_i, the more effective the alternative. The preferences of alternatives can be observed according to the value S_i.

Step 5 computes the level of the alternative utility. To do so, it is necessary to make a comparison of the solutions with is ideal solution (S_0). The calculation of the utility level K_i of an alternative a_i is calculated by Equation (17):

$$K_i = \frac{S_i}{S_0}; i = \overline{0, \, m}; \tag{17}$$

where S_i and S_0 are the optimality criterion values. The calculated values K_i are between 0 and 1.

4. Application of the Hybrid–ARAS Method to 3PL Evaluation and Selection

The previously described methodology was applied to a hypothetical example, and the results are described in this section. The main reason for the hypothetical example was, as aforementioned, because it was hard to obtain real data for 3PL selection given the time constraints of this research, etc. However, three experts helped the authors define the criteria that influence the decision-making process and agreed on their ranking by the entropy and *CRITIC* methods. In this case, the authors selected five 3PL providers as the possible alternatives. According to the experts' opinions, five criteria that should be taken into consideration when evaluating and selecting 3PL providers were selected. These criteria were price (C_1), delivery service (C_2), quality of service (QoS) from customer experience (C_3), territorial coverage of the EU (C_4), and flexibility (C_5). The selected criteria are completely expressed in the following table (Table 2). It is important to point out that the considered criteria were mostly included in the economic aspect. When it comes to the consulted experts' knowledge and expertise, expert 1 held a managerial position in a tire manufacturing company with four years of experience and had a Ph.D. in the field of logistics; expert 2 held a managerial position in a multinational beverage company with five years of experience and had a Master's degree; and Expert 3 was a manager of a cold chain company with more than eight years of experience and a Ph.D. in the field of logistics and supply chains. Because of the COVID-19 outbreak, the authors interviewed the experts by telephone. The experts gave some limited information about themselves but required complete anonymity because of the business policies of their firms. The hypothetical input data used in the ARAS decision-making matrix were formulated in collaboration with the experts as well; the experts agreed that the data should respond to real conditions in the 3PL logistics market.

Table 2. Criteria for 3PL evaluation and selection (Source: Authors).

Price (C_1)	This criterion is expressed as the price that a company pays to the 3PL provider for its service provided. It is expressed in eurocent per km. Different 3PL providers request different prices for their services.
Delivery service (C_2)	This criterion is expressed as the percentage of goods delivered in a promised timeframe.
QoS from customer experience (C_3)	This criterion is expressed on a scale from 1 to 10, where 10 expresses maximal quality from the customer perspective.
Territorial coverage of the EU (C_4)	This criterion is expressed as the percentage of EU territory covered by the 3PL provider.
Flexibility (C_5)	This criterion represents the readiness of the 3PL provider to accommodate changing customer demands and expectations. It is expressed on a scale from 1 to 10, where 10 denotes the maximal degree of flexibility.

After the description of the criteria, the entropy and CRITIC methods were used in order to obtain the criteria weights. First, the weights were obtained by each separate method. Second, hybrid weights were obtained by combining the results of the two methods by applying Equation (7). The entropy method was applied to find criteria importance. After applying the entropy method, the following criteria weights were obtained (Tables 3–5).

Table 3. Initial entropy decision-making matrix (Source: Authors).

	Price (EUR/km)	Delivery Service (%)	QoS from Customer Experience (Scale 1–10)	Territorial Coverage of the EU (%)	Flexibility (Scale 1–10)
3PL-1	0.95	99.98	9	88	9
3PL-2	0.92	99.95	10	92	10
3PL-3	0.99	99.90	10	75	9
3PL-4	0.90	98.98	8	85	8
3PL-5	1.20	99.97	8	95	10
Sum	4.96	498.78	45	435	46

Table 4. Normalization of the entropy decision-making matrix (Source: Authors).

	Price (EUR/km)	Delivery Service (%)	QoS from Customer Experience (Scale 1–10)	Territorial Coverage of the EU (%)	Flexibility (Scale 1–10)
3PL-1	0.1915	0.2004	0.2000	0.2023	0.1957
3PL-2	0.1855	0.2004	0.2222	0.2115	0.2174
3PL-3	0.1996	0.2003	0.2222	0.1724	0.1957
3PL-4	0.1815	0.1984	0.1778	0.1954	0.1739
3PL-5	0.2419	0.2004	0.1778	0.2184	0.2174

Table 5. Computed entropy value (h) and final weights (Source: Authors).

	Price (EUR/km)	Delivery Service (%)	QoS from Customer Experience (Scale 1–10)	Territorial Coverage of the EU (%)	Flexibility (Scale 1–10)	h = 1/ln(m)
3PL-1	−0.3165	−0.3222	−0.3219	−0.3233	−0.3192	
3PL-2	−0.3125	−0.3221	−0.3342	−0.3286	−0.3318	
3PL-3	−0.3216	−0.3221	−0.3342	−0.3031	−0.3192	−0.62133
3PL-4	−0.3097	−0.3209	−0.3071	−0.3190	−0.3042	
3PL-5	−0.3433	−0.3221	−0.3071	−0.3323	−0.3318	
Sum	−1.6037	−1.6094	−1.6045	−1.6062	−1.6061	
e_j	0.9964	1.0000	0.9969	0.9980	0.9979	Sum = 0.0107
$d_j = 1 - e_j$	0.0036	0.0000	0.0031	0.0020	0.0021	$d_j = 1 - e_j$
Weights	0.3323	0.0004	0.2871	0.1861	0.1941	1

According to the entropy method, the highest importance was assigned to price and QoS, while lesser importance was assigned to territorial coverage of the EU, delivery service, and flexibility. By applying the previously described CRITIC method, the following criteria weights were calculated (Tables 6–9).

Table 6. Initial CRITIC decision-making matrix (Source: Authors).

	Price (EUR/km)	Delivery Service (%)	QoS from Customer Experience (Scale 1–10)	Territorial Coverage of the EU (%)	Flexibility (Scale 1–10)
3PL-1	0.95	99.98	9	88	9
3PL-2	0.92	99.95	10	92	10
3PL-3	0.99	99.90	10	75	9
3PL-4	0.90	98.98	8	85	8
3PL-5	1.20	99.97	8	95	10
Sum	4.96	498.78	45	435	46
min/max	min	max	max	max	max
Best	0.90	99.98	10	95	10
Worst	1.20	98.98	8	75	8

Table 7. Initial CRITIC decision-making matrix with standard deviations (Source: Authors).

	Price (EUR/km)	Delivery Service (%)	QoS from Customer Experience (Scale 1–10)	Territorial Coverage of the EU (%)	Flexibility (Scale 1–10)
3PL-1	0.8333	1.0000	0.5000	0.6500	0.5000
3PL-2	0.9333	0.9700	1.0000	0.8500	1.0000
3PL-3	0.7000	0.9200	1.0000	0.0000	0.5000
3PL-4	1.0000	0.0000	0.0000	0.5000	0.0000
3PL-5	0.0000	0.9900	0.0000	1.0000	1.0000
Standard deviation σ_j	0.4037	0.4349	0.5000	0.3857	0.4183

Table 8. $m \times m$ matrix (Source: Authors).

	Price (EUR/km)	Delivery Service (%)	QoS from Customer Experience (Scale 1–10)	Territorial Coverage of the EU (%)	Flexibility (Scale 1–10)
Price (EUR/km)	1.0000	−0.4378	0.3922	−0.3934	−0.5625
Delivery service (%)	−0.4378	1.0000	0.5174	0.2035	0.8135
QoS from customer experience (Scale 1–10)	0.3922	0.5174	1.0000	−0.4213	0.2988
Territorial coverage of the EU (%)	−0.3934	0.2035	−0.4213	1.0000	0.5811
Flexibility	−0.5625	0.8135	0.2988	0.5811	1.0000

Table 9. $m \times m$ matrix with the final weights (Source: Authors).

	Price (EUR/km)	Delivery Service (%)	QoS (Scale 1–10)	Territorial Coverage of the EU (%)	Flexibility (Scale 1–10)	Sum by Rows	σ_j	H_j	W_j
Price (EUR/km)	0.0000	1.4378	0.6078	1.3934	1.5625	5.0015	0.4037	2.0192	0.2642
Delivery service (%)	1.4378	0.0000	0.4826	0.7965	0.1865	2.9035	0.4349	1.2627	0.1652
QoS from customer experience (Scale 1–10)	0.6078	0.4826	0.0000	1.4213	0.7012	3.2130	0.5000	1.6065	0.2102
Territorial coverage of the EU (%)	1.3934	0.7965	1.4213	0.0000	0.4189	4.0302	0.3857	1.5544	0.2034
Flexibility (1–10)	1.5625	0.1865	0.7012	0.4189	0.0000	2.8691	0.4183	1.2002	0.1570
								7.6430	1

According to this method, the highest importance was assigned to price, followed by QoS from customer experience, territorial coverage of the EU, delivery service, and flexibility. The following table (Table 10) compares the criteria weights obtained from both methods and the hybrid weights obtained by applying Equation (7).

Table 10. Obtained hybrid criteria weights (Source: Authors).

Criteria Weights	Entropy Weights	CRITIC Weights	Hybrid Weights
Price (EUR/km)	0.3323	0.2642	0.2983
Delivery service (%)	0.0004	0.1652	0.0828
QoS from customer Experience (1–10)	0.2871	0.2102	0.2486
Territorial coverage of the EU (%)	0.1861	0.2034	0.1947
Flexibility	0.1941	0.157	0.1755
			1

The hybrid method ranked the criteria in a following way: the highest importance was assigned to price (0.2983), second place was related to QoS from the customer perspective (0.2486), third place was related to territorial coverage of the EU (0.1947), and flexibility (0.1755) and delivery service (0.0828) had less importance. Using the criteria weights obtained by the hybrid method, the final ranking of the 3PL providers was obtained by applying the ARAS method. For better clarity, the obtained criteria weights by the hybrid method are presented in Figure 2.

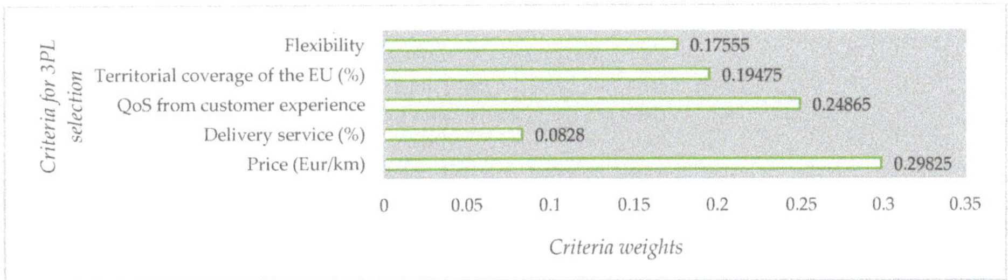

Figure 2. Obtained hybrid criteria weights (Source: Authors).

4.1. Application of the Hybrid–ARAS Method to 3PL Evaluation and Selection Problem

After the criteria weights were determined, the ARAS method was applied to obtain the final ranking of the 3PL providers. The input data used in the ARAS decision-making matrix are presented in Table 11.

Table 11. Initial ARAS decision-making matrix (Source: Authors).

	Price (EUR/km)	Delivery Service (%)	QoS from Customer Experience (Scale 1–10)	Territorial Coverage of the EU (%)	Flexibility (Scale 1–10)
0—optimal value	0.90	99.98	10	95	10
3PL-1	0.95	99.98	9	88	9
3PL-2	0.92	99.95	10	92	10
3PL-3	0.99	99.90	10	75	9
3PL-4	0.90	98.98	8	85	8
3PL-5	1.20	99.97	8	95	10
min/max	min	max	max	max	max
sum	6.2	598.8	55.0	530.0	56.0

The normalization of the input data is presented in Table 12.

Table 12. Normalization of the input data (Source: Authors).

	Price (EUR/km)	Delivery Service (%)	QoS from Customer Experience (Scale 1–10)	Territorial Coverage of the EU (%)	Flexibility (Scale 1–10)
0	0.1791	0.1670	0.1818	0.1792	0.1786
3PL-1	0.1696	0.1670	0.1636	0.1660	0.1607
3PL-2	0.1752	0.1669	0.1818	0.1736	0.1786
3PL-3	0.1628	0.1668	0.1818	0.1415	0.1607
3PL-4	0.1791	0.1653	0.1455	0.1604	0.1429
3PL-5	0.1343	0.1670	0.1455	0.1792	0.1786
min/max	min	max	max	max	max
Hybrid Weights	0.2983	0.0828	0.2487	0.1948	0.1756

The weighted decision-making matrix is described in Table 13.

Table 13. Weighted D–M matrix (Source: Authors).

	Price (EUR/km)	Delivery Service (%)	QoS from Customer Experience (Scale 1–10)	EU Territorial Coverage (%)	Flexibility (Scale 1–10)	S	K	Rank
0	0.0534	0.0138	0.0452	0.0349	0.0314	0.1787	Preference	
3PL-1	0.0506	0.0138	0.0407	0.0323	0.0282	0.1657	0.9270	2
3PL-2	0.0523	0.0138	0.0452	0.0338	0.0314	0.1765	0.9873	1
3PL-3	0.0486	0.0138	0.0452	0.0276	0.0282	0.1634	0.9141	3
3PL-4	0.0534	0.0137	0.0362	0.0312	0.0251	0.1596	0.8930	4
3PL-5	0.0401	0.0138	0.0362	0.0349	0.0314	0.1563	0.8747	5

As shown in Figure 3, by applying the hybrid–ARAS method, the best alternative was shown to be 3PL-2, with the preference value of 0.9873, followed by 3PL-1, with the preference of 0.9270; 3PL-3, with the preference of 0.9141; 3PL-4, with the preference of 0.8930; and in the last place 3PL-5, with the preference of 0.8747.

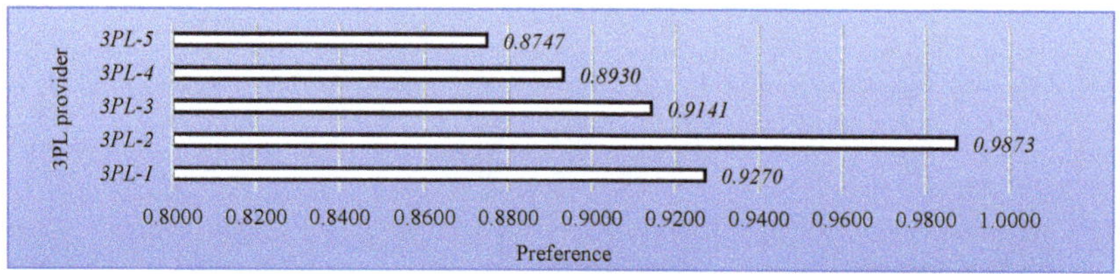

Figure 3. Final ranking of 3PL providers obtained by the hybrid–ARAS method (Source: Authors).

4.2. Sensitivity Analysis

To test the stability of the proposed hybrid–ARAS method, a sensitivity analysis was conducted. The main purpose of the analysis was to examine how the change in trade-off parameter ξ affected the final ranking of the alternatives. In this regard, the parameter ξ was changed within the interval of [0, 1] with an increment value of 0.1 (Figure 4).

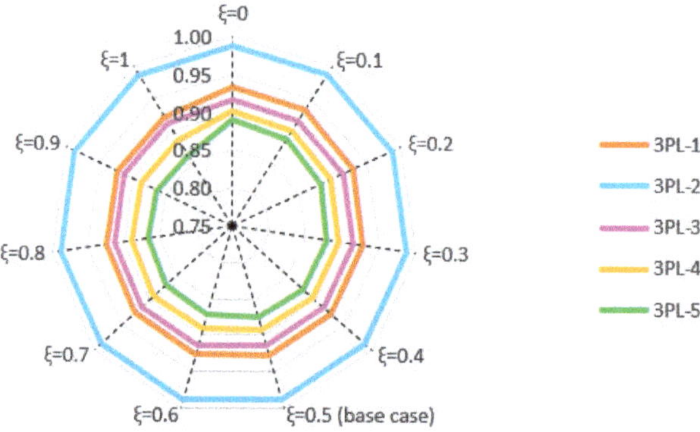

Figure 4. The sensitivity analysis to changes in the trade-off parameter ξ (Source: Authors).

When $\xi = 1$, only the entropy ARAS method was applied to prioritize the 3PL providers. When $\xi = 0$, only the CRITIC ARAS method was used to evaluate the 3PL providers. Therefore, in the base case scenario, ξ was set to 0.5 to equally appraise both methods and generate hybrid criteria importance. According to Figure 4, 3PL-2 was the best alternative under all ξ values. In addition, there was no change in the ranks of any 3PL provider in all 10 new test cases; i.e., the 3PL service provider ranking order was 3PL-2

> 3PL-1 > 3PL-3 > 3PL-4 > 3PL-5. The sensitivity analysis revealed that the proposed model has a high level of stability.

4.3. Comparative Analysis

A comparative analysis was performed to check the reliability of the results obtained through the hybrid–ARAS method for 3PL selection. The 3PL selection process was solved with two state-of-the art MCDM approaches, WASPAS [72] and EDAS [73]. The result of the comparative analysis is presented in Figure 5.

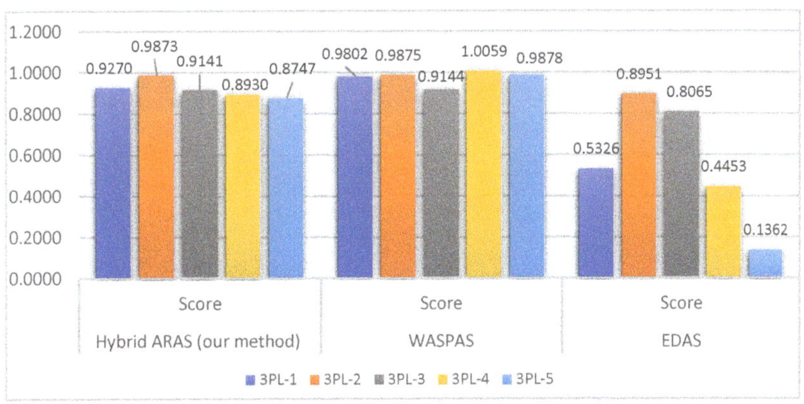

Figure 5. The comparative analysis of the hybrid–ARAS approach with WASPAS and EDAS.

The proposed hybrid–ARAS and EDAS methods ranked 3PL-2 as the best solution and 3PL-5 as the worst one. When it comes to 3PL-4, it was ranked as the fourth best option according to hybrid–ARAS and EDAS, but the WASPAS method ranked it as the best solution. According to the results, the model is reliable.

5. Managerial Insights

Since third-party logistics play an important role in the logistics market, managers all around the world should carefully monitor and identify the most important parameters that may affect this field. Not all the parameters are equally important, so sorting out the most important ones may greatly influence the 3PL business. When identifying the criteria for 3PL selection, each company should consider the whole picture of its business environment and select the most important criteria to attract the best collaboration partner according to its expectations. In addition, collaboration between experts is of crucial importance for 3PL selection. The wrong choice of a 3PL provider can have negative consequences for a company such as loss of profit, business image, reputation, customer loyalty, etc. The recommendation for managers should be to eliminate subjectivity and combine one or more objective methods to identify the priority of the criteria with MCDM methods in order to obtain the best possible 3PL provider to collaborate with. The results of using more objective methods are more robust and confident, which leads to a more stable managerial solution.

6. Discussion and Conclusions

This paper aimed to propose a possible hybrid–ARAS approach to the 3PL evaluation and selection process. The main reason for the methodology proposed was to give a theoretical contribution in the 3PL logistics field that could help managers and scientists think about the possible approaches for evaluating and selecting 3PL service providers. Three methods were combined in order to obtain a ranking of the best alternatives. The first method was the entropy method, applied to obtain objective criteria importance. The second was the CRITIC method, utilized to obtain objective criteria importance. The third was the ARAS method combined with the entropy and CRITIC methods in order to obtain a ranking of the possible alternatives. Furthermore, the combination of the entropy and CRITIC methods resulted in obtaining *hybrid* criteria weights, which were further coupled with the ARAS method to rank the best alternative. In each case, either separate (entropy–ARAS; CRITIC–ARAS) or coupled (hybrid–ARAS), the final ranking results were not changed. The main idea of coupling the former two methods was to obtain objective criteria weights, which should eliminate subjectivity and give more robust results.

To examine the stability of the proposed hybrid–ARAS method, a sensitivity analysis was conducted. The results of the analysis revealed that there was no changing in the ranking alternatives when 10 variations in criteria were taken. In addition, a comparative analysis was carried out to compare the obtained results. The results of the comparative analysis revealed that the model is reliable.

The major contributions of this paper are: (i) for the first time, a new combination of methods was introduced to rank 3PL service providers; (ii) a combination CRITIC–entropy (hybrid) method was employed to prioritize the criteria that influence the 3PL evaluation and selection process—a more robust solution was obtained by coupling two objective methods into a hybrid one, and subjectivity was eliminated; (iii) a new hybrid–ARAS method was developed to rank 3PL service providers; (iv) the methodology was very simply described and easy to implement and should be beneficial for the 3PL logistics field.

In our hypothetical case, the hybrid–ARAS MCDM approach for 3PL evaluation and selection process generated the following ranking order: *A2* (3PL-2) > *A1* (3PL-1) > *A3* (3PL-3) > *A4* (3PL-4) > *A5* (3PL-5). This approach identified 3PL-2 as the best possible alternative. On the other hand, the worst-ranked alternative was 3PL-5. As a result, it would be strongly recommended to select 3PL-2, since it was shown as the best alternative according to the methodology.

Limitations of this paper can indicate possible areas for its extension. The limitations included: (1) the methodology was not applied to a real-life case, but only to a hypothetical example; (2) the criteria influencing the decision-making process were filtered not only according to a review of the literature, but by the consideration of experts' opinions— indeed, filtering was mostly performed according to discussion with experts in the field. However, the methodology is general and can be applied with any other criteria influencing the 3PL evaluation and selection process. The main point of the paper was to show the applicability of the methodology in order to respond to the aforementioned research questions; (3) there were only considered criteria from the economic pillar. However, there is space to include any other aspects that should matter in the decision-making process, such as environmental, technical, and social aspects; (4) the authors did not consider how 3PL logistics may fit with the industry 4.0.

This study can be seen as an important trigger for future papers in the field. The future directions inspired by this paper should be to (a) apply the methodology to real-life cases; (b) include the criteria most often used by other authors in the field; (c) apply the methodology in the fuzzy and picture fuzzy environment, since fuzzy logic deals with uncertainty, which often factors into the decision-making process; (d) examine how 3PL logistics may fit with the industry 4.0.

Author Contributions: Conceptualization, S.J. and P.P.; methodology: S.J.; validation: S.J. and P.P.; investigation: S.J. and P.P.; writing—original draft preparation, S.J. and P.P.; writing—review and editing: S.J. and P.P.; visualization, S.J. and P.P.; supervision, S.J. and P.P. All authors have read and agreed to the published version of the manuscript.

Funding: This research received no external funding.

Institutional Review Board Statement: Not applicable.

Informed Consent Statement: Not applicable.

Data Availability Statement: Not applicable.

Conflicts of Interest: The authors declare no conflict of interest.

References

1. Simić, V.; Lazarević, D.; Dobrodolac, M. Picture fuzzy WASPAS method for selecting last-mile delivery mode: A case study of Belgrade. *Eur. Transp. Res. Rev.* **2021**, *13*, 43. [CrossRef]
2. Lazarević, D.; Dobrodolac, M.; Švadlenka, L.; Stanivuković, B. A model for business performance improvement: A case of the postal company. *J. Bus. Econ. Manag.* **2020**, *21*, 564–592. [CrossRef]
3. Wang, C.-N.; Nguyen, N.-A.-T.; Dang, T.-T.; Lu, C.-M. A Compromised Decision-Making Approach to Third-Party Logistics Selection in Sustainable Supply Chain Using Fuzzy AHP and Fuzzy VIKOR Methods. *Mathematics* **2021**, *9*, 886. [CrossRef]
4. Wang, C.-N.; Dang, T.-T.; Nguyen, N.-A.-T.; Le, T.-T.-H. Supporting Better Decision-Making: A Combined Grey Model and Data Envelopment Analysis for Efficiency Evaluation in E-Commerce Marketplaces. *Sustainability* **2020**, *12*, 10385. [CrossRef]
5. Wang, G.; Dou, W.; Zhu, W.; Zhou, N. The effects of firm capabilities on external collaboration and performance: The moderating role of market turbulence. *J. Bus. Res.* **2015**, *68*, 1928–1936. [CrossRef]
6. Soh, S. A decision model for evaluating third-party logistics providers using fuzzy analytic hierarchy process. *Afr. J. Bus. Manag.* **2009**, *4*, 339–349.
7. Hsu, C.H.; Wang, F.K.; Tzeng, G.H. The best vendor selection for conducting the recycled material based on a hybrid MCDM model combining DANP with VIKOR. *Resour. Conserv. Recycl.* **2012**, *66*, 95–111. [CrossRef]
8. Jovčić, S.; Průša, P.; Dobrodolac, M.; Švadlenka, L. A Proposal for a Decision-Making Tool in Third-Party Logistics (3PL) Provider Selection Based on Multi-Criteria Analysis and the Fuzzy Approach. *Sustainability* **2019**, *11*, 4236. [CrossRef]
9. Saaty, T.L. *The Analytic Hierarchy Process*; McGraw-Hill: New York, NY, USA, 1980.
10. Korpela, J.; Tuominen, M. A decision aid in warehouse site selection. *Int. J. Prod. Econ.* **1996**, *45*, 169–180. [CrossRef]
11. Yahya, S.; Kingsman, B. Vendor rating for an entrepreneur development program: A case study using the Analytic Hierarchy Process Method. *J. Oper. Res. Soc.* **1999**, *50*, 916–930. [CrossRef]
12. Akarte, M.M.; Surendra, N.V.; Ravi, B.; Rangaraj, N. Web based casting supplier evaluation using analytical hierarchy process. *J. Oper. Res. Soc.* **2001**, *52*, 511–522. [CrossRef]
13. Muralidharan, C.; Anantharaman, N.; Deshmukh, S.G. A multi-criteria group decision-making model for supplier rating. *J. Supply Chain Manag.* **2002**, *38*, 22–33. [CrossRef]
14. Liu, F.H.F.; Hai, H.L. The voting analytic hierarchy process method for selecting supplier. *Int. J. Prod. Econ.* **2005**, *97*, 308–317. [CrossRef]
15. So, S.H.; Kim, J.J.; Cheong, K.J.; Cho, G. Evaluating the service quality of third-party logistics service providers using the analytic hierarchy process. *J. Inf. Syst. Technol. Manag.* **2006**, *3*, 261–270.
16. Göl, H.; Catay, B. Third-party logistics provider selection: Insights from a Turkish automotive company. *Supply Chain Manag.* **2007**, *12*, 379–384. [CrossRef]
17. Chan, F.T.S.; Chan, H.K.; Ip, R.W.L.; Lau, H.C.W. A decision support system for supplier selection in the airline industry. *Proc. Inst. Mech. Eng. Part B–J. Eng. Manuf.* **2007**, *221*, 741–758. [CrossRef]
18. Hou, J.; Su, D. EJB–MVC oriented supplier selection system for mass customization. *J. Manuf. Technol. Manag.* **2007**, *18*, 54–71. [CrossRef]
19. Gómez, J.C.O.; Herrera, M.F.U.; Vinasco, M.A. Modelo para la evaluación del desempeño de los proveedores utilizando AHP. *Ing. Y Desarro.* **2008**, *23*, 43–58.
20. Hudymáčová, M.; Benková, M.; Pócsová, J.; Škovránek, T. Supplier selection based on multi-criterial AHP method. *Acta Montan. Slovaca* **2010**, *15*, 249.
21. Asamoah, D.; Annan, J.; Nyarko, S. AHP approach for supplier evaluation and selection in a pharmaceutical manufacturing firm in Ghana. *Int. J. Bus. Manag.* **2012**, *7*, 49. [CrossRef]
22. Hruška, R.; Průša, P.; Babić, D. The use of AHP method for selection of supplier. *Transport* **2014**, *29*, 195–203. [CrossRef]
23. Jayant, A.; Singh, P. Application of AHP-VIKOR Hybrid MCDM Approach for 3pl Selection: A Case Study. *Int. J. Comput. Appl.* **2015**, *125*, 4–11.
24. Tuljak-Suban, D.; Bajec, P. Integration of AHP and GTMA to Make a Reliable Decision in Complex Decision-Making Problems: Application of the Logistics Provider Selection Problem as a Case Study. *Symmetry* **2020**, *12*, 766. [CrossRef]
25. Aguezzoul, A.; Pache, G. An AHP-ELECTRE I Approach for 3PL-Provider Selection. *Int. J. Transp. Econ.* **2020**, *47*, 37–50.
26. Meade, L.; Sarkis, J. A conceptual model for selecting and evaluating third-party reverse logistics providers. *Supply Chain Manag. Int. J.* **2002**, *7*, 283–295. [CrossRef]
27. Sarkis, J.; Talluri, S. A model for strategic supplier selection. *J. Supply Chain Manag.* **2002**, *38*, 18–28. [CrossRef]
28. Bayazit, O. Use of Analytic Network Process in vendor selection decisions. *Benchmarking Int. J.* **2006**, *13*, 566–579. [CrossRef]
29. Jharkharia, S.; Shankar, R. Selection of logistics service provider: An analytic Network process (ANP) approach. *Omega: Int. J. Manag. Sci.* **2007**, *35*, 274–289. [CrossRef]
30. Zareinejad, M.; Javanmard, H. Evaluation and selection of a third-party reverse logistics provider using ANP and IFG-MCDM methodology. *Life Sci. J.* **2013**, *10*, 350–355.
31. Chen, Z.; Yang, W. An MAGDM based on constrained FAHP and FTOPSIS and its application to supplier selection. *Math. Comput. Model.* **2011**, *54*, 2802–2815. [CrossRef]
32. Zeydan, M.; Çolpan, C.; Çobanoğlu, C. A combined methodology for supplier selection and performance evaluation. *Expert Syst. Appl.* **2011**, *38*, 2741–2751. [CrossRef]

33. Singh, R.; Rajput, H.; Chaturvedi, V.; Vimal, J. Supplier selection by technique of order preference by similarity to ideal solution (TOPSIS) method for automotive industry. *Int. J. Adv. Technol.* **2012**, *2*, 157–160.
34. Jayant, A.; Gupta, P.; Garg, S.K.; Khan, M. TOPSIS-AHP Based Approach for Selection of Reverse Logistics Service Provider: A Case Study of Mobile Phone Industry. *Procedia Eng.* **2014**, *97*, 2147–2156. [CrossRef]
35. Laptate, V.R. Fuzzy Modified TOPSIS for Supplier Selection Problem in Supply Chain Management. *Int. J. Innov. Res. Comput. Sci. Technol.* **2015**, *3*, 22–28.
36. Aguezzoul, A.; Pires, S. 3PL performance evaluation and selection: A MCDM method. *Supply Chain Forum Int. J.* **2016**, *17*, 87–94. [CrossRef]
37. Govindan, K.; Grigore, M.C.; Kannan, D. Ranking of third-party logistics provider using Fuzzy-Electre II. In Proceedings of the 40th International Conference on Computers & Industrial Engineering, Awaji, Japan, 25–28 July 2010. [CrossRef]
38. Cheng, C.H. Evaluating Naval Tactical Missile System by Fuzzy AHP Based on the Grade Value of Membership Function. *Eur. J. Oper. Res.* **1997**, *96*, 343–350. [CrossRef]
39. Cheng, C.H.; Yang, L.L.; Hwang, C.L. Evaluating Attack Helicopter by AHP Based on Linguistic Variable Weight. *Eur. J. Oper. Res.* **1999**, *116*, 423–435. [CrossRef]
40. Ayhan, M.B. A Fuzzy-AHP approach for supplier selection problem: A case study in a Gear-motor company. *Int. J. Manag. Value Supply Chain.* **2013**, *4*, 12–23. [CrossRef]
41. Kilincci, O.; Onal, S.A. Fuzzy AHP approach for supplier selection in a washing machine company. *Expert Syst. Appl.* **2011**, *38*, 9656–9664. [CrossRef]
42. Shaw, K.; Shankar, R.; Yadav, S.S.; Thakur, L.S. Supplier Selection Using Fuzzy AHP and Fuzzy Multi Objective Linear programming for developing low carbon supply chain. *Expert Syst. Appl.* **2012**, *39*, 8182–8192. [CrossRef]
43. Zhang, H.; Xiu, L.; Wenhuang, L.; Bing, L.; Zhang, Z. An application of the AHP in 3PL vendor selection of a 4PL system. In Proceedings of the IEEE International Conference on Systems, Man and Cybernetics (IEEE Cat. No.04CH37583), The Hague, The Netherlands, 10–13 October 2004; pp. 1255–1260.
44. Zhang, Y.; Feng, Y. A selection approach of reverse logistics provider based on fuzzy AHP. In Proceedings of the Fourth International Conference on Fuzzy Systems and Knowledge Discovery (FSKD 2007), Haikou, China, 24–27 August 2007; pp. 479–482.
45. Cheng, J.-H.; Chen, S.-S.; Chuang, Y.-W. An application of fuzzy-Delphi and Fuzzy AHP formulation-criteria evaluation model of fourth party logistics. *WSEAS Trans. Syst.* **2008**, *7*, 466–478.
46. Arikan, F. An interactive solution approach for multiple-objective supplier selection problem with fuzzy parameters. *J. Intell. Manuf.* **2015**, *26*, 989–998. [CrossRef]
47. Jagannath, R.; Pamučar, D.; Kar, S. Evaluation and selection of third-party logistics provider under sustainability perspectives: An interval valued fuzzy-rough approach. *Ann. Oper. Res.* **2019**, *293*, 669–714. [CrossRef]
48. Rezaeisaray, M.; Ebrahimnejad, S.; Khalili-Damghani, K. A novel hybrid MCDM approach for outsourcing supplier selection. *J. Model. Manag.* **2016**, *11*, 536–559. [CrossRef]
49. Sremac, S.; Stević, Ž.; Pamučar, D.; Arsić, M.; Matić, B. Evaluation of a Third-Party Logistics (3PL) Provider Using a Rough SWARA–WASPAS Model Based on a New Rough Dombi Aggregator. *Symmetry* **2018**, *10*, 305. [CrossRef]
50. Zarbakhshnia, N.; Wu, Y.; Govindan, K.; Soleimani, H. A novel hybrid multiple attribute decision-making approach for outsourcing sustainable reverse logistics. *J. Clean. Prod.* **2019**, *242*, 118461. [CrossRef]
51. Özcan, E.; Ahiskali, M. 3PL Service Provider Selection with a Goal Programming Model Supported with Multi-criteria Decision-Making Approaches. *Gazi Univ. J. Sci.* **2020**, *33*, 413–427. [CrossRef]
52. Lai, K.; Ngai, E.W.T.; Cheng, T.C.E. Measures for evaluating supply chain performance in transport logistics. *Transp. Res. Part E Logist. Transp. Rev.* **2002**, *38*, 439–456. [CrossRef]
53. Sheen, G.L.; Tai, C.T. A study on decision factors and third party selection criterion of logistics outsourcing: An exploratory study of direct selling industry. *J. Am. Acad. Bus.* **2006**, *9*, 331–337.
54. Yeung, C.L.A. The impact of third-party logistics performance on the logistics and export performance of users: An empirical study. *Marit. Econ. Logist.* **2006**, *8*, 121–139. [CrossRef]
55. Lai, K.-h. Service capability and performance of logistic service providers. *Transp. Res. Part E Logist. Transp. Rev.* **2004**, *40*, 385–399. [CrossRef]
56. Sinkovics, R.R.; Roath, S.A. Strategic orientation, capabilities, and performance in manufacturer-3PL relationships. *J. Bus. Logist.* **2004**, *25*, 43–64. [CrossRef]
57. Knemeyer, M.A.; Murphy, R.P. Evaluating the performance of third-party logistics arrangements: A relationship marketing perspective. *J. Supply Chain Manag.* **2004**, *40*, 35–51. [CrossRef]
58. Falsini, D.; Fondi, F.; Schiraldi, M.M. A logistics provider evaluation and selection methodology based on AHP, DEA and linear programming integration. *Int. J. Prod. Res.* **2012**, *50*, 4822–4829. [CrossRef]
59. Zhou, G.; Min, H.; Xu, C.; Cao, Z. Evaluating the comparative efficiency of Chinese third-party logistics providers using data envelopment analysis. *Int. J. Phys. Distrib. Logist. Manag.* **2008**, *38*, 262–279. [CrossRef]
60. Hamdan, A.; Rogers, K.J. Evaluating the efficiency of 3PL logistics operations. *Int. J. Prod. Econ.* **2008**, *113*, 235–244. [CrossRef]
61. Kumar, M.; Vrat, P.; Shankar, R. A multi-objective 3PL allocation problem for fish distribution. *Int. J. Phys. Distrib. Logist. Manag.* **2006**, *36*, 702–715. [CrossRef]

62. Tsai, C.-M.; Lee, H.-S.; Gan, G.-Y. A New Fuzzy DEA Model for Solving the MCDM Problems in Supplier Selection. *J. Mar. Sci. Technol.* **2021**, *29*, 7. [CrossRef]
63. Liu, G.; Fan, S.; Tu, Y.; Wang, G. Innovative Supplier Selection from Collaboration Perspective with a Hybrid MCDM Model: A Case Study Based on NEVs Manufacturer. *Symmetry* **2021**, *13*, 143. [CrossRef]
64. Hoseini, S.A.; Fallahpour, A.; Wong, K.Y.; Mahdiyar, A.; Saberi, M.; Durdyev, S. Sustainable Supplier Selection in Construction Industry through Hybrid Fuzzy-Based Approaches. *Sustainability* **2021**, *13*, 1413. [CrossRef]
65. Kurniawan, V.R.B.; Puspitasari, F.H. A Fuzzy BWM Method for Evaluating Supplier Selection Factors in a SME Paper Manufacturer. *Int. Conf. Adv. Sci. Technol.* **2021**. [CrossRef]
66. Bošković, S.; Radonjić-Djogatović, V.; Ralević, P.; Dobrodolac, M.; Jovčić, S. Selection of Mobile Network Operator Using the Critic-Aras Method. *Int. J. Traffic Transp. Eng.* **2021**, *11*, 17–29. [CrossRef]
67. Diakoulaki, D.; Mavrotas, G.; Papayannakis, L. Determining objective weights in multiple criteria problems: The critic method. *Comput. Oper. Res.* **1995**, *22*, 763–770. [CrossRef]
68. Ghorabaee, K.M.; Amiri, M.; Zavadskas, E.K.; Antuchevičienė, J. Assessment of third-party logistics providers using a CRITIC–WASPAS approach with interval type-2 fuzzy sets. *Transport* **2017**, *32*, 66–78. [CrossRef]
69. Zhang, H. Application on the Entropy Method for Determination of Weight of Evaluating Index in Fuzzy Mathematics for Wine Quality Assessment. *Adv. J. Food Sci. Technol.* **2015**, *7*, 195–198. [CrossRef]
70. Ranđelović, M.; Nedeljković, S.; Jovanović, M.; Čabarkapa, M.; Stojanović, V.; Aleksić, A.; Ranđelović, D. Use of Determination of the Importance of Criteria in Business-Friendly Certification of Cities as Sustainable Local Economic Development Planning Tool. *Symmetry* **2020**, *12*, 425. [CrossRef]
71. Zavadskas, E.K.; Turskis, Z. A new additive ratio assessment (ARAS) method in multi-criteria decision-making. *Technol. Econ. Dev. Econ.* **2010**, *16*, 159–172. [CrossRef]
72. Alinezhad, A.; Khalili, J. WASPAS Method. In *New Methods and Applications in Multiple Attribute Decision Making (MADM)*; International Series in Operations Research & Management Science; Springer: Cham, Switzerland, 2019; pp. 93–98. [CrossRef]
73. Ghorabaee, M.K.; Zavadskas, E.K.; Olfat, L.; Turskis, Z. Multi-Criteria Inventory Classification Using a New Method of Evaluation Based on Distance from Average Solution (EDAS). *Informatica* **2015**, *26*, 435–451. [CrossRef]

mathematics

Article

Design and Numerical Implementation of V2X Control Architecture for Autonomous Driving Vehicles

Piyush Dhawankar [1,2], Prashant Agrawal [2], Bilal Abderezzak [3], Omprakash Kaiwartya [4], Krishna Busawon [2] and Maria Simona Raboacă [5,6,*]

1 Department of Computer Science, University of York, York YO10 5GH, UK; Piyush.Dhawankar@york.ac.uk
2 Department of Mathematics, Physics and Electrical Engineering, Northumbria University, Newcastle Upon Tyne NE1 8ST, UK; Prashant.Agrawal@northumbria.ac.uk (P.A.); Krishna.Busawon@northumbria.ac.uk (K.B.)
3 Laboratoire de l'Énergie et des Systèmes Intelligent (LESI), University of Khemis Miliana, Road of Theniet El Had, Khemis Miliana 44225, Algeria; b.abderezzak@univ-dbkm.dz
4 School of Science and Technology, Nottingham Trent University, Nottingham NG11 8NS, UK; Omprakash.Kaiwartya@ntu.ac.uk
5 Faculty of Electrical Engineering and Computer Science, Stefan cel Mare University of Suceava, 720229 Suceava, Romania
6 National Research and Development Institute for Cryogenic and Isotopic Technologies—ICSI Rm. Vâlcea, Uzinei Street, No. 4, P.O. Box 7 Râureni, 240050 Rm. Vâlcea, Romania
* Correspondence: Simona.Raboaca@icsi.ro

Abstract: This paper is concerned with designing and numerically implementing a V2X (Vehicle-to-Vehicle and Vehicle-to-Infrastructure) control system architecture for a platoon of autonomous vehicles. The V2X control architecture integrates the well-known Intelligent Driver Model (IDM) for a platoon of Autonomous Driving Vehicles (ADVs) with Vehicle-to-Infrastructure (V2I) Communication. The main aim is to address practical implementation issues of such a system as well as the safety and security concerns for traffic environments. To this end, we first investigated a channel estimation model for V2I communication. We employed the IEEE 802.11p vehicular standard and calculated path loss, Packet Error Rate (PER), Signal-to-Noise Ratio (SNR), and throughput between transmitter and receiver end. Next, we carried out several case studies to evaluate the performance of the proposed control system with respect to its response to: (i) the communication infrastructure; (ii) its sensitivity to an emergency, inter-vehicular gap, and significant perturbation; and (iii) its performance under the loss of communication and changing driving environment. Simulation results show the effectiveness of the proposed control model. The model is collision-free for an infinite length of platoon string on a single lane road-driving environment. It also shows that it can work during a lack of communication, where the platoon vehicles can make their decision with the help of their own sensors. V2X Enabled Intelligent Driver Model (VX-IDM) performance is assessed and compared with the state-of-the-art models considering standard parameter settings and metrics.

Keywords: autonomous driving vehicles; vehicular communication; intelligent driver model; data-driven control model

Citation: Dhawankar, P.; Agrawal, P.; Abderezzak, B.; Kaiwartya, O.; Busawon, K.; Raboacă, M.S. Design and Numerical Implementation of V2X Control Architecture for Autonomous Driving Vehicles. *Mathematics* **2021**, *9*, 1696. https://doi.org/10.3390/math9141696

Academic Editor: Bo-Hao Chen

Received: 9 June 2021
Accepted: 14 July 2021
Published: 19 July 2021

Publisher's Note: MDPI stays neutral with regard to jurisdictional claims in published maps and institutional affiliations.

1. Introduction

The automotive industry has recently shifted from developing advanced vehicles to smart transportation, which focuses on the evolution of new intelligent vehicles with autonomous driving and control capabilities [1]. Autonomous Driving Vehicles (ADVs) are highly complex multidisciplinary products, which integrate sensors, automotive control, information processing, artificial intelligence and ultra-fast communication capabilities. The ADVs must have a precise knowledge of the locations of other vehicles in the vicinity. They should be able to determine how to reach the destination optimally without any human intervention [2]. They should also comprehensively sense the surrounding environment for other road users and weather conditions to avoid collisions and accidents. Furthermore, it should detect road signs as well as other static road infrastructure details such as traffic lights, lanes, crosswalks, and speed bumps. In existing technologies, for detecting the surrounding environment in different driving conditions, the sensing systems use a range of cameras, radar, lidar, laser range finders and advanced autonomous driving algorithms [3].

The vehicles on the road with some common interests can cooperatively form a platoon-based driving pattern whereby one vehicle follows another vehicle and maintains a small and nearly constant distance to the preceding vehicle. In this context, several mathematical models on car-following (CF) behaviour have been proposed by traffic engineers and traffic psychologists based on empirical examination and conceptual principles. The foundation of these models is based on simulations and traffic control theories. One of the early linear CF models has been proposed by Chandler et al. [4] and Herman et al. [5], while non-linear models proposed by Reuschel [6], Pipes [7], Gazis et al. [8], Helbing and Tilch [9], and Jiang et al. [10]. Some of these models provided a steady-flow equation for traffic flow from the collected observational and experimental data. Moreover, they discussed several properties of actual traffic flow, such as unstable traffic flow, rise in traffic congestion and formulation of stop-and-go traffic scenarios.

However, these types of models are not suitable for predicting accurate gaps between the vehicles in dense traffic situations as the gap between the vehicles does not relax to an equilibrium value. For example, suppose the relative velocity between two vehicles is zero. In that case, the smallest value of the bumper-to-bumper (actual) gap will not necessarily initiate deceleration, which will result in an accident. Also, these models do not consider any speed limit restriction for vehicles speeding on the free road. To address the abovementioned challenges, Newell [11] and Bando et al. [12] proposed an Optimal Velocity Model (OVM) with an acceleration function with more improvements to provide accident-free driving. These CF models combine the vehicles' desired velocity on a single road and free driving condition with implementing a new breaking strategy to avoid any collisions or accidents.

Moreover, these CF models play a crucial role in describing how one vehicle uninterruptedly follows another vehicle by maintaining its relative velocities with respect to the desired distance between them [13,14]. The follower vehicles would require an automatic control of velocity to the other platoon vehicles as well as their inter-vehicular distance to operate efficiently [15]. Therefore, the CF model and vehicle platooning can be envisioned as the next progressive step towards the more advanced driver assistance systems such as dynamic platoon detection or avoidance of front collision, lane departure warning, autonomous acceleration, and emergency braking [16]. We adopted one such CF model from the OVM family, called Intelligent Driver Model (IDM), for this research work.

Regarding platooning coordination, vehicular communication technologies can be used to develop cooperation among vehicles in traffic environments [17]. The IEEE 802.11p and IEEE 1609-family standards are developed for vehicular communications in telematics. The Federal Communications Commission and European Telecommunications Standard Institute have mandated the use of 5.9 GHz frequency spectrum for Dedicated Short-Range Communication (DSRC) in the vehicular environment and safety-related application in Intelligent Transportation Systems (ITS). Technology-enabled vehicular communication includes Vehicle-to-Vehicle (V2V) and Vehicle-to-Infrastructure (V2I) communication, commonly known as Vehicle-to-Everything (V2X) communication [18]. Moreover, the vehicular networking performance can be improved significantly by the platoon-based driving pattern, as the relatively fixed position of the vehicles in the single platoon can facilitate better cooperative communication [19]. V2X communication can help detect surrounding environments more precisely and help promote a safe and accident-free driving environment for the vehicle platoon [20]. The V2X communication can serve as an extra layer of protection in platoon based autonomous driving with vehicles periodically broadcasting Cooperative Awareness Messages (CAMs) to the neighbouring vehicles and infrastructure. It helps other vehicles accurately map their surroundings and improve decision-making capability [21]. The coordination of vehicular communication technology with platoon-based traffic modelling can support and boost the development of safe and efficient semi-autonomous driving environments.

Considering the above remarks, in this work, we propose an integrated design of a V2X control architecture for a platoon of vehicles focusing on safety in urban and highway traffic environments. For this, we consider the well-known IDM as it emulates human driver's physiological behaviours [22]. It is to be noted that, in the context of this work, any other model such as a cooperative control-based model or a consensus control-based model could be used. Our focus is not on designing or improving CF models but rather investigating the numerical implementation of the V2X control architecture incorporating any kind of CF model. The contributions of the paper can be summarised in the following major folds:

(1)   Firstly, an extension of the IDM is derived considering V2V and V2I communication together called V2X.
(2)   Secondly, a channel estimation model for V2X communication is developed, optimising signal level altering physical layer parameters settings.
(3)   Thirdly, six case studies are designed for integrating vehicular communication-centric coordination within different on-road vehicles in the platoon.
(4)   Finally, the V2X Enabled Intelligent Driver Model (VX-IDM) model performance is assessed and compared with the state-of-the-art models considering standard vehicular parameter settings and metrics.

The rest of this paper is organised as follows. Section 2 presents related work on autonomous vehicles. Section 3 presents the details of the proposed framework VX-IDM for cooperative autonomous driving. The validity of the findings and results are discussed in Section 4, followed by the conclusion presented in Section 5.

2. Related Works

In this section, an overview of platoon-based vehicular communication and their implementation on vehicle platoon and implementation challenges are discussed and presented. Several empirical studies have been performed in the SARTRE project from 2009–2012 to evaluate and demonstrate vehicle platooning's performance by implementing IEEE 802.11p-based communication systems [23]. They successfully deployed a platoon of two trucks and three cars driven autonomously at a speed of 90 km/h with approximately 5–7 m inter-vehicular gap. In Japan's national ITS project named Energy ITS, three fully autonomous trucks were successfully tested on an expressway at 80 km/h with 10 m of an inter-vehicular gap. They used 76 GHz radar and lidar to control longitudinal manoeuvres, while inter-vehicular communication is facilitated by the 5.8 GHz DSRC. For the European Truck Platooning Challenge, several major truck vendors—such as DAF, MAN, and Daimler—drove their trucks in platoons on public roads from various European Cities to Rotterdam in The Netherlands [24]. The published work from these vendors indicates that they used IEEE 802.11p communication modules for inter-vehicular communication. Another study, called the PATH program, led by UC Berkeley and Volvo, successfully demonstrated the platoon of three IEEE 802.11p-equipped trucks driving on the busy 110 Interstate Freeway in Los Angeles with a 15 m inter-vehicular gap [25–27].

These field trials certainly provide valuable information with a thorough analysis of platooning performance under a realistic radio propagation environment for V2X communication with actual vehicle dynamics. However, they do not account for emergency scenarios where the platoons on the highway would require more information about traffic, accidents, weather conditions further away from them on the road as they would not know how to refrain themselves from such an emergency situation if their information database only consists of the local driving environment rather than global driving environment. Therefore, V2I communication is necessary along with platooning, and therefore simulations are an essential tool to study the ability of different V2I technologies to meet the requirements of the ITS applications [28,29]. A comprehensive simulation framework to investigate the Adaptive Cruise Control (ACC) performance in noisy communication conditions is presented by [30]. The results show that the ACC algorithm's performance significantly depends on the broadcast frequency and loss ratio of the CAM messages.

In [31,32], a consensus-based study on platoon with multiple wireless control communication topologies are proposed in the presence of interference, delay and fading conditions. The results show the improvement in comfort level and safety of platoon drivers. Another study to investigate the inter platoon communication facilitated by IEEE 802.11p on control, safety, and dedicated service channels is proposed by [33]. The results show that the guarantee of timely channel access for all packets within a specified deadline is achieved while still providing reasonable dissemination delay. The multiple adaptive CAM beaconing schemes are implemented in IEEE 802.11p for the ACC to manage and maintain the platoons on a freeway is evaluated in [34–36]. The results show that this system met the stringent requirement regarding update frequency and communication reliability for ACC.

Furthermore, few more studies that evaluate ACC's performance implemented with IEEE 802.11p are provided in [37]. The Block Error Rate (BER) versus Signal-to-Noise Ratio (SNR) for wireless communication link and Packet Reception Rate (PRR) versus Distance for the system stability performances for Cellular-Vehicle-to-Everything (C-V2X) and IEEE 802.11p are compared and evaluated [38]. To achieve the 10% BER target, the C-V2X provides 4–5 dB and 1.3 dB SNR gains over IEEE 802.11p in Line-of-Sight (LOS) condition and Non-Line-of-Sight (NLOS) condition, respectively. To achieve the 90% PRR target, the C-V2X provides a 95% and 55% gain in coverage over IEEE 802.11p in highway and urban scenarios, respectively [39].

Similar models investigating the ACC performance in terms of complex traffic scenarios with V2V and V2I communication are suggested in the studies [40,41]. They designed a suitable vehicle driving strategy and proposed an improved consensus-based control algorithm for the Cooperative Driving System considering V2X communications for CF models. Network and traffic simulators are employed for studying platooning scenarios to support the simulations for wireless communication and complex traffic scenarios. The effects of V2X communication on system performance—such as transmission delay, transmission coverage, and measurement noise—is studied theoretically and verified by numerical simulations. However, none of these studies shows how wireless communication link and system stability gains of C-V2X would affect platoon performance. There is no mapping between the implementation of the wireless technologies and the achievable inter-vehicle distance or how the platoon would react and follow the policies enforced by the V2I communication.

3. A Framework for Cooperative Autonomous Vehicles

This section presents an IDM and channel estimation model for V2X communication in a platoon. Its focus is on the control mechanism of IDM and implementing IEEE 802.11p vehicular standards for ADVs in urban and highway driving environments.

3.1. IDM for a Platoon of ADVs

In this work, IDM is used with an 802.11p V2I channel for modelling communication for future ADVs [22]. The IDM is a deterministic, autonomous, and microscopic CF model from the OVM family. This traffic flow model is used for the simulation of highway and urban traffic scenarios. The IDM considers the quadratic relation between speed and braking distance of the vehicles. It also considers the relative velocity $\Delta \dot{x}_n$ between the preceding and the following vehicle, which makes it accident-free.

The IDM uses two microscopic measures mainly, safe time gap or distance headway T, and actual distance (gap) s_n to calculate longitudinal spacing between the vehicles, which are dependent on the length of vehicles and the actual gap between the vehicles.

Figure 1 depicts the general layout of an IDM platoon with their notations as follows: the vehicles in a platoon are marked with an index $n \in N$, where $n - 1$ represents the first vehicle in platoon followed by vehicle n, and a denotes the maximum acceleration. Thus, the leader vehicle's position and velocity at time t are characterised by $x_{n-1}(t)$ and $\dot{x}_{n-1}(t)$ respectively. The actual distance (gap) between two vehicles is denoted by s_n at time t. Due to the presence of safe time gap T and considering the vehicle's length l_n, the actual gap between vehicles is expressed by

$$s_n = x_{n-1} - x_n - l_n. \tag{1}$$

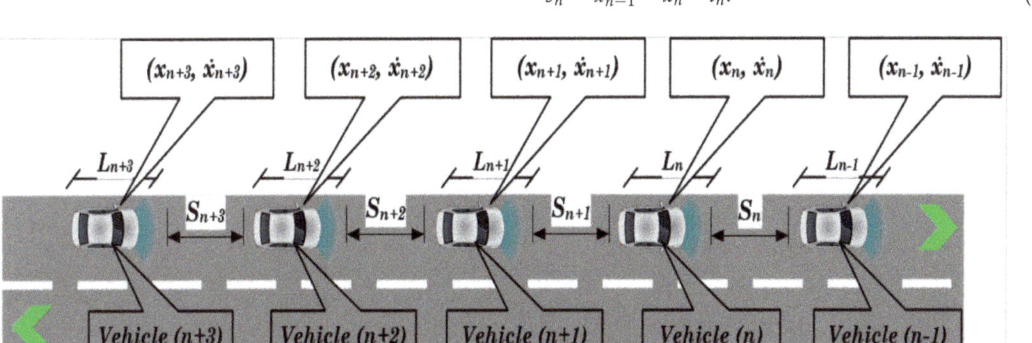

Figure 1. The working of the IDM in physical world environments, incorporating V2V communication.

The relative velocity is a function of time t, and for the vehicle n, it can be described by

$$\Delta \dot{x}_n = \dot{x}_n - \dot{x}_{n-1}. \tag{2}$$

With this definition, the acceleration equation of IDM for any vehicle n is given by

$$\ddot{x}_n\left(s_n, \dot{x}_n, \Delta \dot{x}_n\right) = a\left[1 - \left(\frac{\dot{x}_n}{\dot{x}_{0,n}}\right)^{\delta} - \left(\frac{s^*\left(\dot{x}_n, \Delta \dot{x}_n\right)}{s_n}\right)^2\right], \tag{3}$$

where the s^* is the desired gap and is determined by the equation

$$s^*\left(\dot{x}_n, \Delta \dot{x}_n\right) = s_{0,n} + T\dot{x}_n + \frac{\left(\dot{x}_n \Delta \dot{x}_n\right)}{2\sqrt{ab}}. \tag{4}$$

From the above equation, the free road acceleration and deceleration strategy of the vehicle n on the road within IDM is symbolized by the terms

$$Acceleration = \left[a\left(1 - \left(\frac{\dot{x}_n}{\dot{x}_{0,n}}\right)\right)^{\delta}\right];\ Deceleration = \left[-a\left(\frac{s^*\left(\dot{x}_n, \Delta \dot{x}_n\right)}{s_n}\right)\right]^2 \tag{5}$$

However, the deceleration strategy is only effective if the actual gap s_n between vehicles is not significantly larger than the desired gap $s^*\left(\dot{x}_n, \Delta \dot{x}_n\right)$. The follower vehicle's acceleration is reduced from initial acceleration a to zero when vehicle n is approaching the leader vehicle $n - 1$ with the desired velocity $\dot{x}_{0,n}$. The desired velocity, $\dot{x}_{0,n}$ for all the vehicles is kept constant as given in Table 1. The computed acceleration from Equation (3) generates the velocity and displacement profile of the follower vehicles at time t based on the velocity and displacement of the platoon leader at time t. In our presented model, initially, the platoon leader moves with a constant velocity and then adopts to change in velocity when instructed by the Road Side Unit (RSU), i.e., V2I infrastructure.

Table 1. Summary of parameters used for IDM.

Notation	Description	Realistic Bound	Realistic Values
$\dot{x}_{0n}$	Desired velocity	[0, 50]	33.33 m/s
T	Safe Time Headway	[1, 3]	1.6 m/s
a	Maximum Acceleration	[0.5, 2]	2 m/s^2
b	Comfortable Deceleration	[0.5, 2]	2 m/s^2
δ	Acceleration Exponent	[0 − 4]	4 m/s
l	Length of Vehicle	[5, 10]	5 m
s_0	Linear Jam Distance	[0, 5]	2 m

CF models are primarily defined in acceleration functions. The function defined for the IDM has second-order continuous partial derivatives. The only point at which this is not continuous is when $(x_{n-1} - x_n - l_n) = 0$, which means the actual gap s_n between two vehicles is zero. In order to examine how each component of the model controls the results, this function was reordered as directly dependent on $\dot{x}_n$ and transformed to an Ordinary Differential Equations (ODE-45) system. The position and velocity in these simulations were calculated using ODE-45 in MATLAB (version R2020a, MathWorks, Inc., Natick, MA, USA), which uses Runge–Kutta methods to solve systems of ordinary differential equations.

The parameters that were used to attain the results were in the form $IDM = (x_{0n}, T, a, b, \delta, l, s_0, x_{n-1}, \dot{x}_{n-1})$. When running a simulation, the values in IDM are chosen ahead of time, i.e., the simulation runtime initialised with zero seconds and increased it by a step function every one second with 600-time steps in each second. Then this function is used to calculate the parameterised coefficients, and an ODE-45 solver can interpolate the function to find the position and velocity at any time. After initialising all vehicles' positions and velocities (including the platoon leader) in the platoon, the second-order IDM equations are solved using an ODE-45 solver for the acceleration, displacement, and velocity for all vehicles. Table 1 lists the parameters used for determining the velocity and displacement profile for the vehicle platoon.

The essential characteristics of the IDM Model, which makes it collision-free, are as follows:

(1)   The model is collision-free with a unique dependency on $\Delta \dot{x}_n$.
(2)   All the model parameters are known to be relevant and can be easily interpreted.
(3)   All the model parameters are also empirically measurable and within the expected order of magnitude.
(4)   The model is stable, and it can be calibrated using empirical data.
(5)   The model functions can be numerically/analytically simulated.
(6)   An identical macroscopic model is known for IDM, which simplifies the model's calibration [42].

3.2. Channel Estimation for V2X Communication

In this section, we investigated and proposed a channel estimation model for V2I communication. We employ the IEEE 802.11p vehicular standard and calculated path loss, PER, SNR, and throughput between transmitter and receiver end. The model is divided into three different parts, (i) path loss model, (ii) 802.11p model, and (iii) throughput model. The path loss model considers the effect of the signal's path loss during transmission from the transmitter (RSU) and the receiver (platoon leader). The 802.11p transmitter and receiver model is shown in Figure 2, and it calculates the PER and SNR at the receiver.

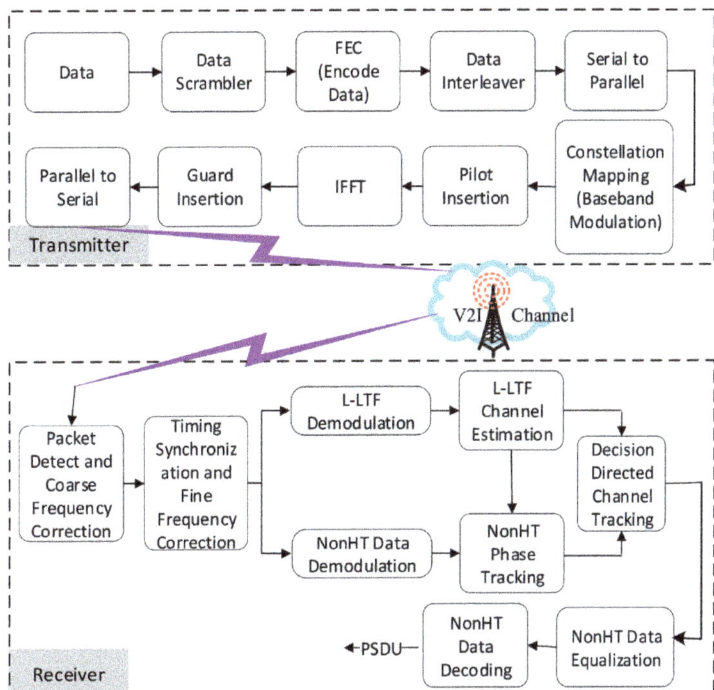

Figure 2. Transmitter and receiver design of IEEE 802.11p for V2I communication.

In contrast, the throughput model calculates the time lag during a platoon leader is entering the RSU's range, and a data communication link is established. The attenuation of radio signal during propagation is termed path loss, and it includes the propagation losses due to the free space, absorption, and diffraction. The path loss is a function of the surrounding environment and the distance between the transmitter and the receiver (d), given by d^{γ}, where γ is the path loss exponent, which depends on the surrounding environment. An increase in the path loss leads to attenuation of radio signal (SNR), limiting the transmission range and data rates of the wireless channel between the receiver and the transmitter. Therefore, in our model, the path loss will determine the transmission range of the RSUs. Table 2 lists the parameters for calculating path loss. We estimate the path loss using the free-space path loss model, given by

$$PL(d) = 20log_{10}\left(\frac{4\pi}{\lambda}\right) + 10\gamma log_{10}(d), \tag{6}$$

where, $PL(d)$ is the path loss at distance d, λ is the free space wavelength defined as the ratio of the velocity of radio signal c to the carrier frequency f of the radio signal. The SNR, without considering fading, can be calculated using the Equation (7).

$$SNR = P_t + G_t + G_r - PL(d) - N_{Power} - I_m, \tag{7}$$

where P_t is the transmitted power, G_t is the antenna gain at the transmitter, G_r is the antenna gain at receiver, N_{Power} is the noise power, and I_m is the implementation noise. N_{Power} is the additional power attenuation per meter distance and is given by the Equation (8)

$$N_{Power} = 10log_{10}(KTB) + N_{Figure}. \tag{8}$$

Table 2. Parameters to calculate path loss of wireless channel for V2I communication [43].

Parameters	V2I Parameters
Channel Bandwidth	10 MHz
Channel Frequency	5.860 GHz
Transmit Power	24 dBm
Transmit Antenna Gain	9 dBi
Transmit EIRP	33 dBm
Receive Antenna Gain	5 dBi
Average Path Loss	2–3
Noise Power	10 dB
Implementation Margin	5 dB

In Equation (8), k, T, B, and N_{Figure} denotes Boltzmann constant, noise temperature, bandwidth, and noise figure, respectively. In our model, we consider SNR of 15 db, and it is obtained at the maximum distance between the transmitter and receiver at which transmission occurs. Therefore, the transmission range of the RSU (denoted as RSU_{range}) is derived from Equations (6) and (7) for a threshold SNR of 15 dB. Equations (6) and (7) are also used to obtain the SNR at each distance d from the transmitter. This SNR value is used to calculate the packet error over the transmitter's distance in the next section. As the transmitter's signal strength weakens depending on the distance or obstacle, a packet loss is probable during transmission. Therefore, we model the receiver's internal working to identify the packet error from the SNR values obtained from the path loss model. Tables 2 and 3 detail the values of all the parameters used in the simulation to calculate the path loss and SNR at a certain distance over the V2I channel in MATLAB.

Table 3. Communication parameters of 802.11p PHY layer used for simulations [44].

Parameters	Notations	V2I Parameters
Channel Width	$OFDM_{bw}$	10 MHz
Symbol Duration	T_{Symbol}	8 µs
Guard Time	T_g	1.6 µs
Signal Field Duration	T_{Signal}	8 µs
Sub-Carrier Spacing	Δ_f	0.15625 MHz
Frequency Range	F_{range}	$USA : 5.86 - 5.92$ GHz $EU : 5.87 - 5.92$ GHz
Maximum EIRP	$EIRP_{max}$	$USA : 30$ W (44.8 dBm) $EU : 2$ W (33 dBm)

The BER and PER affect the data rate and throughput received on the wireless channel. The BER of 10^{-5} is acceptable for the wireless LANs, while PER values depend on the various transmission parameters such as transmission power, the modulation scheme (transmission rate) and packet size [45]. If BER and PER's value is more than their threshold, then the transmitter and receiver cannot establish the connectivity. In IEEE 802.11 wireless standards, the transmitter transmits multiple bits, and they are grouped into packets to share the physical medium. The receiver receives these packets using the checksum in every packet to detect and discard the packets with any bit errors. We obtain the packet error loss at the receiver end. This packet error loss reduces the net data rate (Throughput) received at the receiver. This resulting throughput B_{th} of the wireless channel can be estimated using the Equation (9).

$$B_{th} = \frac{(1 - PER)B_r}{n_{rec}},\qquad(9)$$

where, B_r denotes the bit rate of the modulation scheme. The throughput obtained at each receiver will also depend on the number of receivers (n_{rec}) in communication with the RSU. A non-zero value of data packet size (P_s) is transmitted between transmitter and receiver with the finite throughput value, creates communication lag (t_c) between the transmitter and the receiver, which can be obtained as

$$t_c = \frac{P_s}{B_{th}}. \tag{10}$$

3.3. Development and Working of V2X Enabled IDM Model

This section integrates the above discussed platoon-based CF model and the V2I communication scheme between the platoon leader and an RSU. By integrating these two, we demonstrate how a V2I communication scheme can control a platoon of vehicles' motion while communicating only with a leader vehicle. The presence of a CF scheme within the platoon always ensures vehicles' safety in the platoon. This, in turn, reduces the excess data generation and decision-making load on the RSU and provides an extra layer of control and safety. The V2I communication enables the transmission of periodic safety information, such as weather changes, traffic information and accident knowledge, to an entire platoon. At the same time, the IDM ensures that vehicles within a platoon are safe.

In our model, IDM controls the longitudinal motion of ADVs with sensors' help and controls the acceleration, deceleration, safety time headway, and inter-distance (safety gap) between the ADVs. The IDM determines all platoon vehicles' initial positions and velocities in the proposed model, along with their fixed safety gap (environment dependent) and maximum velocities (road type dependent) for all ADVs. The ADVs follow each other safely, maintaining the safety gap between them and adopting the platoon leader's behaviour. To simulate the platoon of ADVs driving safely, without colliding with each other, the IDM uses the system of ODE-45 as mentioned in Section 3.1 and Figure 3 below shows the IDM integration in MATLAB using the ODE45 solver.

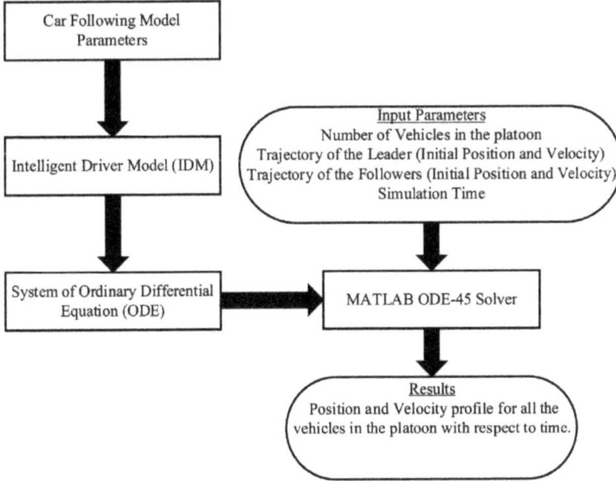

Figure 3. Integration of IDM in MATLAB.

The proposed data-driven hybrid model integrates the 802.11p V2I communication channel with the IDM in MATLAB using ODE-45 and 802.11p simulation toolbox for vehicular channels. For the simulation, we used the IDM and 802.11p parameters from Tables 1–3. The V2I communication is facilitated with the help of RSUs, and the RSUs transmit periodic information about traffic, weather conditions to the platoon leader of ADVs. Therefore, the vehicle's driving behavior is controlled by the RSUs for V2I communication. However, when the vehicle is not in the communication range of the RSUs, the IDM plays a vital role to govern the safety of the vehicle platoon. Although both models work independently with their own limitations, their integration will facilitate the efficient and simultaneous operation of V2V and V2I communication for ADVs. Here, we identify the key parameters that interlink these two models, and they are depicted in Figure 4. It also shows the simulation process using these parameters and how effectively they work together to govern the safety of ADVs.

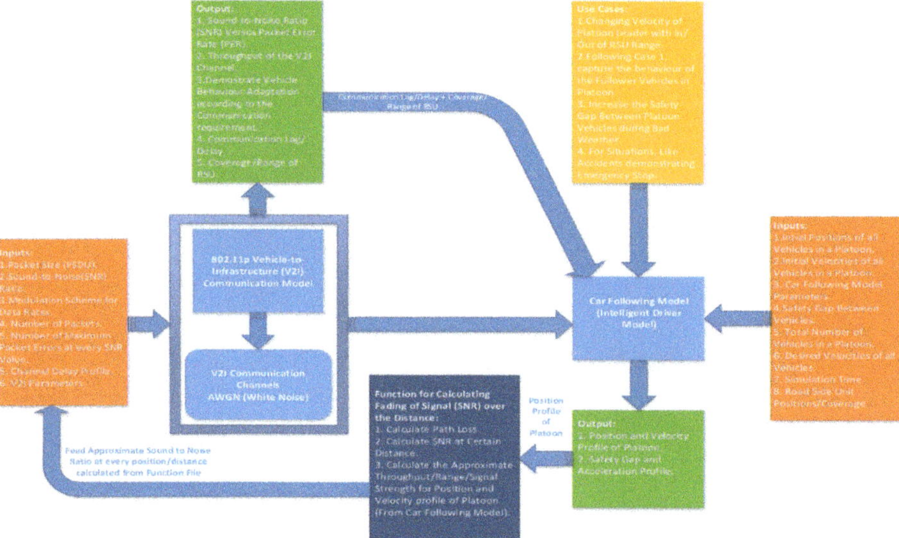

Figure 4. Integration of IDM and 802.11p V2I communication in MATLAB.

The simplified modelling of the path loss mentioned in Section 3.2 is very accurate and suitable for the vehicular environment. The path loss and SNR threshold will define the efficient range of the nearest available RSU. This model calculates the average path loss between ADVs and RSU in real-time for every distance instant from the IDM position profile data. It also determines if the ADVs are in the communication range of the RSU. We employed MATLAB's IEEE 802.11p simulation tool for configuring the vehicular channel. This includes selecting waveform, channel, SNR, urban and highway LOS and NLOS configuration. The V2I wireless channel is configured using the time and frequency selective multipath Rayleigh fading channel specified in [46].

The created channel object filters the real and complex input signal through the multipath V2I wireless channel to obtain the channel-impaired signal. For each SNR point, multiple packets (1000) of 1000 bytes are transmitted through the V2I channel, with the Binary Phase Shift Keying (BPSK) modulated Scheme and Modulation Coding Scheme (MCS) index of 2. Moreover, Root-Mean-Square (RMS) delay spread, Average path gains and RMS Doppler spread for different driving conditions are adopted from the real-world field trials mentioned in [46,47]. The V2I channel estimation is evaluated using SNR ranges from SNR threshold of 15 decibels (dB) to Maximum SNR of 60 decibels. For every SNR point, a Physical Layer Convergence Procedure (PLCP) and PLCP Service Data Unit (PSDU) is created and encoded to create a single packet waveform.

3.4. Case Studies for V2X Centric Cooperation in Autonomous Vehicles

The simulation layout of our integrated model is shown in Figure 5. Here, we consider a road of length $L_{road} = 15$ km with 3 RSU's arranged at equal distances, and their transmission range is also depicted in Figure 5. We considered a platoon consisting of five vehicles (including the leader) to be present on the road with given initial positions and velocities. As the platoon leader enters the transmission range of any RSU, communication is established between the platoon leader and the RSU after a certain time lag (t_c). The leader relays information regarding the platoon, and the RSU (as an observer) provides new control parameters for the car-following model to the platoon leader. Presently, the distance between the RSU's is kept such that their transmission ranges do not overlap. This is done primarily to avoid deciding between the controls of RSU when the platoon leader is in the overlap region and to show that in the loss of communication, how IDM controls the platoon trajectory for accident-free driving.

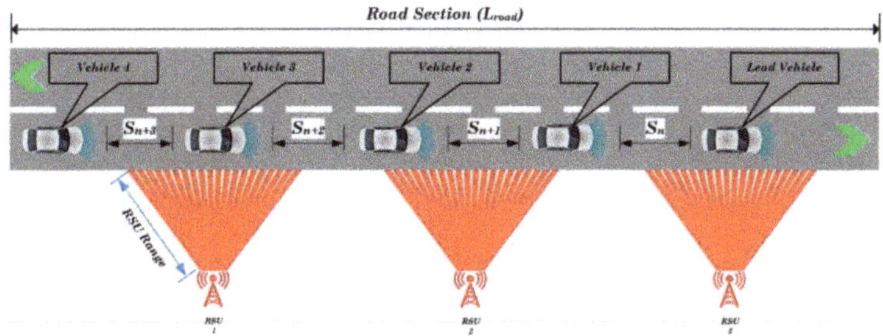

Figure 5. V2X enabled platoon vehicles in physical world environments.

The general control layout of the integrated communication and IDM model is shown in Figure 6. The platoon leader is given a constant velocity in the current model, controlled by the RSU's. The other vehicles follow the platoon leader based on IDM, as discussed in Section 3.1. As the platoon leader moves, the model compares the platoon leader's position with the communication range of each RSU. The path loss model discussed in Section 3.2 is used to approximate the range of RSUs. When the platoon leader enters the communication range, the RSU detects the platoon leader's presence; however, message transmission is not established instantly. There is a marginal communication lag depending on the available data rate, interference, and the amount of data to be transferred, provided by the communication model discussed in Section 3.2.

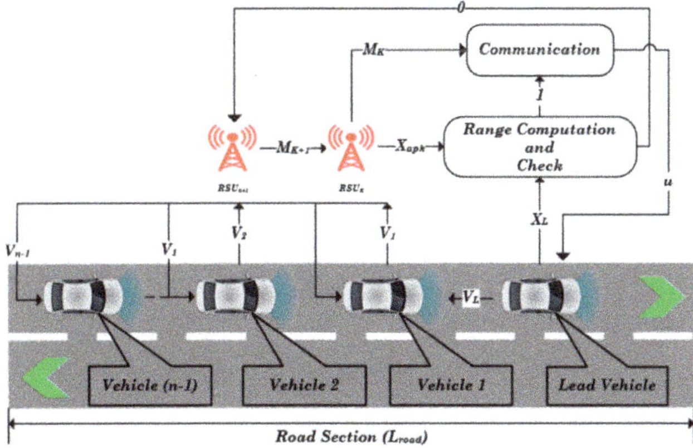

Figure 6. V2X oriented Intelligent Driver Control Model.

When a message transmission is established, the RSU provides new instructions to the platoon leader to control the vehicles' behaviour in the platoon following IDM. These instructions comprise controlling the velocity, safety gap, platoon length, and vehicle's acceleration based on the current safety information relayed to the RSU's from our V2X controller model depicted in Figure 6. The data shared by the platoon leader comprise information such as the number of vehicles in the platoon, position, and velocity of each vehicle in the platoon, condition of each vehicle, and emergency messages relayed from each vehicle. The RSUs periodically update this information and forward it to the platoon leader while remaining in the range of the RSU. Once the platoon leader leaves the RSU range, it retains the driving instructions provided by the preceding RSU while the simulation code compares their positions with the following RSU's range. During this entire time, IDM controls the position of the other vehicles in the platoon. Therefore, the RSUs communicate only with the platoon leader, while IDM manages the other vehicles' trajectories in the platoon. We designed six test cases simulating a few different driving scenarios to demonstrate this integrated communication and car-following model working. Table 4 summarises the parameters' initial and altered values for all test cases during simulation. The cases are described as follows:

Table 4. Test cases used for simulation.

Case No.	Changing Parameters	Parameters Value (Before)	Parameters Value (After)
1	v_1 m/s	33.33 m/s	$13.88; 6.94; 5.55$ m/s
	γ	2.02	*No Change*
2	v_1 m/s	33.33 m/s	$27.77; 27.77; 27.77$ m/s
	γ	2.02	*No Change*
3	v_1 m/s	33.33 m/s	$13.88; 9.25; 6.94$ m/s
	γ	2.02	2.96
4	v_1 m/s	33.33 m/s	$27.77; 27.77; 27.77$ m/s
	γ	2.02	2.56
	n	1	$1; 10; 100$
5	v_1 m/s	25 m/s	$27.77; 27.77; 27.77$ m/s
	γ	2.02	2.56
	n	1	$1; 10; 100$
6	v_1 m/s	27.78 m/s	$25; 30; 30$ m/s
	γ	2.02	2.56
	n	1	$1; 10; 100$

Case I—response to multiple RSUs instructions: In this case, controlling the vehicles platoon's velocities by multiple RSU's is demonstrated. Each RSU enforces a different driving velocity on the platoon leader; the out-of-range velocity and the RSU imposed velocities are given in Table 4. The v_1 denotes the velocity of the leader, γ denotes the path loss exponent depending on the driving environment, and n represents the number of platoons. The platoon is in Highway LOS condition; therefore, the communication parameters—such as average path gains, Doppler shift, and path loss exponent—are used as considered in [48].

Case II—sensitivity to an emergency: This case simulates how a platoon reacts to a dynamic change in instruction from RSU's, such as in an emergency of an accident. At the 400 s, the RSU broadcast an emergency safety message reporting about an accident ahead on the road and instructs all platoon leaders to reduce their velocities. Additionally, the RSU's broadcast another message at 450 s, relaying that the road is cleared and instructs the platoon leader to increase the speed again.

Case III—performance under the loss of communication: In this case, we show the effect of a change in LOS condition between RSU and the platoon leader due to an obstacle's appearance. At the 800 s, we consider that a truck blocks the direct view between the platoon leader and RSU, affecting communication. The change in this condition from Highway LOS to Highway NLOS changes the path loss exponent, which would alter the transmission range (as discussed in Section 3.2).

Case IV—performance under changing environmental conditions: In this case, we simulate a change in the platoon's driving environment, considering the scenario where the platoon enters into a city from a highway. The RSU 1, as shown in Figure 5, is considered to be in a Highway LOS environment while handling only a single platoon. The RSU 2 is also considered to be in Highway LOS environment and is placed between RSU 1 and 3 to mimic slip road leaving the highway and entering urban (city) environment. Moreover, we wanted to demonstrate the slip road that would create bottleneck traffic conditions due to physical differences in the road infrastructure and considered 10 platoons in the RSU 2 region. The RSU 3 is deemed to be in a city environment with high traffic density, where the RSU operates 100 platoons. Other communication parameters such as path loss exponent, average path gains, channel delay profile, and Doppler spread also change in this Urban LOS environment, and their values are adopted from [48].

Case V—sensitivity to single large perturbation: In this case, we demonstrated stable traffic flow wherein all vehicles drive at the same constant speed of 25 m/s. The single large perturbation occurs where the leading vehicle first experiences a deceleration phase with 4 m/s^2, and reduce the velocity from 25 m/s to 5 m/s, after that the vehicle maintains the lower speed of 5 m/s for a time of 160 s, and finally accelerates with 2 m/s^2 increasing its velocity from 5 m/s to 25 m/s.

Case VI—sensitivity to inter-vehicular gap: In this case, a platoon leader starts with a velocity of 27.78 m/s. The platoon's leader's speed is changed to 25 m/s and 30 m/s at the time 200 and 400 s, respectively. Moreover, after 60 s, an instruction to increase the inter-vehicular gap is sent to all the vehicles in the platoon. The default time headway is set to 0.5 s. After receiving the 'increase gap' command, our controller modifies the gap parameter to 20 m and time headway to 1 s.

4. Empirical Results and Discussion

The results of the six test cases for the integrated model proposed in the previous section are discussed in this section. The list of test cases was initially compiled, projecting onto the future technologies; however, they are adapted from existing transportation requirements and functionalities.

Case I—Figure 7 illustrates a basic control test conducted to show the effect of an RSU control over a vehicle platoon's position and velocity, adhering to a CF model. As detailed in the previous section, as the platoon leader starts communicating with an RSU, the RSU relays a message that changes the platoon leader's speed.

This velocity change forces the follower vehicles to adjust their motion under the influence of the ever-present IDM model, as shown in Figure 7a,b. It can be seen from Figure 7(ai), the follower vehicles adopt the behaviour of the platoon leader while maintaining a safe gap determined by the IDM.

Furthermore, each RSU imposes a different speed condition on the platoon leader (values are given in Table 4). The lightly shaded red regions represent instances when the leader vehicle is in an RSU's coverage. Similarly, Figure 7(bi,bii) shows the velocity profile of the vehicles in the platoon with respect to time, reflecting their CF behaviour. It is also visible that the platoon retains the driving conditions imposed by the respective RSU as it leaves their coverage area. The communication lag obtained at each RSU is about 0.0135 s, which is relatively insignificant in this case's driving conditions. This pilot case shows the basic functionality of the V2I communication while controlling the movement of platoon vehicles as they move in and out of the coverage range of RSUs. The following cases in this section will demonstrate some simple exemplary scenarios of how and where such an integrated V2I communication and CF model can be used.

Case II—Figure 8 illustrates the vehicle platoon's functioning with V2I communication in an emergency situation, such as an accident. We imitate an accident scenario by simulating that a message is relayed by all the RSU's to the platoon leaders in their coverage range to reduce their velocities. In our model, this instruction is provided at the 400 s. As shown in Figure 8, at this time, the platoon is in the range of RSU 3, when the platoon leader suddenly decreases the velocity, and the follower vehicles adapt accordingly. To further demonstrate the dynamic functioning of the V2I communication with car behaviour, we relay another message at 450 s, simulating that the road is clear, to instruct the leader vehicle to increase its velocity. The follower vehicles again adapt and increase their velocities while adhering to the IDM. Due to this dynamic message relay, the platoon leader and following vehicles have a displacement of only 29 m during this emergency. Therefore, this case shows the platoon motion can react and adopt different driving conditions by responding to dynamic instructions from the RSUs.

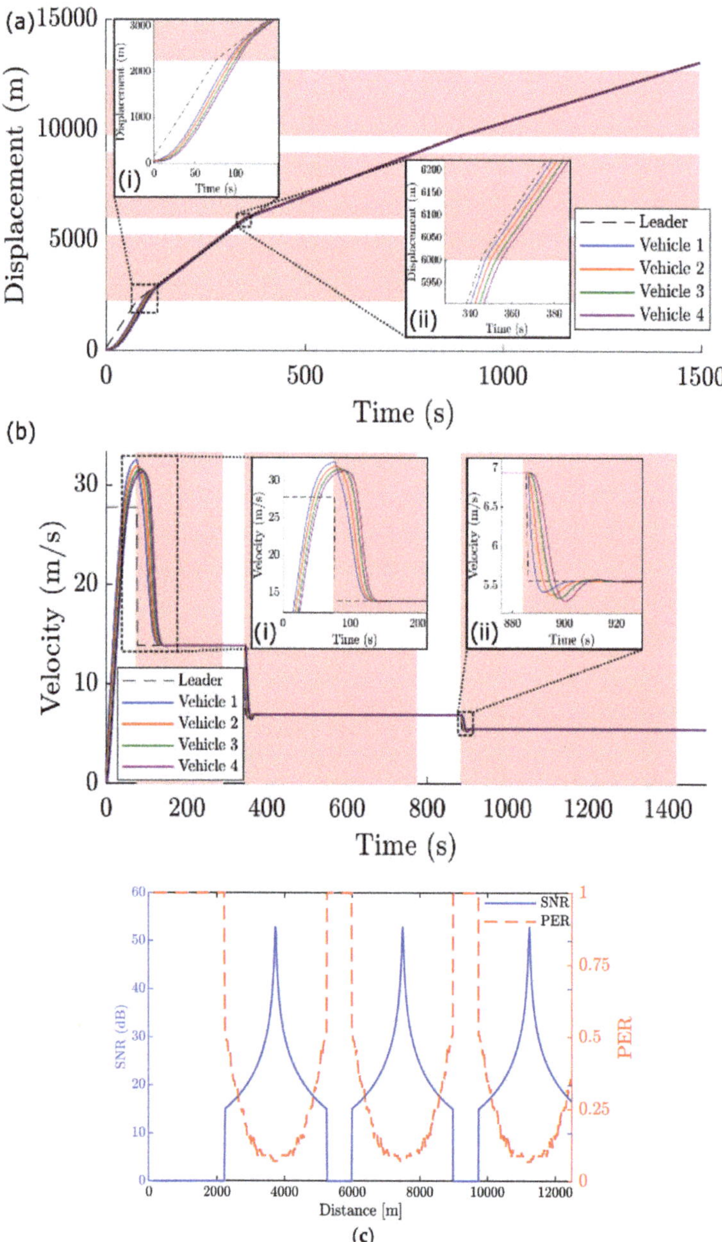

Figure 7. Case I vehicle platoon motion as described in Section 3.4: (**a**) Displacement of the vehicles over the simulation time. The inset shows a magnified image of the vehicle movement as the platoon enters the RSU coverage; (**b**) Velocity of the vehicles over the simulation time. The inset shows a magnified image of the vehicle velocities as the platoon enters the RSU coverage; (**c**) Variation of SNR and PER as the platoon moves along the road. Leader is the platoon leader.

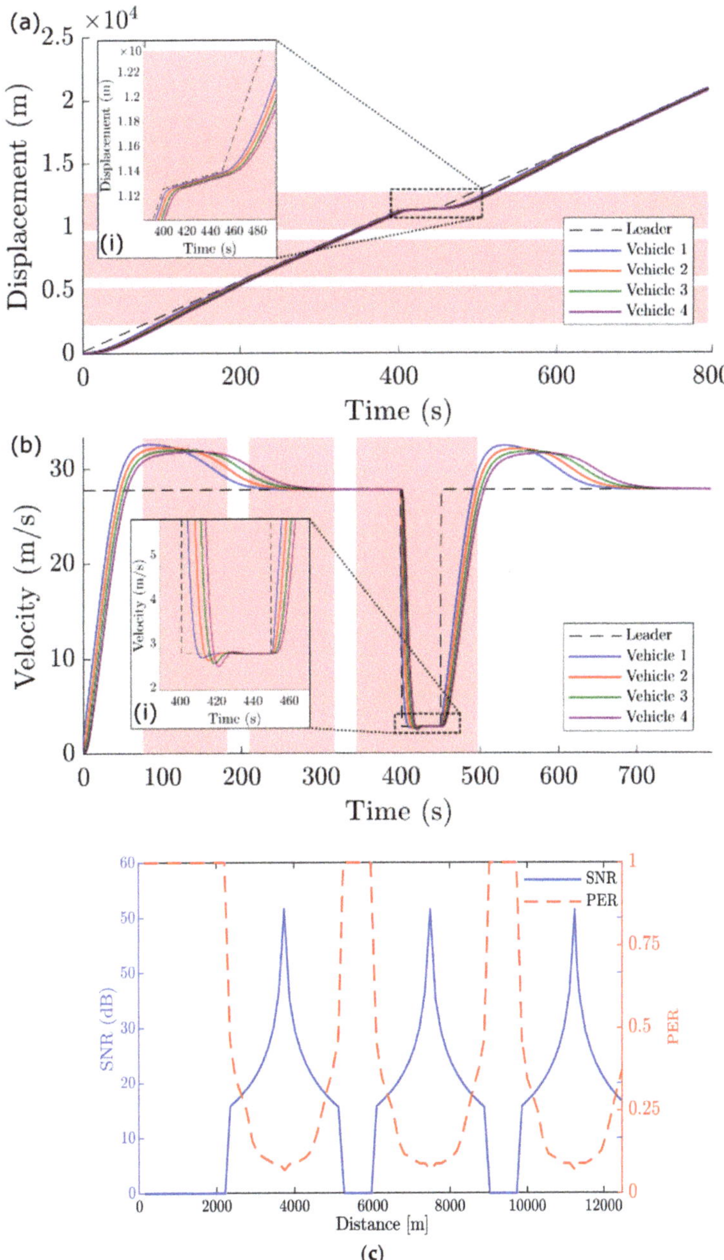

Figure 8. Case II vehicle platoon motion as described in Section 3.4: (**a**) Displacement of the vehicles over the simulation time. The inset shows a magnified image of the vehicle displacements as the platoon encounters a dynamic message from the RSU's; (**b**) Velocity of the vehicles over the simulation time. The inset shows a magnified image of the car velocities as the platoon encounters a dynamic message from the RSU's; (**c**) Variation of SNR and PER as the platoon moves along the road. Leader is the platoon leader.

Case III—Figure 9 illustrates the vehicle platoon's functioning with V2I communication when an obstacle such as a truck or trailer blocks the direct LOS path between the platoon leader and the RSUs. We imitate the scenario by considering a barrier that blocks the direct LOS path between RSU 3 and platoon leader at 800 s of simulation time. As seen from Figure 9a,b, the platoon enters the coverage range of the RSU 3 at 750 s, till which point the platoon was traversing under LOS conditions. Under these conditions, the communication range of the RSU's was about 3000 m.

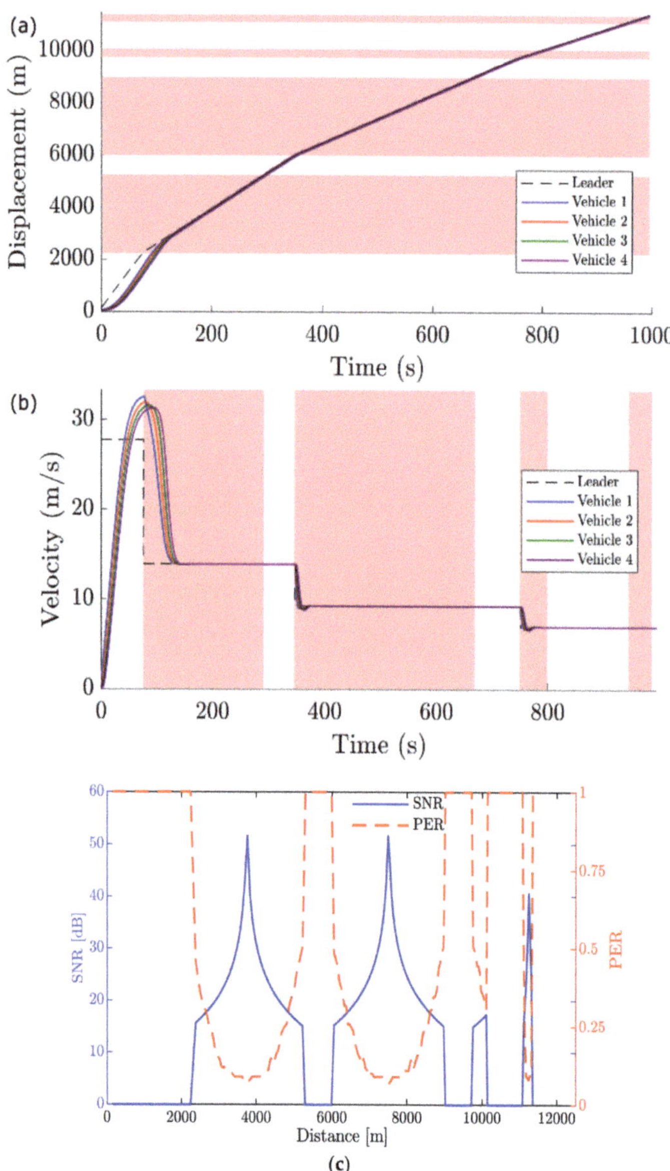

Figure 9. Case III vehicle platoon motion as described in Section 3.4: (**a**) Displacement of the vehicles over the simulation time; (**b**) Velocity of the vehicles over the simulation time; (**c**) Variation of SNR and PER as the platoon moves along the road. Leader is the platoon leader.

As the obstacle blocks the line of sight between the RSU and the platoon leader (at 800 s), there is an increase in the transmitted signal's path loss. This path loss is observed as a decrease in the RSU's communication range, which depletes its range to 1500 m. In this NLOS condition, the platoon leader falls out of range of RSU and has to travel some more distance to re-connect with the RSU; at around 1000 s, as can be seen in Figure 9a.

Although, in this case, we assumed the line of sight between the platoon leader and the RSU 3 to be consistently blocked after 800 s, the example shows the ability of this model to predict how a platoon would react to dynamic changes in signal attenuation due to the environment and surrounding obstacles. This simple example can also be expanded to a more complex scenario of intermittent changes in the line-of-sight due to multiple trucks overtaking the platoon.

Case IV—In Figure 10, we simulate our integrated model's working in a scenario where a vehicle platoon enters an urban environment from a highway environment. Here, we show the effect of multiple isolated vehicle platoons communicating with the RSU's in driving environments changing from a Highway LOS to Urban LOS. As mentioned in Section 3.4, RSU 1 and RSU 2 are considered to be in a highway environment, and RSU 3 is regarded as an urban environment.

All the RSU's are assumed to handle a different number of platoons, as mentioned in Table 4; we simulate only one platoon's motion assuming all platoons are isolated. We imitate the scenario where the platoon crosses the RSU 2's coverage (Highway LOS condition) and follows the slip road into the city (Urban LOS condition). The RSU 3 is in an Urban LOS condition and due to multiple scatters in this environment, path loss exponent changes, that reduces the RSU range (3000 m) as opposed to that on the highway (640 m), shown in Figure 10a–c.

Additionally, we assume that due to a high vehicle density in the urban environment, the number of platoons in communication with the RSU increases, decreasing the data throughput received by each platoon leader. This decreasing throughput increases the communication lag as a platoon enters the RSU range; the communication or time lag t_{lag} of 0.01, 0.02, and 0.031 s is observed at RSU 1, RSU2, and RSU 3, respectively. This communication lag implies that, for any instance, even though the platoon leader is in the coverage range of an RSU 3, there is no effective communication established, and it moved 0.85 m, termed as the distance lag d_{lag} in Figure 10c.

This distance lag may become necessary for relaying safety messages in conditions when the coverage range of an RSU changes dynamically and is of a similar order of magnitude to the distance lag. Therefore, this parameter is essential in designing an appropriate RSU or V2I infrastructure layout, given any environmental and traffic conditions. This case shows that the coupled system works dynamically and shows the platoon vehicles can adapt to changes in range and communication ability of RSUs in a LOS and NLOS condition on the highway and urban environment. This shows the effect of traffic density on communication lag, and it can help RSUs implementation design in a transition from a highway to an urban environment.

Another aspect that can be included in communication is the change in path loss due to weather change. We found no concrete evidence of weather change significantly affecting path loss exponent, which varies at frequencies below 10 GHz. However, this integrated scheme can be used to simulate path loss changes due to weather alterations at other frequency schemes as well.

Corresponding changes in vehicular movement such as velocity decrement, safety gap increment, change in maximum acceleration, and deceleration exponents can be adapted and included as shown above accordingly.

Case V and VI—These two cases are discussed separately in the following subsection, as they are mainly employed for comparison purposes with existing methods.

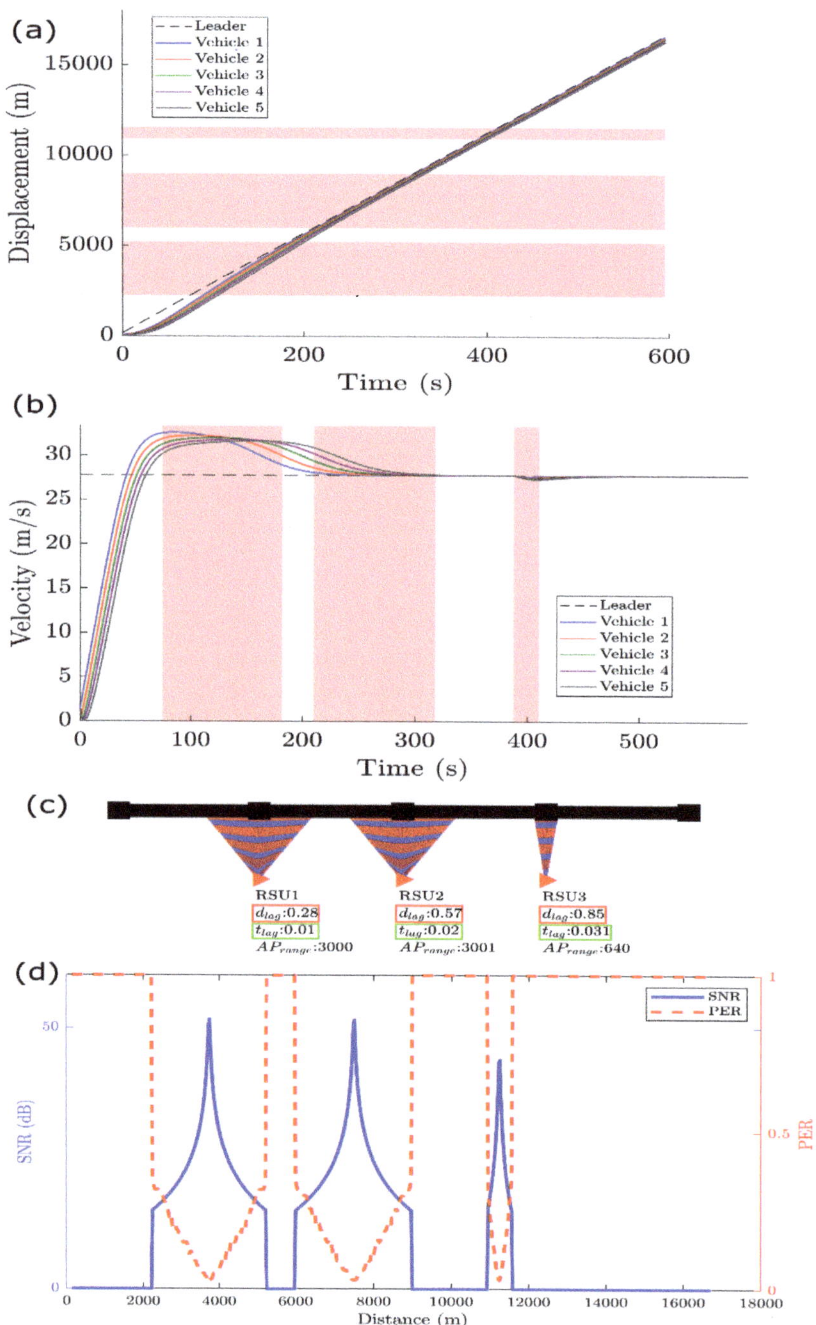

Figure 10. Case IV vehicle platoon motion for case IV, as described in Section 3.4: (**a**) Displacement of the vehicles over the simulation time; (**b**) Velocity of the vehicles over the simulation time; (**c**) Depiction of the coverage range of RSU's (red) and the range during which the car communicates with the RSU (blue). The distance lag (red), time lag (green), and the RSU range are also mentioned; (**d**) Variation of SNR and PER as the platoon moves along the road. Leader is the platoon leader.

Comparison with the State-of-the-Art Models

We implemented two state-of-the-art models for cooperative intelligent transport systems, such as [40,41] shown in Figures 11 and 12 as Case V and Case VI, respectively, and compared with our VX-IDM model described as Case IV in Figure 10. The test cases from [40,41] are used to compare with our model test cases for simulation defined in Table 4. The relevant results are compared in terms of dynamic change in velocity and gap between the platoons. Figure 10 shows the VX-IDM model where a platoon's behaviour is captured in terms of their displacement and velocity in urban and highway driving conditions. Moreover, the distance lag (d_{lag}), time lag (t_{lag}) and AP range are also calculated from RSU to Vehicle Platoon communication.

In comparison to the state-of-the-art models in [40,41], the d_{lag} in Figure 10c, at RSU 1, 2, and 3 are 0.28, 0.57, and 0.85 m, respectively. While in Figure 11c, the d_{lag} at RSU 1, 2, and 3 are 0.38, 0.15 and 1.13 m, respectively. In Figure 12, the d_{lag} at RSU 1, 2, and 3 are 0.34, 0.67, and 0.91 m, respectively. The distance lag d_{lag} in all the three cases are compared, and it can be seen that overall, our model performs better than the compared models, where the distance lag (longitudinal spacing) is more compared to our model except in one case of RSU 2 in Figure 11c. The d_{lag} signifies that the platoon leader travels a certain distance in the communication range without connecting or establishing two-way communication with the RSUs. Therefore, the lower value of d_{lag} is very important during a real-world driving scenario. Suppose the platoon leader's communication is not established with RSU even when it is in the communication range. In that case, the platoon leader will fail to receive important updates or information. For example, safety messages about an accident on the road ahead are not communicated to the platoon leader and failing to act on the information relayed by the RSU, the platoon vehicles will collide with the vehicles in the front. Moreover, the stopping distance of these platoon vehicles at a velocity of 29 m/s to 33 m/s is about 58.77 to 67.8 m, respectively [49]. Therefore, the lower d_{lag} is very crucial for the safety of platoon vehicles.

Moreover, the time lag (t_{lag}) calculates the time delay between instances where the platoon leader enters the RSUs range and establishes data communication. Figure 10c shows that the time lag at RSU 1, 2, and 3 are 0.01, 0.02, and 0.031 s, respectively. While in Figure 11c for RSU 1, 2, and 3 is, 0.015, 0.03, and 0.045 s respectively. In addition, Figure 12c depicts the t_{lag} of 0.012, 0.024, and 0.036 s for RSU 1, 2 and 3, respectively. The proposed VX-IDM model shows that the time lag required to establish communication with the RSUs is substantially lower than the compared model. Such low latency and seamless communication capabilities of the proposed model would help reduce the number of casualties and fatalities in road accidents. They would also improve traffic flow by tracking the vehicular network and mapping the least congested routes. Therefore, the t_{lag} is an important aspect of vehicular communication for their safety, where platoon vehicles can communicate seamlessly with the RSUs and establish communication links even during a high-density traffic scenario such as an urban (city) traffic environment. The result attests that our model performed well compared to the referenced models.

Overall, the results clearly illustrate that incorporating IDM with V2I communication improves the fidelity of our proposed model. Moreover, our model shows consistent improvement in terms of distance and time lag compared with the referenced frameworks. These results are heavily focused on the value of spacing between the platoon vehicles, distance lag, and time delay for establishing reliable communication. Moreover, these mentioned variable values are considered very important for several applications, such as safety considerations, wireless network connectivity, and various traffic systems. Accurate simulation of the spacing and time lag is essential for all other physical attributes for platoon vehicles' position, velocity, and acceleration. The results attested that the compared models' longitudinal spacing between the platoon vehicles in motion is higher than our model, restricting relaying safety-related messages in a dynamically changing communication topology in real-time. Moreover, the communication delay (time lag), essential for vehicles to establish communication with the infrastructures and starts relaying safety-related messages, was substantially lower than compared state-of-the-art models in [40,41]. This makes the VX-IDM model collision-free for a platoon of vehicles on a single road for highway and urban driving environments.

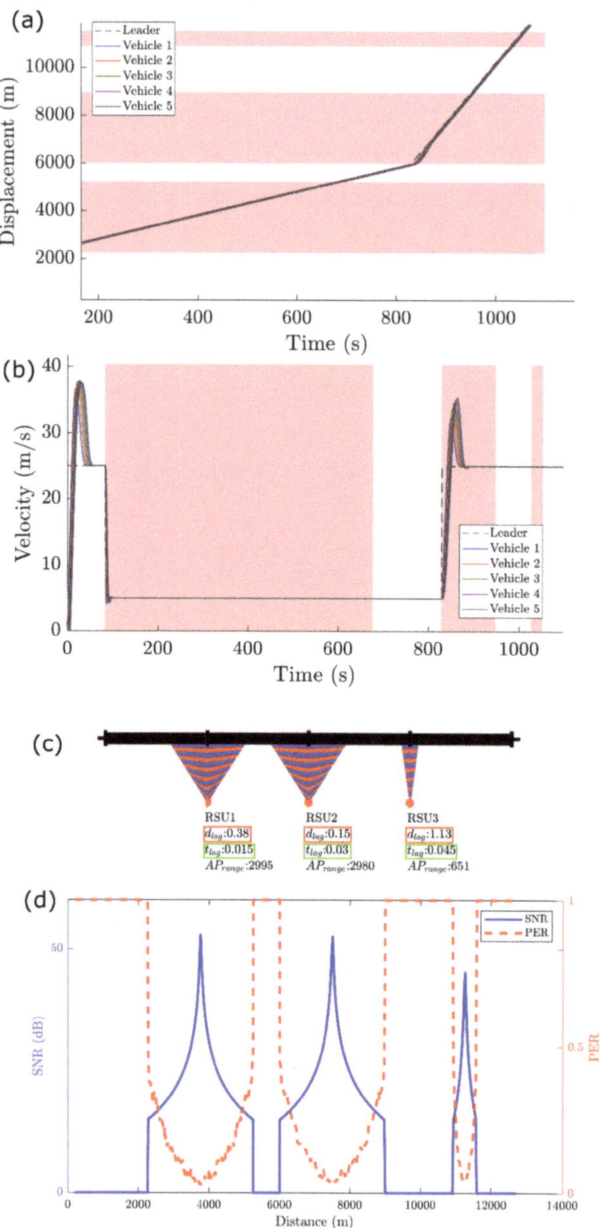

Figure 11. Case V vehicle platoon motion as described in Section 3.4: (**a**) Displacement of the vehicles over the simulation time; (**b**) Velocity of the vehicles over the simulation time; (**c**) Depiction of the coverage range of RSU's (red) and the range during which the car communicates with the RSU (blue). The distance lag (red), time lag (green) and the RSU range are also mentioned; (**d**) Variation of SNR and PER as the platoon moves along the road. Leader is the platoon leader.

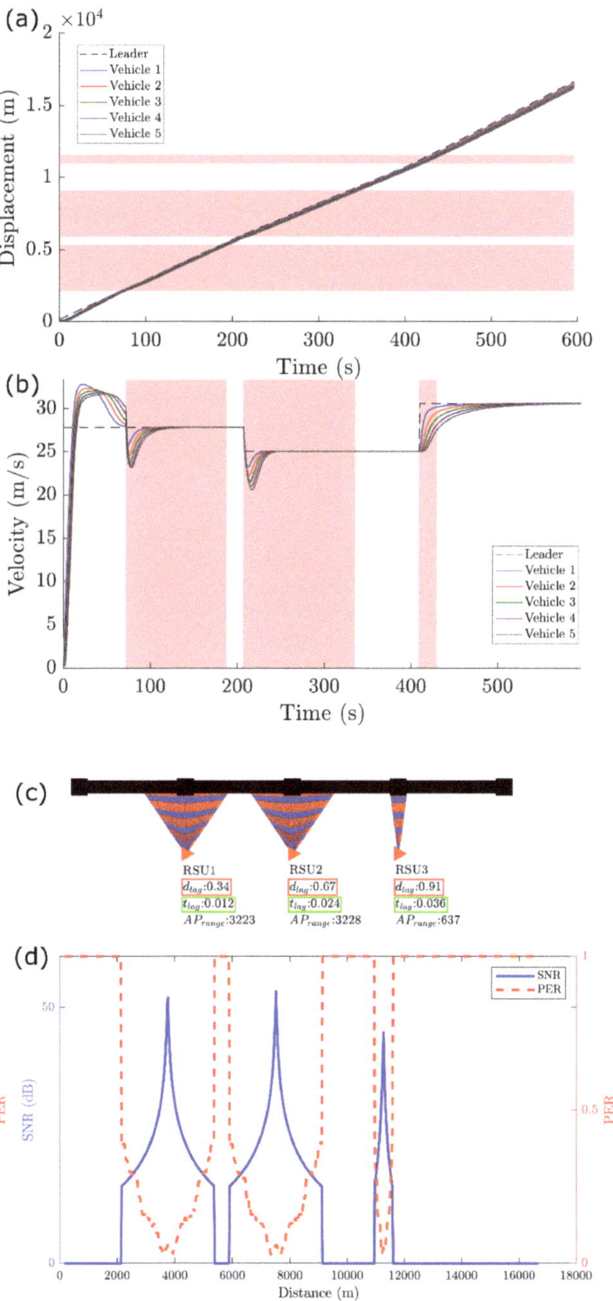

Figure 12. Case VI vehicle platoon motion as described in Section 3.4: (**a**) Displacement of the vehicles over the simulation time; (**b**) Velocity of the vehicles over the simulation time; (**c**) Depiction of the coverage range of RSU's (red) and the range during which the car communicates with the RSU (blue). The distance lag (red), time lag (green) and the RSU range are also mentioned; (**d**) Variation of SNR and PER as the platoon moves along the road. Leader is the platoon leader.

5. Conclusions and Future Work

This paper has presented a novel design and numerical implementation of a V2X control system architecture for a platoon of autonomous vehicles. More precisely, we have demonstrated some aspects of platoon-based driving with vehicular network architecture and their effects on platoon dynamics and control. This involves an IDM controller jointly operating with 802.11p communication architecture. We have provided a clear insight into the impact of the communication topology on the controller performance to define platoon vehicles' control strategy. We have systematically elaborated on the effects of cooperative platoon driving and platoon based vehicular communication. Simulation results show the effectiveness of the implemented control architecture in its sensitivity to an emergency, inter-vehicular gap, large perturbation, and the loss of communication and changing driving environment. Moreover, the proposed simulation platform proves that it can integrate any CF or consensus control model with any communication system.

Future work will explore integrating the United Kingdom's Road networks statistical traffic data provided by the Department for Transport (DfT) with the V2X enabled autonomous vehicles, considering big traffic data analytics. Moreover, the lane-changing strategies will be considered along with various platoons in a multi-lane road driving environment. The communication within the lane changing vehicles and their association with the other platoons will be exploited using the V2I communication employing software-defined networking. Moreover, we will also include the data collected from a real traffic system, which will consider the platoon behaviour in very dense traffic situations such as rush hours on weekdays and weekend traffic scenarios.

Author Contributions: Conceptualisation, P.D.; Methodology, P.D.; Validation, P.D.; Formal analysis P.D.; Investigation P.D.; Resources, P.D.; Data curation, P.D., P.A., B.A., O.K. and K.B.; Writing—original draft preparation, P.D.; Writing—review and editing, P.D., P.A., O.K. and K.B.; Visualisation, P.D., P.A., O.K., K.B. and M.S.R.; Supervision, P.A., B.A., M.S.R., O.K., and K.B.; Project administration, P.D., O.K., K.B. and P.A.; Funding acquisition P.D., O.K., B.A., M.S.R., K.B. and P.A. All authors have read and agreed to the published version of the manuscript.

Funding: This research received no external funding.

Institutional Review Board Statement: Not applicable.

Informed Consent Statement: Not applicable.

Data Availability Statement: More help on experimental data and associated settings will be made available to researchers and practitioners on individual requests to this research team with the restrictions that it will solely be used for further research in literature progress. The associated research data is being utilised for further development by the research team.

Acknowledgments: This work was supported by a grant from the Romanian Ministry of Research and Innovation, CCCDI—UEFISCDI, project number PN-III-P1-1.2-PCCDI-2017-0776/No. 36 PC-CDI/15.03.2018, within PNCDI III and project number PN-III-P1-1.2-PCCDI-2017-0194/25 PCCDI within PNCDI III.

Conflicts of Interest: The authors declare no conflict of interest.

References

1. Kichun, K.; Junsoo, K.; Dongchul, J.C.; Myoungho, S. Development of autonomous car—Part II: A case study on the implementation of an autonomous driving system based on distributed architecture. *IEEE Trans. Ind. Electr.* **2015**, *62*, 5119–5132.
2. Choi, J.L.; Kim, D.; Soprani, G.; Cerri, P.; Broggi, A.; Yi, K. Environment-detection-and-mapping algorithm for autonomous driving in rural or off-road environment. *IEEE Trans. Intell. Transp. Syst.* **2012**, *13*, 974–982. [CrossRef]
3. Morisio, C.; Morisio, M. Connected car: Technologies, issues, future trends. *ACM Comput. Surv.* **2016**, *49*, 46.
4. Chandler, R.E.; Herman, R.; Montroll, E.W. Traffic dynamics: Studies in car following. *Oper. Res.* **1958**, *6*, 165–184. [CrossRef]
5. Herman, R.; Montroll, E.W.; Potts, R.B.; Rothery, R.W. Traffic dynamics: Analysis of stability in car following. *Oper. Res.* **1959**, *7*, 86–106. [CrossRef]
6. Gazis, D.C.; Herman, R.; Rothery, R.W. Analytical methods in transportation: Mathematical car-following theory of traffic flow. *J. Eng. Mech. Div.* **1963**, *89*, 29–46. [CrossRef]

7. Pipes, L.A. An operational analysis of traffic dynamics. *J. Appl. Phys.* **1953**, *24*, 274–281. [CrossRef]
8. Gazis, D.C.; Herman, R.; Rothery, R.W. Nonlinear follow-the-leader models of traffic flow. *Oper. Res.* **1961**, *9*, 545–567. [CrossRef]
9. Helbing, D.; Tilch, B. Generalized force model of traffic dynamics. *Phys. Rev. E* **1998**, *58*, 133. [CrossRef]
10. Jiang, R.; Wu, Q.; Zhu, Z. Full velocity difference model for a car-following theory. *Phys. Rev. E* **2001**, *64*, 017101. [CrossRef]
11. Newell, G.F. Nonlinear effects in the dynamics of car following. *Oper. Res.* **1961**, *9*, 209–229. [CrossRef]
12. Bando, M.; Hasebe, K.; Nakayama, A.; Shibata, A.; Sugiyama, Y. Dynamical model of traffic congestion and numerical simulation. *Phys. Rev. E* **1995**, *51*, 1035. [CrossRef] [PubMed]
13. Van Arem, B.; Van Driel, C.J.; Visser, R. The impact of cooperative adaptive cruise control on traffic-flow characteristics. *IEEE Trans. Intell. Transp. Syst.* **2006**, *7*, 429–436. [CrossRef]
14. Rahman, M.; Chowdhury, M.; Khan, T.; Bhavsar, P. Improving the efficacy of car-following models with a new stochastic parameter estimation and calibration method. *IEEE Trans. Intell. Transp. Syst.* **2015**, *16*, 2687–2699. [CrossRef]
15. Besselink, A.B.; Turri, V.; Martensson, J.; Johansson, K.H. Heavy-duty vehicle platooning for sustainable freight transportation: A cooperative method to enhance safety and efficiency. *IEEE Control Syst. Mag.* **2015**, *35*, 34–56.
16. Fleming, B. New innovative ICs [automotive electronics]. *IEEE Veh. Technol. Mag.* **2011**, *6*, 4–8. [CrossRef]
17. Jia, D.; Lu, K.; Wang, J.; Zhang, X.; Shen, X. A survey on platoon-based vehicular cyber-physical systems. *IEEE Commun. Surv. Tutor.* **2016**, *18*, 263–284. [CrossRef]
18. Kaiwartya, O.; Abdullah, A.H.; Cao, Y.; Altameem, A.; Prasad, M.; Lin, C.T.; Liu, X. Internet of vehicles: Motivation, layered architecture, network model, challenges, and future aspects. *IEEE Access* **2017**, *4*, 5356–5373. [CrossRef]
19. Axelsson, J. Safety in vehicle platooning: A systematic literature review. *IEEE Trans. Intell. Transp. Syst.* **2017**, *18*, 1033–1045. [CrossRef]
20. Dhawankar, P.; Raza, M.; Le-Minh, H.; Aslam, N. Software-defined approach for communication in autonomous transportation systems. *EAI Endorsed Trans. Energy Web.* **2017**, *4*, 7. [CrossRef]
21. Sasaki, K.; Suzuki, N.; Makido, S.; Nakao, A. Vehicle control system coordinated between cloud and mobile edge computing. In Proceedings of the 55th Annual Conference of the Society of Instrument and Control Engineers of Japan (SICE), Tsukuba, Japan, 20–23 September 2016; pp. 1122–1127.
22. Kesting, A.; Treiber, M.; Helbing, D. Enhanced intelligent driver model to access the impact of driving strategies on traffic capacity. *Philos. Trans. R. Soc. A Math. Phys. Eng. Sci.* **2010**, *368*, 4585–4605.
23. Chan, E. *Overview of the SARTRE Platooning Project: Technology Leadership Brief*; SAE Convergence, International; SAE International: Warrendale, PA, USA, 2012. [CrossRef]
24. Volvo. European Truck Platooning Challenge. 2016. Available online: https://www.eutruckplatooning.com (accessed on 15 May 2019).
25. Volvo. Volvo Trucks Successfully Demonstrates on-Highway TRUCK Platooning in California. 2017. Available online: https://www.volvotrucks.us/news-and-stories/press-releases/2017/march/volvo-trucks-successfully-demonstrates-on-highway-truck-platooning-in-california/ (accessed on 15 May 2019).
26. Umair, S. Significance of smart grid in electric power systems: A brief overview. *J. Electr. Eng. Electron. Control Comput. Sci.* **2020**, *6*, 7–12.
27. Sorlei, I.-S.; Bizon, N.; Thounthong, P.; Varlam, M.; Carcadea, E.; Culcer, M.; Iliescu, M.; Raceanu, M. Fuel cell electric vehicles—A brief review of current topologies and energy management strategies. *Energies.* **2021**, *14*, 252. [CrossRef]
28. Raboaca, M.S.; Bizon, N.; Phatiphat, T. Intelligent charging station in 5G environment: Challenges and perspective. *Int. J. Energy Res.* **2021**. [CrossRef]
29. Raboaca, M.S.; Bizon, N.; Grosu, O.V. Optimal energy management strategies for the electric vehicles compiling bibliometric maps. *Int. J. Energy Res.* **2021**, *45*, 10129–10172. [CrossRef]
30. Lei, C.; Eenennaam, E.M.; Wolterink, W.K.; Karagiannis, G.; Heijenk, G.; Ploeg, J. Impact of packet loss on CACC string stability performance. In Proceedings of the 11th International Conference on ITS Telecommunications, St. Petersburg, Russia, 23–25 August 2011; pp. 381–386.
31. Santini, S.; Salvi, A.; Valente, A.S.; Pescapé, A.; Segata, M.; Lo Cigno, R. A consensus-based approach for platooning with inter vehicular communications and its validation in realistic scenarios. *IEEE Trans. Veh. Technol.* **2017**, *66*, 1985–1999. [CrossRef]
32. Gautam, M.K.; Pati, A.; Mishra, S.K.; Appasani, B.; Kabalci, E.; Bizon, N.; Thounthong, P. A Comprehensive Review of the Evolution of Networked Control System Technology and Its Future Potentials. *Sustainability* **2021**, *13*, 2962. [CrossRef]
33. Böhm, A.; Jonsson, M.; Uhlemann, E. Performance comparison of a platooning application using the IEEE 802.11p MAC on the control channel and a centralised MAC on a service channel. In Proceedings of the 2013 IEEE 9th International Conference on Wireless and Mobile Computing, Networking and Communications (WiMob), Lyon, France, 7–9 October 2013; pp. 545–552.
34. Segata, M.; Bloessl, B.; Joerer, S.; Sommer, C.; Gerla, M.; Lo Cigno, R.; Dressler, F. Toward communication strategies for platooning: Simulative and experimental evaluation. *IEEE Trans. Veh. Technol.* **2015**, *64*, 5411–5423. [CrossRef]
35. Bizon, N.; Thounthong, P. A simple and safe strategy for improving the fuel economy of a fuel cell vehicle. *Mathematics* **2021**, *9*, 604. [CrossRef]
36. Yodwong, B.; Thounthong, P.; Guilbert, D.; Bizon, N. Differential flatness-based cascade energy/current control of battery/supercapacitor hybrid source for modern e–vehicle applications. *Mathematics* **2020**, *8*, 704. [CrossRef]

37. Vinel, A.; Lan, L.; Lyamin, N. Vehicle-to-vehicle communication in c-acc/platooning scenarios. *IEEE Commun. Mag.* **2015**, *53*, 192–197. [CrossRef]

38. Hu, J.; Chen, S.; Zhao, L.; Li, Y.; Fang, J.; Li, B.; Shi, Y. Link level performance comparison between LTE V2X and DSRC. *J. Commun. Inf. Netw.* **2017**, *2*, 101–112. [CrossRef]

39. Blasco, R.; Do, H.; Shalmashi, S.; Sorrentino, S.; Zang, Y. 3GPP LTE enhancements for v2v and comparison to IEEE 802.11 p. In Proceedings of the 11th ITS European Congress, Glasgow, Scotland, 6–9 June 2016.

40. Aramrattana, M.; Larsson, T.; Jansson, J.; Nåbo, A. A simulation framework for cooperative intelligent transport systems testing and evaluation. *Transp. Res. Part F Traffic Psychol. Behav.* **2019**, *61*, 268–280. [CrossRef]

41. Rata, M.; Rata, G.; Filote, C.; Raboaca, M.S.; Graur, A.; Afanasov, C.; Felseghi, A.-R. The electricalvehicle simulator for charging station in mode 3 of IEC 61851-1 standard. *Energies* **2020**, *13*, 176. [CrossRef]

42. Helbing, D.; Hennecke, A.; Shvetsov, V.; Treiber, M. Micro-and macro-simulation of freeway traffic. *Math. Comput. Model.* **2002**, *35*, 517–547. [CrossRef]

43. Vandenberghe, W.; Moerman, I.; Demeester, P. Approximation of the IEEE 802.11 p standard using commercial off-the-shelf IEEE 802.11 a hardware. In Proceedings of the 11th International Conference on ITS Telecommunications IEEE, Graz, Austria, 15–17 June 2011; pp. 21–26.

44. Alexander, P.; Haley, D.; Grant, A. Cooperative intelligent transport systems: 5.9-GHz field trials. *Proc. IEEE* **2011**, *99*, 1213–1235. [CrossRef]

45. Purandare, R.; Kshirsagar, S.; Koli, S. Analysis of various parameters for link adaptation in wireless transmission. In *Innovations in Computer Science and Engineering*; Springer: Berlin/Heidelberg, Germany, 2016; pp. 9–19.

46. Bernado, L.; Zemen, T.; Tufvesson, F.; Molisch, A.F.; Mecklenbräuker, C.F. Delay and doppler spreads of nonstationary vehicular channels for safety-relevant scenarios. *IEEE Trans. Veh. Technol.* **2013**, *63*, 82–93. [CrossRef]

47. Raboaca, M.S.; Dumitrescu, C.; Manta, I. Aircraft trajectory tracking using radar equipment with fuzzy logic algorithm. *Mathematics* **2020**, *8*, 207. [CrossRef]

48. Shivaldova, V.; Paier, A.; Smely, D.; Mecklenbräuker, C.F. On roadside unit antenna measurements for vehicle-to-infrastructure communications. In Proceedings of the IEEE 23rd International Symposium on Personal, Indoor and Mobile Radio Communications-(PIMRC), Sydney, Australia, 9–12 September 2012; pp. 1295–1299.

49. Dhawankar, P.; Le Minh, H.; Aslam, N.; Raza, M. Communication infrastructure and data requirements for autonomous transportation. In Proceedings of the 2nd International Workshop on Sustainability and Green Technologies, Da Nang, Vietnam, 6–8 March 2017.

mathematics

Article

Optimal Parameter Estimation Methodology of Solid Oxide Fuel Cell Using Modern Optimization

Hesham Alhumade [1,2,*], Ahmed Fathy [3,4], Abdulrahim Al-Zahrani [1], Muhyaddin Jamal Rawa [2,5] and Hegazy Rezk [6,7]

1   Department of Chemical and Materials Engineering, Faculty of Engineering, King Abdulaziz University, Jeddah 21589, Saudi Arabia; azahrani@kau.edu.sa
2   Center of Research Excellence in Renewable Energy and Power systems, King Abdulaziz University, Jeddah 21589, Saudi Arabia; mrawa@kau.edu.sa
3   Electrical Engineering Department, Faculty of Engineering, Jouf University, Sakaka 72388, Saudi Arabia; afali@ju.edu.sa
4   Electrical Power and Machine Department, Faculty of Engineering, Zagazig University, Zagazig 44519, Egypt
5   Department of Electrical and Computer Engineering, Faculty of Engineering, King Abdulaziz University, Jeddah 21589, Saudi Arabia
6   College of Engineering at Wadi Addawaser, Prince Sattam Bin Abdulaziz University, Al-Kharj 11911, Saudi Arabia; hr.hussien@psau.edu.sa
7   Electrical Engineering Department, Faculty of Engineering, Minia University, Minia 61517, Egypt
*   Correspondence: halhumade@kau.edu.sa

Citation: Alhumade, H.; Fathy, A.; Al-Zahrani, A.; Rawa, M.J.; Rezk, H. Optimal Parameter Estimation Methodology of Solid Oxide Fuel Cell Using Modern Optimization. *Mathematics* **2021**, *9*, 1066. https://doi.org/10.3390/math9091066

Academic Editor: Nicu Bizon

Received: 27 March 2021
Accepted: 4 May 2021
Published: 10 May 2021

Abstract: An optimal parameter estimation methodology of solid oxide fuel cell (SOFC) using modern optimization is proposed in this paper. An equilibrium optimizer (EO) has been used to identify the unidentified parameters of the SOFC equivalent circuit with the assistance of experimental results. This is presented via formulating the modeling process as an optimization problem considering the sum mean squared error (SMSE) between the observed and computed voltages as the target. Two modes of the SOFC-based model are investigated under variable operating conditions, namely, the steady-state and the dynamic-state based models. The proposed EO results are compared to those obtained via the Archimedes optimization algorithm (AOA), Heap-based optimizer (HBO), Seagull Optimization Algorithm (SOA), Student Psychology Based Optimization Algorithm (SPBO), Marine predator algorithm (MPA), Manta ray foraging optimization (MRFO), and comprehensive learning dynamic multi-swarm marine predators algorithm. The minimum fitness function at the steady-state model is obtained via the proposed EO with value of 1.5527×10^{-6} at 1173 K. In the dynamic based model, the minimum SMSE is 1.0406. The obtained results confirmed the reliability and superiority of the proposed EO in constructing a reliable model of SOFC.

Keywords: solid oxide fuel cell; parameter identification; optimization

1. Introduction

There is a growing demand for energy to meet the requirements of continuous industrial development and modern civilization. In parallel, there is a growing concern about the depletion of traditional energy sources such as fossil fuel and drawbacks of continuous consumption of fossil fuel such as climate change [1]. Indeed, a recent study expected that future energy demands might exceed the limits of current energy systems [2]. Moreover, the increasing global energy demands and consumption of fossil fuel will escalate the emissions of greenhouse gases and other toxic air pollutants. Therefore, alternative sources of energy such as renewable energy have earned significant attention in the recent decades. In particular, fuel cell is among the power generation systems that can deliver environmentally friendly quality energy with great energy conversion efficiency. Furthermore, fuel cell has a great potential in power delivery for stationary and movable applications compared to other storage technologies [3–7]. Other remarkable features of fuel cell over other energy alternatives include lower fuel oxidation temperature and reduced emissions [8]. Solid oxide fuel cell (SOFC) and polymer composites-based electrolyte fuel cell represent the most attractive types of fuel cell for a wide range of applications. A growing effort is made to deliver a model that can predict the performance of fuel cell over steady state or dynamics operating environmental conditions [9–12]. In fact, appropriate model identification requires feeding accurate input parameters to the governing equations that encompass physical and chemical properties of the cell, where the modelling methodologies can be empirical, semi-empirical, or theoretical [13–16].

The parameter extraction of the fuel cell model plays an important role in the simulation, evaluation, control, and optimization of a fuel cell system. The voltage drops in SOFC are mainly reliant on the parameters associated with the chemical processes inside SOFC [17,18]. Several methods were used to identify the accurate parameters of SOFC. Among these methods, the metaheuristic optimization-based methodologies were superior in resolving the SOFC parameter estimation problem due to their reliability, robustness, and simplicity. Shi et al. [19] proposed a strategy, Converged Grass Fibrous Root, to determine the best parameters of the SOFC model. Both temperature and pressure variation are considered. During the optimization process, seven parameters are assigned to be decision variables: the standard potential, the current limitation density, the Tafel line slope, a constant depends on the operating state of SOFC, the area-specific resistance, the anode exchange current density, and the cathode exchange current density. El-Hay et al. [10] suggested a methodology based on an interior search algorithm to estimate the steady state and transient parameters of SOFC. A proportional-integral controller is integrated with the dynamic model to enhance its performance throughout transient disturbances. A similar study also carried by the same authors based on Satin Bowerbird Optimizer was conducted [9]. During the optimization process, the decision variables are represented by the unknown parameters of SOFC, whereas the cost function is represented by mean squared deviations between experimental data and estimated SOFC voltages. In the same direction, Yousri et al. [20] proposed a modified algorithm called comprehensive learning dynamic multi-swarm marine predators to determine both static and dynamic parameters of SOFC. During the optimization process, the mean squared error between the experimental data and estimated SOFC voltage is used as the objective function that is required to be minimum. Nassef et al. [21] used the radial movement optimization algorithm (RMOA) to determine the best parameters of the SOFC model. The model of SOFC was created using a neural network. During the optimization process, four parameters, including electrolyte thickness, cathode interlayer thickness, anode porosity, and anode support layer thickness, are used as decision variables; in contrast, the objective function is represented by the power density of SOFC. By using the RMOA, the power density was increased by 17.28% compared with the genetic algorithm. In the same direction, Fathy et al. [22] suggested a methodology based on the moth-flame optimization algorithm (MFOA). The power density of SOFC using the MFOA was improved. It was increased by 18.92% and 5.56% compared to the genetic algorithm and RMOA, respectively.

The contribution of the current research work can be summarized as follows:

- A novel approach based on Equilibrium Optimizer (EO) is suggested to determine the optimal parameters of the SOFC-based model.
- The suggested methodology is validated through both steady-state and dynamic-state models of SOFC with the changing of the operational conditions.
- A comprehensive comparison with previous works and other programs of the Archimedes optimization algorithm (AOA), Heap-based optimizer (HBO), Seagull Optimization Algorithm (SOA), Student Psychology Based Optimization Algorithm (SPBO), Marine predator algorithm (MPA), and Manta ray foraging optimization (MRFO).
- The superiority and reliability of the suggested EO-based strategy in solving the SOFC parameter determination problem is verified.

The rest of the paper is organized as follows: The mathematical model of SOFC is illustrated in Section 2. Section 3 presents an overview about main aspects of the equilibrium optimizer. Then, the suggested optimization problem and solution methodology are explained in Section 4. Section 5 presents a detailed discussion of the obtained results and a comparative study with other methods. Finally, the main findings and the future work are outlined in Section 6.

2. SOFC Mathematical Model

In this section, the authors illustrate the static- and dynamic-based models of SOFC.

2.1. Steady-State Model

The output voltage of SOFC can be estimated using the following relation considering the activation loss, ohmic loss, and concentration loss [23]:

$$V_{cell} = E_n - V_a - V_o - V_c \tag{1}$$

where E_n denotes the reversible voltage of the cell, and V_a, V_o, and V_c denote the activation, ohm, and concentration voltage drops, respectively.

The activation loss, concentration loss and ohmic loss can be estimated using the following relations:

$$V_a = a \cdot \ln\left(\frac{J}{2J_0}\right) \tag{2}$$

$$V_o = \sum_k i \times R_k \tag{3}$$

$$V_c = -b \cdot \ln\left(1 - \frac{J}{J_{max}}\right) \tag{4}$$

where a and b are constants; J denotes the current density; J_0 denotes the exchange current density; J_{max} is the maximum current density; R_k is the sum of ionic (electrolyte) and electronic resistances; and i denotes the output SOFC current.

To increase the rating voltage of SOFC, the number of cells are connected in series. Therefore, the total stack output voltage can be estimated using the following relation.

$$V_s = n_c \times V_{cell} = n_c \times (E_n - V_a - V_o - V_c) \tag{5}$$

where vs. is the stack voltage and n_c is the number of cells. The reversible voltage can be written as follows [24]:

$$E_n = E_0 + \frac{RT}{2F} \ln\left(\frac{P_{H_2}\sqrt{P_{O_2}}}{P_{H_2O}}\right) \tag{6}$$

where E_0 denotes the reference voltage at unit activity and atmospheric pressure; T denotes operating temperature (K); P_{H_2}, P_{O_2}, and P_{H_2O} denote the hydrogen, oxygen, and water partial pressures, respectively; R is the universal gas constant with a value of 8.314 kJ $(\text{kmol K})^{-1}$; and F is the Faraday constant.

2.2. SOFC Dynamic Model

The gas molar flow in SOFC is reliant on hydrogen and oxygen partial pressures as follows [24]:

$$\frac{q_{H_2}}{P_{H_2}} = \frac{k_{an}}{\sqrt{M_{H_2}}} = K_{H_2} \tag{7}$$

$$\frac{q_{O_2}}{P_{O_2}} = \frac{k_{an}}{\sqrt{M_{O_2}}} = K_{O_2} \tag{8}$$

where q_{H_2} is the molar flow of hydrogen; q_{O_2} is the molar flow of oxygen; K_{H_2} is the hydrogen molar constant; K_{O_2} is the oxygen molar constant; k_{an} is the anode valve constant; M_{H_2} is the molar masses of hydrogen; and M_{O_2} is the molar masses of oxygen.

The partial pressure derivative can be estimated using the following relation:

$$\frac{dP_{H_2}}{dt} = \frac{RT}{V_{an}}\left(q_{H_2}{}^{in} - q_{H_2}{}^{out} - q_{H_2}{}^{r}\right) \tag{9}$$

where V_{an} denotes the anode volume; $q_{H_2}{}^{in}$ is the hydrogen input flow rate; $q_{H_2}{}^{out}$ is the hydrogen output flow rate; and $q_{H_2}{}^{r}$ denotes the reacted hydrogen flow rate.

The reacted hydrogen flow rate can be estimated based on the following relation:

$$q_{H_2}{}^{r} = \frac{n_c i}{2F} = 2K_r i \tag{10}$$

where K_r is constant.

Considering the above equations and the Laplace transform, hydrogen and oxygen partial pressures can be formulated as follows.

$$P_{H_2}(s) = \frac{1/K_{H_2}}{1 + \tau_{H_2}}\left(q_{H_2}{}^{in} - 2K_r i\right) \tag{11}$$

$$P_{H_2O}(s) = \frac{1/K_{H_2O}}{1 + \tau_{H_2O}}(2K_r i) \tag{12}$$

$$P_{O_2}(s) = \frac{1/K_{O_2}}{1 + \tau_{O_2}}\left(q_{O_2}{}^{in} - K_r i\right) \tag{13}$$

where τ_{H_2}, τ_{H_2O}, and τ_{O_2} denote the flow time constants of hydrogen, water, and oxygen, respectively.

Ultimately, the dynamic model of SOFC voltage is given by the following relation.

$$V_s = n_c\left(E_0 + \frac{RT}{2F}\left(\ln\frac{P_{H_2}\sqrt{P_{O_2}}}{P_{H_2}}\right)\right) - \left(a\cdot\ln\left(\frac{J}{2J_0}\right) + r\times i - b\cdot\ln\left(1 - \frac{J}{J_{max}}\right)\right) \tag{14}$$

3. Overview of Equilibrium Optimizer

Equilibrium optimizer (EO) is a recent algorithm that was proposed by Faramarzi et al. in 2020 [25]. The core idea of the EO is extracted from the control volume mass balance models. During the optimization process of the EO, the particles and positions are assigned to the solutions and concentrations, respectively. The details explanations about the inspiration, mathematical model, and algorithm of the EO can be found in [25]. The mass balance formula is expressed as follows.

$$V\frac{dC}{dt} = QC_{eq} - QC - G \tag{15}$$

where C denotes the concentration of the control volume; $V\frac{dC}{dt}$ denotes the changing rate of the mass; Q denotes the flow rate; C_{eq} denotes the concentration at the balance state; and G denotes the mass generation rate.

By integration over time and rearranged with the above formula, the following relation can be used to express the concentration of the control volume.

$$C = C_{eq} + (C_0 - C_{eq})F + \frac{G}{\lambda V}(1 - f) \tag{16}$$

where λ denotes the turnover rate ($\lambda = Q/V$) and $f = e^{-\lambda(t - t_0)}$, and t_0 and C_0 denote the initial time and concentration, respectively.

The time t is decreasing with increasing the number of iterations as follows:

$$t = (1 - I/z)^{\left(a_2 \frac{I}{z}\right)} \tag{17}$$

$$\vec{t_0} = \frac{1}{\vec{\lambda}} \ln(-a_1 sign(\vec{r} - 0.5)[1 - e^{\vec{\lambda}t}]) + t \tag{18}$$

$$\vec{F} = a_1 sign(\vec{r} - 0.5)[1 - e^{\vec{\lambda}t}] \tag{19}$$

where I and z are the current iteration and maximum number of iterations, respectively. a_1 and a_2 are constants. $\vec{r}$ is a random vector in range $[0, 1]$.

Considering Equation (16), there are three sections describing the updating process for particles. The first section represents the equilibrium concentration. It represents the optimum solutions arbitrarily chosen from a pool. The second section is related to the concentration variations between a particle and the equilibrium state. The last section is related to the generation rate. It is mainly performing the role of an exploiter. The equilibrium state is the final convergence state of the EO optimization process. The equilibrium pool can be represented as follows.

$$\vec{C}_{eq.pool} = \left\{ \vec{C}_{eq(1)}, \vec{C}_{eq(2)}, \vec{C}_{eq(3)}, \vec{C}_{eq(4)}, \vec{C}_{eq(ave)} \right\} \tag{20}$$

During the first iteration, the particle modifies the concentration using $\vec{C}_{eq(1)}$ whereas it is uses $\vec{C}_{eq(ave)}$ with other iterations. The generation rate is defined as follows.

$$\vec{G} = \vec{G_0} e^{-\vec{k}(t-t_0)} \tag{21}$$

where G_0 is the initial value, and k denotes a decay constant ($k = \lambda$).

$$\vec{G} = \vec{G_0} e^{-\vec{\lambda}(t-t_0)} = \vec{G_0}\vec{F} \tag{22}$$

$$\vec{G_0} = G\vec{C}P(\vec{C}_{eq} - \lambda\vec{C}) \tag{23}$$

$$G\vec{C}P = \begin{cases} 0.5r_1 & r_2 \geq GP \\ 0 & r_2 < GP \end{cases} \tag{24}$$

where r_1 and r_2 are random variables in range $[0, 1]$, and GCP is a parameter that controls the generation rate.

The addition of memory saving helps each particle to save its coordinates in the search space. Moreover, it informs its fitness function value. The fitness function related to a particular particle in the ongoing iteration is compared with the previous one; then, the updating process is placed if it reaches better fit. This action enhances the exploitation phase. The optimization process of EO is illustrated in Figure 1.

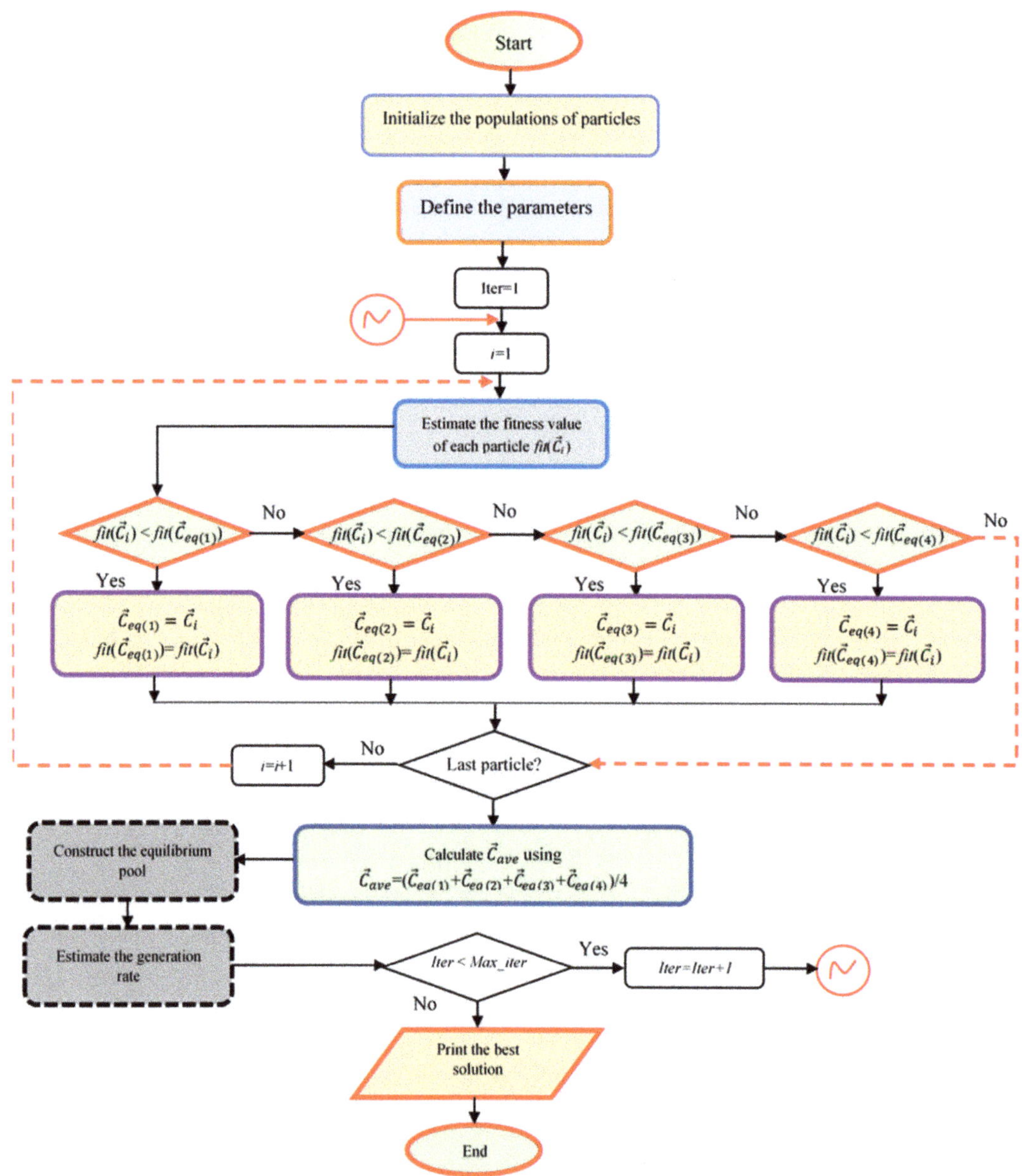

Figure 1. Optimization process of EO.

4. The Proposed Methodology

In this section, the proposed methodology is explained via formulating the problem of SOFC parameter estimation as an optimization problem, and an objective function and corresponding constraints are also introduced. Moreover, the proposed solution methodology is also presented.

4.1. The Proposed Objective Function

In this section, the formulation of the identification process of SOFC circuit's parameters as an optimization problem is explained via specifying the fitness function, the corresponding constraints, and the proposed methodology of the solution incorporated by the EO. The main target of the work is to construct a reliable equivalent circuit of SOFC by identifying six parameters, E_0, a, J_o, r, b, and J_{max}; this is achieved with the aid of experimental data of the current and voltage of FC. The fitness function represented in this work is the sum mean squared error between the measured and calculated terminal voltages of SOFC, and this can be described as follows:

$$Minimize \quad SMSE = \sum_{k=1}^{N} \frac{1}{N} \left(V_{meas,k} - V_{cal,k} \right)^2 \tag{25}$$

where $V_{meas,k}$ and $V_{cal,k}$ are the kth observed and computed voltages, respectively, and N is the number of measured datasets. The constraints related to the variables to be designed can be described as follows:

$$\begin{aligned}
E_0^{min} &\leq E_0 \leq E_0^{max} \\
a^{min} &\leq a \leq a^{max} \\
J_o^{min} &\leq J_o \leq J_o^{max} \\
r^{min} &\leq r \leq r^{max} \\
b^{min} &\leq b \leq b^{max} \\
J_{max}^{min} &\leq x_6 \leq J_{max}^{max}
\end{aligned} \tag{26}$$

where min denotes the minimum limit and max denotes the maximum limit. In this work the authors only considered the disturbance on the SOFC output voltage. However, in future work, they will consider the disturbance not only on the SOFC output but also on the SOFC input such as those given in [26–29].

4.2. The Proposed EO Based Methodology

The equilibrium optimizer is selected due to many advantages: It is simple in implementation; it achieves balance between the exploration and exploitation phases; and there is diversity between the population individuals. These features enable the algorithm to be applicable for many optimization problems. Six parameters are required to be identified such that the SMSE is minimized. The proposed methodology incorporating the EO begins by defining the specifications of SOFC and the recorded measured data of the terminal voltage. Then, an initial population with a dimension of $n_{pop} \times dim$, n_{pop} is the population size, and dim is the problem dimension, which is constructed with the aid of the minimum and maximum limits defined by the user. The initial corresponding fitness function (SMSE) is calculated, and the iterative process is implemented by calculating the fitness function of each particle. The obtained fitness function is checked with those of the equilibrium pool to decide the updating action of each particle. After that, the condition of the last particle is investigated, and the average of the equilibrium pool is calculated, which helps in estimating the generation rate. The process is continued until the constraint of maximum iteration is achieved. At this moment, the optimal results are obtained and can be printed. The proposed methodology incorporating EO is shown in Figure 2.

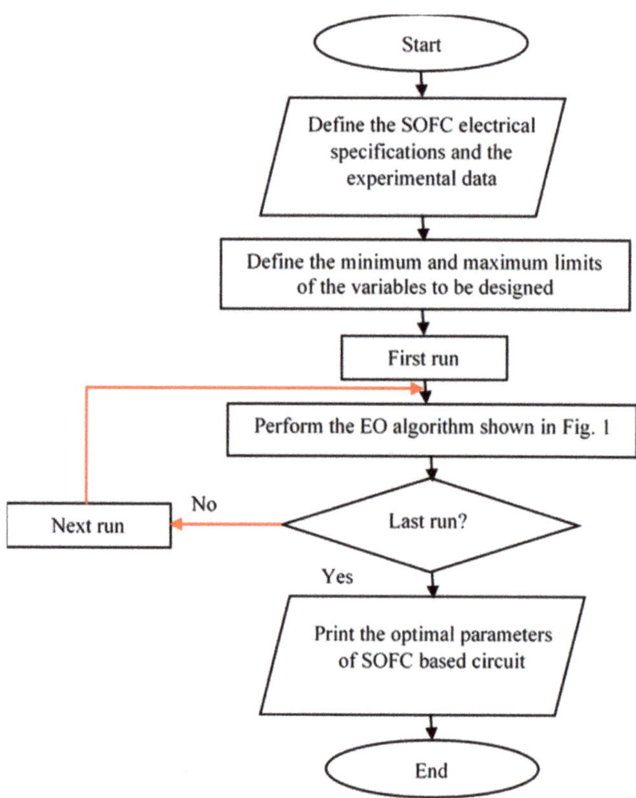

Figure 2. The proposed steps incorporating EO.

5. Numerical Analysis

The analysis is performed on two modes of the SOFC operation which are steady-state and dynamic-state. Both of them are investigated under variable-operating conditions. The commercial SOFC, which is manufactured by Siemens [30], is employed in steady-state analysis. In such a case, four measured datasets are recorded at temperatures of 1073, 1173, 1213, and 1273 K, where the proposed EO size of population is assigned as 50, and the number of iterations is selected as 100. The population-based approach presented in this work has some difficulty, such as getting premature and local optima, and the authors take into consideration this problem by performing the approach with 50 independent runs, and the best one is selected as global optima. This action minimizes the problem of falling in local optima. Other metaheuristic approaches are implemented and compared to the proposed EO; these algorithms are the Archimedes optimization algorithm (AOA), Heap-based optimizer (HBO), Seagull Optimization Algorithm (SOA), Student Psychology Based Optimization Algorithm (SPBO), Marine predator algorithm (MPA), Manta ray foraging optimization (MRFO), and comprehensive learning dynamic multi-swarm marine predators algorithm CLDMMPA [20]. Table 1 shows the obtained optimal parameters of SOFC operated at 1073 K via the proposed EO and the others. The proposed approach succeeded in achieving a fitness function of 2.6906×10^{-6} which is the same obtained via CLDMMPA. However, the CLDMMPA is complex in construction; moreover, the proposed EO consumes only 272.198102 s, which is the best compared to the others. The measured and calculated polarization curves obtained via the proposed EO are shown in Figure 3. Both curves are closely converged. Moreover, Figure 4 shows the estimated polarization curves obtained via the other approaches and the measured ones. Furthermore, the performance of each optimizer during the iterative process is shown in Figure 5. It is clear that the EO performance is the best compared to the others.

Table 1. The optimal parameters of SOFC steady-state based model operated at 1073 K.

	CLDMMPA [20]	MPA	HBO	SOA	MRFO	The Proposed EO
E_0 (V)	0.90754	0.9127	0.91214	0.91101	0.91827	0.91056
a (V)	0.010741	0.011058	0.010929	0.020502	0.0116	0.010724
J_o (A/cm^2)	0.098627	0.059994	0.063321	0.048127	0.035918	0.074522
r (k$\Omega\cdot$cm^2)	1.0	1.0	1.0	1.0	0.99919	1.0
b (V)	0.044104	0.042784	0.043502	0.0	0.036886	0.044165
J_{max} (A/cm^2)	1.0	1.0	1.0	0.64297	0.9183	1.0
Elapsed time (sec.)	NA	468.526	403.607	278.254	560.753	272.198102
SMSE	2.6906×10^{-6}	2.692×10^{-6}	2.7003×10^{-6}	4.123×10^{-6}	2.7213×10^{-6}	2.6906×10^{-6}

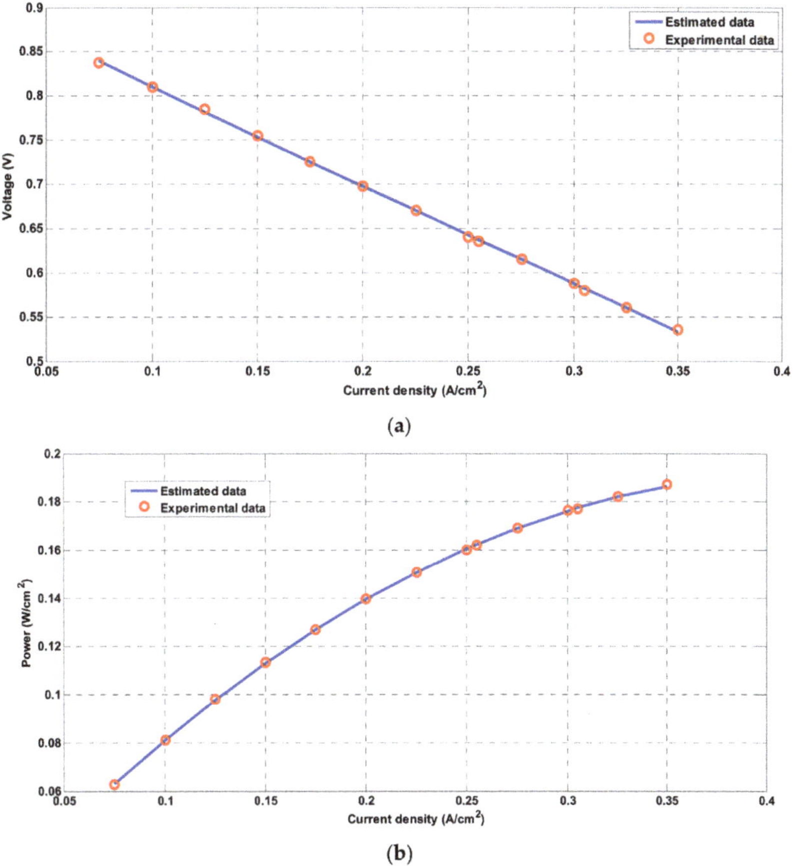

Figure 3. The measured and calculated polarization curves of SOFC operated at 1073 K obtained via EO at (**a**) current density-voltage, (**b**) current density-power.

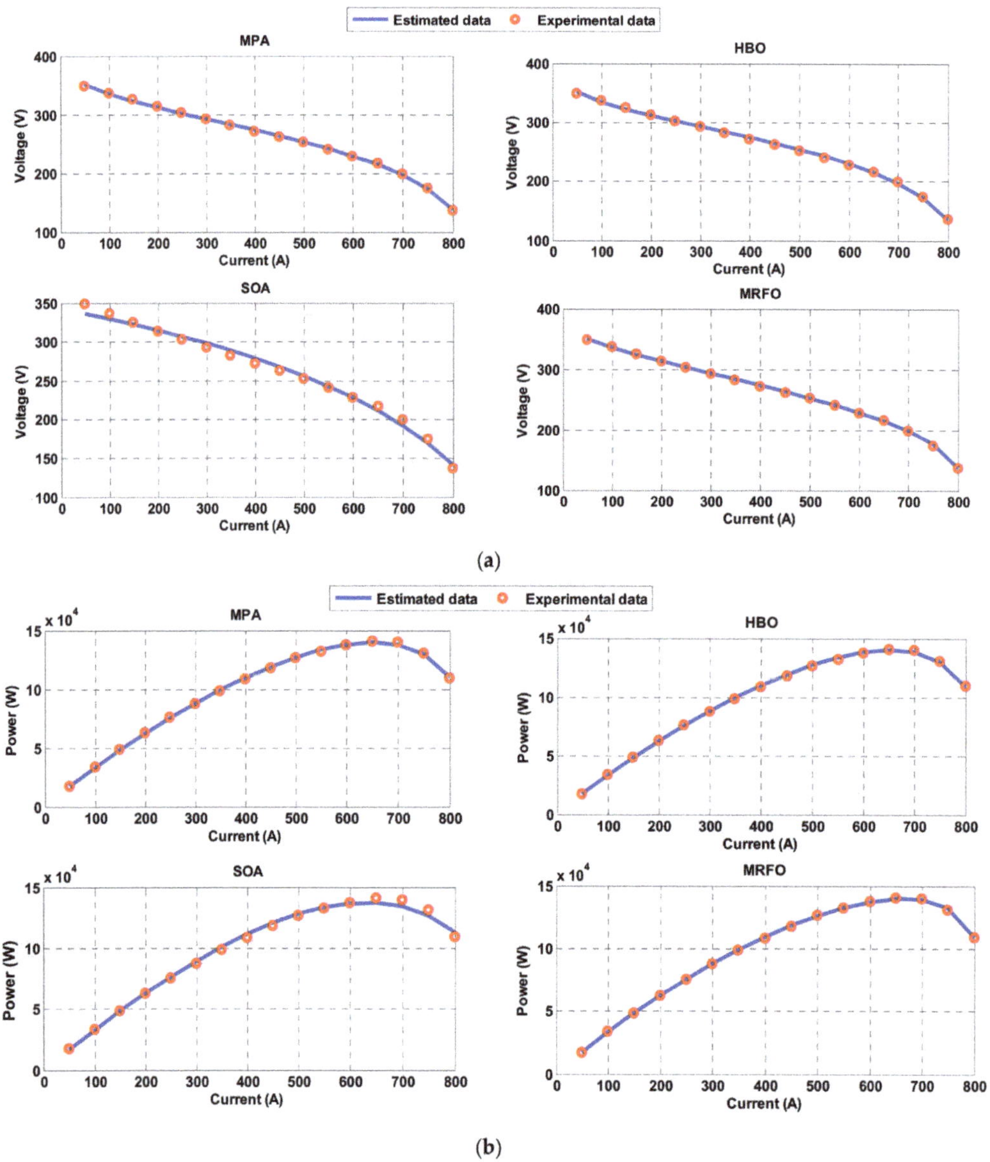

Figure 4. (**a**) Current-voltage curve, (**b**) Current-power curve of SOFC operated at 1073 K obtained via other approaches.

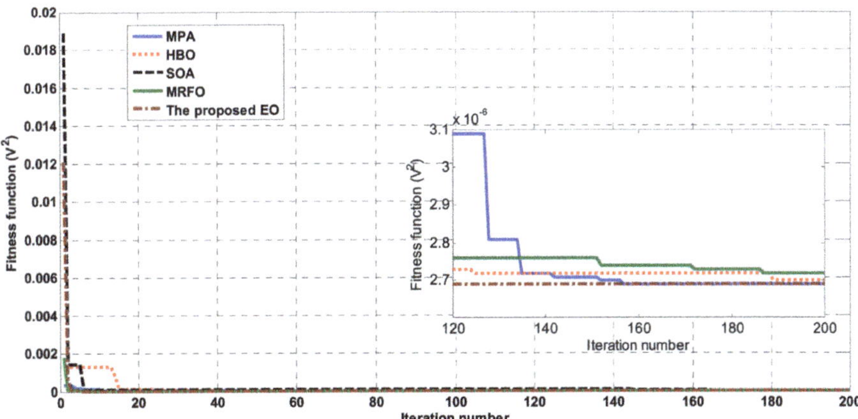

Figure 5. The variation of fitness function during iterative process for all employed optimizers applied for steady-state SOFC model.

The optimal parameters of the SOFC steady-state based model at 1173, 1213, and 1273 K obtained via the proposed EO and the others are tabulated in Tables 2–4. Regarding the obtained results at 1173 K, the best fitness function is 1.5527×10^{-6} obtained via the proposed EO while CLDMMPA comes in the second rank with a SMSE of 1.5529×10^{-6}. On the other hand, the worst approach is SOA with a fitness function of 3.1657×10^{-6}. Moreover, during the operation at 1213 K, the EO outperformed the others in terms of elapsed time and fitness function. The reader can see that during operation at 1273 K, the proposed EO achieved a SMSE of 2.2995×10^{-6}, which is the best compared to the others.

Table 2. The optimal parameters of SOFC steady-state based model operated at 1173 K.

	CLDMMPA [20]	MPA	HBO	SOA	MRFO	The Proposed EO
E_0 (V)	0.89103	0.89129	0.89083	0.87956	0.89087	0.89108
a (V)	3.671×10^{-13}	6.1748×10^{-13}	3.23334×10^{-13}	0.0077567	9.8137×10^{-6}	3.14568×10^{-8}
J_o (A/cm^2)	0.095127	0.018358	0.041194	0.087363	0.068684	0.09998
r (kΩ·cm^2)	0.40473	0.41593	0.39952	0.21139	0.4027	0.40610
b (V)	0.18841	0.17741	0.19212	0.28497	0.18885	0.18741
J_{max} (A/cm^2)	1.0	0.98862	1.0	0.9465	0.9968	0.99999
Elapsed time (sec.)	NA	495.653992	396.181972	303.170716	612.841366	303.185040
SMSE	1.5529×10^{-6}	1.5594×10^{-6}	1.5557×10^{-6}	3.1657×10^{-6}	1.5563×10^{-6}	1.5527×10^{-6}

Table 3. The optimal parameters of SOFC steady-state based model operated at 1213 K.

	CLDMMPA [20]	MPA	HBO	SOA	MRFO	The Proposed EO
E_0 (V)	0.86169	0.86189	0.86134	0.85622	0.86176	0.86164
a (V)	5.0588×10^{-13}	4.2805×10^{-28}	3.4456×10^{-30}	3.1223×10^{-29}	3.3451×10^{-5}	7.12575×10^{-8}
J_o (A/cm^2)	0.08871	0.054568	0.011975	0.047885	0.063468	0.07768
r (kΩ·cm^2)	0.15982	0.16633	0.14858	0.000124	0.16629	0.15873
b (V)	0.28529	0.28032	0.29351	0.4012	0.27918	0.28603
J_{max} (A/cm^2)	1.0	0.99946	1.0	1.0	0.99732	0.99999
Elapsed time (sec.)	NA	464.968854	367.890819	236.154633	571.548052	273.530
SMSE	2.6811×10^{-6}	2.6846×10^{-6}	2.6887×10^{-6}	4.3804×10^{-6}	2.6896×10^{-6}	2.6809×10^{-6}

Table 4. The optimal parameters of SOFC steady-state based model operated at 1273 K.

	CLDMMPA [20]	MPA	HBO	SOA	MRFO	The Proposed EO
E_0 (V)	0.8478	0.84782	0.84802	0.84014	0.84802	0.8478
a (V)	2.887×10^{-14}	3.128×10^{-21}	5.1251×10^{-6}	2.238×10^{-20}	3.2243×10^{-5}	6.7797×10^{-12}
J_o (A/cm^2)	0.061816	0.014523	0.086283	0.017285	0.024437	0.0160
r (k$\Omega{\cdot}$cm^2)	0.21564	0.21634	0.22323	0.1563	0.22021	0.2169
b (V)	0.20575	0.20524	0.20009	0.36046	0.20226	0.2047
J_{max} (A/cm^2)	1.0	1.0	0.99984	1.00	0.99888	1.0
Elapsed time (sec.)	NA	465.757529	352.634515	199.242260	555.636660	317.980527
SMSE	2.2997×10^{-6}	2.2996×10^{-6}	2.3031×10^{-6}	5.4888×10^{-6}	2.3058×10^{-6}	2.2995×10^{-6}

The polarization curves of the measured data and calculated data obtained via the proposed EO for the steady-state SOFC based model at 1173, 1213, and 1273 K are shown in Figure 6. The curves confirm the matching between the experimental and calculated data.

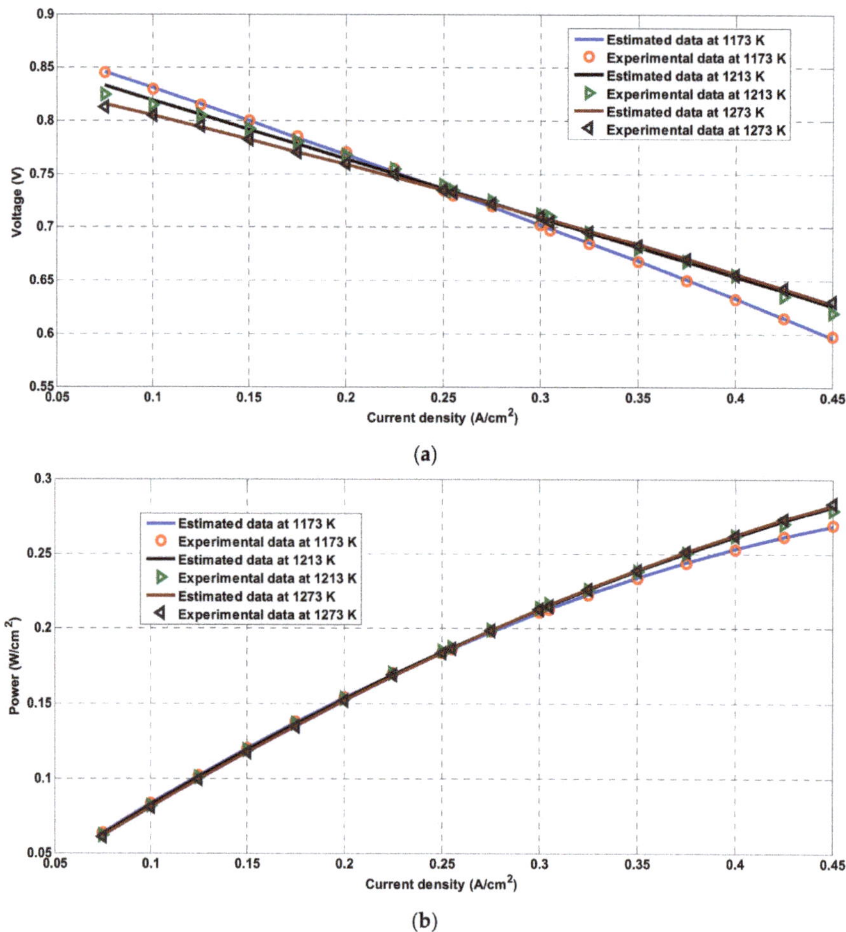

Figure 6. The measured and calculated (**a**) current density-voltage, (**b**) current density-power of SOFC operated at 1173 K, 1213 K, and 1273 K obtained via the proposed EO.

Mathematics **2021**, 9, 1066

It is important to investigate the performance of each optimizer via calculating the statistical parameters, which include the best, worst, mean, median, variance, and standard deviation after 50 independent runs. These data are calculated and tabulated in Table 5. The proposed EO gives acceptable statistical parameters compared to the others.

Table 5. Statistical parameters (best, worst, mean, median, variance, and standard deviation) of all optimizers used for steady-state SOFC model.

	At T = 1073 K					
	Cldmmpa [20]	MPA	HBO	SOA	MRFO	The Proposed EO
Best	2.6906×10^{-6}	2.69203×10^{-6}	2.70035×10^{-6}	4.12297×10^{-6}	2.7213×10^{-6}	2.6906×10^{-6}
Worst	2.696×10^{-6}	1.23064×10^{-5}	4.49932×10^{-6}	0.00034	4.8504×10^{-6}	6.8151×10^{-6}
Mean	2.6926×10^{-6}	4.75170×10^{-6}	3.11043×10^{-6}	9.38355×10^{-5}	3.1121×10^{-6}	3.3645×10^{-6}
Median	2.6917×10^{-6}	3.61749×10^{-6}	2.96160×10^{-6}	9.33996×10^{-5}	2.9395×10^{-6}	2.6906×10^{-6}
Variance	4.1347×10^{-18}	6.38059×10^{-12}	1.75759×10^{-13}	9.20085×10^{-9}	2.2231×10^{-13}	8.38802×10^{-13}
Std. deviation	2.0334×10^{-9}	2.52598×10^{-6}	4.19236×10^{-7}	9.59210×10^{-5}	4.7150×10^{-7}	9.15673×10^{-7}
	At T = 1173 K					
	CLDMMPA [20]	MPA	HBO	SOA	MRFO	The proposed EO
Best	1.5529×10^{-6}	1.5594×10^{-6}	1.5556×10^{-6}	3.1656×10^{-6}	1.5563×10^{-6}	1.55279×10^{-6}
Worst	1.5752×10^{-6}	5.9028×10^{-6}	2.1420×10^{-6}	0.01102	2.6603×10^{-6}	6.00407×10^{-6}
Mean	1.5593×10^{-6}	3.0125×10^{-6}	1.6979×10^{-6}	0.00026	1.7105×10^{-6}	2.31364×10^{-6}
Median	1.5572×10^{-6}	2.4861×10^{-6}	1.6532×10^{-6}	1.26093×10^{-5}	1.6493×10^{-6}	1.58630×10^{-6}
Variance	3.8987×10^{-17}	1.9116×10^{-12}	1.8382×10^{-14}	2.42265×10^{-6}	3.7587×10^{-14}	2.66121×10^{-12}
Std. deviation	6.244×10^{-9}	1.3826×10^{-6}	1.3558×10^{-7}	0.00155	1.9387×10^{-7}	1.63132×10^{-6}
	At T = 1213 K					
	CLDMMPA [20]	MPA	HBO	SOA	MRFO	The proposed EO
Best	2.6811×10^{-6}	2.6846×10^{-6}	2.6887×10^{-6}	4.3804×10^{-6}	2.68961×10^{-6}	2.68099×10^{-6}
Worst	2.7269×10^{-6}	1.2325×10^{-5}	3.4295×10^{-6}	0.00032	3.22001×10^{-6}	1.30287×10^{-5}
Mean	2.6921×10^{-6}	4.6204×10^{-6}	2.8357×10^{-6}	2.37574×10^{-5}	2.85206×10^{-6}	4.02438×10^{-6}
Median	2.6867×10^{-6}	3.3119×10^{-6}	2.7703×10^{-6}	1.31251×10^{-5}	2.81486×10^{-6}	2.78147×10^{-6}
Variance	1.9891×10^{-16}	9.4614×10^{-12}	2.7256×10^{-14}	3.77968×10^{-9}	1.72574×10^{-14}	1.12951×10^{-11}
Std. deviation	1.4103×10^{-8}	3.0759×10^{-6}	1.6509×10^{-7}	6.14791×10^{-5}	1.3136×10^{-7}	3.36081×10^{-6}
	At T = 1273 K					
	CLDMMPA [20]	MPA	HBO	SOA	MRFO	The proposed EO
Best	2.2997×10^{-6}	2.2995×10^{-6}	2.30310×10^{-6}	5.4888×10^{-6}	2.30579×10^{-6}	2.2995×10^{-6}
Worst	2.3392×10^{-6}	7.4728×10^{-6}	3.0608×10^{-6}	0.000247	2.84248×10^{-6}	7.6115×10^{-6}
Mean	2.3156×10^{-6}	3.6859×10^{-6}	2.4184×10^{-6}	1.758508×10^{-5}	2.40213×10^{-6}	3.4890×10^{-6}
Median	2.3141×10^{-6}	3.1707×10^{-6}	2.3647×10^{-6}	7.71186×10^{-6}	2.37664×10^{-6}	2.3085×10^{-6}
Variance	1.3448×10^{-16}	2.1494×10^{-12}	1.9816×10^{-14}	2.25858×10^{-9}	8.64055×10^{-15}	4.8928×10^{-12}
Std. deviation	1.1597×10^{-8}	1.466×10^{-6}	1.4077×10^{-7}	4.7524×10^{-5}	9.29545×10^{-8}	2.2119×10^{-6}

The obtained results confirmed the superiority and reliability of the proposed methodology incorporating EO in identifying the optimal parameters of the SOFC steady-state based model.

It is important to confirm the availability of the presented approach in a dynamic/transient-based model of SOFC. Therefore, a 100 kW stack with specifications given in Table 6 is modeled in a dynamic-state model subjected to variable load disturbances. At the beginning, the proposed EO is applied to identify the optimal parameters of a 100 kW SOFC stack operated at 1273 K; the obtained parameters are tabulated in Table 7 in comparison to those obtained by the others. Regarding the obtained results, the proposed EO outperformed the others, achieving the minimum SMSE with a value of 1.0406. MRFO comes in the second rank with a fitness function of 1.0775, and then CLDMMPA achieves an SMSE of 1.3204 and comes in the third rank. Figure 7 shows the measured and calculated polarization curves obtained via the EO, and both curves are closely converged. However, Figure 8 shows the polarization curves obtained via MPA, HBO, SOA, and MRFO. The statistical parameters of all optimizers in such cases are calculated and tabulated in Table 8, where the best parameters are obtained by the proposed EO. Figure 9 shows the performance of each optimizer during implementing the iterative process. The performance of the proposed EO is confirmed to be better than the others.

Table 6. SOFC stack specifications [30,31].

Parameter	Value
P_{rated} (W)	100 kW
n_c	384
E_0 (V)	1.18
T (K)	1273
K_{H_2} (kmol/s/atm)	8.43×10^{-4}
K_{O_2} (kmol/s/atm)	2.52×10^{-3}
K_{H_2O} (kmol/s/atm)	2.81×10^{-4}
r_{H-O}	1.145
τ_{H_2} (s)	26.1
τ_{O_2} (s)	2.91
τ_{H_2O} (s)	78.3
T_e (s)	0.8

Table 7. The optimal parameters of SOFC dynamic-state based model operated at 1273 K.

	CLDMMPA [20]	MPA	HBO	SOA	MRFO	The Proposed EO
E_0 (V)	1.1405	1.199847	1.194939	0.89461271	1.113337	1.144398
a (V)	0.037449	0.04882335	0.04756176	0.000	0.02729001	0.02982692
J_o (A/cm^2)	0.095442	0.09987029	0.0939504	0.090655585	0.03143729	0.0202932
r (kΩ·cm^2)	0.0001829	4.3684×10^{-5}	4.873756×10^{-5}	0.000	0.0002672047	0.0002547476
b (V)	0.10386	0.1551425	0.1515045	0.32672895	0.08200661	0.08350225
J_{max} (A/cm^2)	0.8367	865.6602	859.6317	1000	825.4696	825.7578
Elapsed time (sec.)	NA	433.946266	331.304354	222.830499	593.725886	327.35802
SMSE	1.3204	3.0887	3.3486	34.1692	1.0775	1.0406

Table 8. Statistical parameters (best, worst, mean, median, variance, and standard deviation) of all optimizers used for dynamic-state SOFC model.

	CLDMMPA [20]	MPA	HBO	SOA	MRFO	The Proposed EO
Best	1.3204	3.08867	3.3486	34.1692	1.0775	1.04061
Worst	3.9835	25.1135	10.473	5401.8725	3.4164	2.00140
Mean	3.2462	8.54964	5.6209	1025.8099	1.7115	1.1352
Median	3.5241	6.07593	4.8835	34.3473	4.5141	1.0816
Variance	0.51027	36.3547	2.7528	3,915,661.957	0.23360	0.02264
Std. deviation	0.71434	6.02948	1.6592	1978.803	0.15284	0.15048

After identifying the parameters of the transient-state based model of SOFC, the model of fuel cell is implemented in Simulink/Matlab, and two load disturbances are applied to the model. The first disturbance is shown in Figure 10 (1st graph), the power is changed from 30 kW to 60 kW at a time of 300 sec., and given the identified parameters via the proposed EO and the stack output power plotted with the load disturbance, it is clear that they are closely matched, this means that the EO succeeded in extracting the correct parameters of the SOFC dynamic model. Moreover, the load current of the constructed model is closely converged to the disturbance current (3rd graph, Figure 10). Moreover, the terminal voltage (2nd graph) and the voltage drops occur inside the stack (4th graph), and they are shown in Figure 10. The constructed model via the proposed EO succeeded in tracking the changes in the load power. Moreover, a second disturbance is applied on the dynamic model in which the load power has two variations; as shown in Figure 11 (1st graph), the load power changes from 20 kW to 40 kW at 200 s and then changes again to 60 kW at 400 s on the same graph. The output power from the constructed model with identified parameters via the proposed EO is given, and both curves are converged. The terminal voltage, the corresponding current, and the voltage drops are shown in Figure 11 (2nd graph, 3rd graph, and 4th graph, respectively).

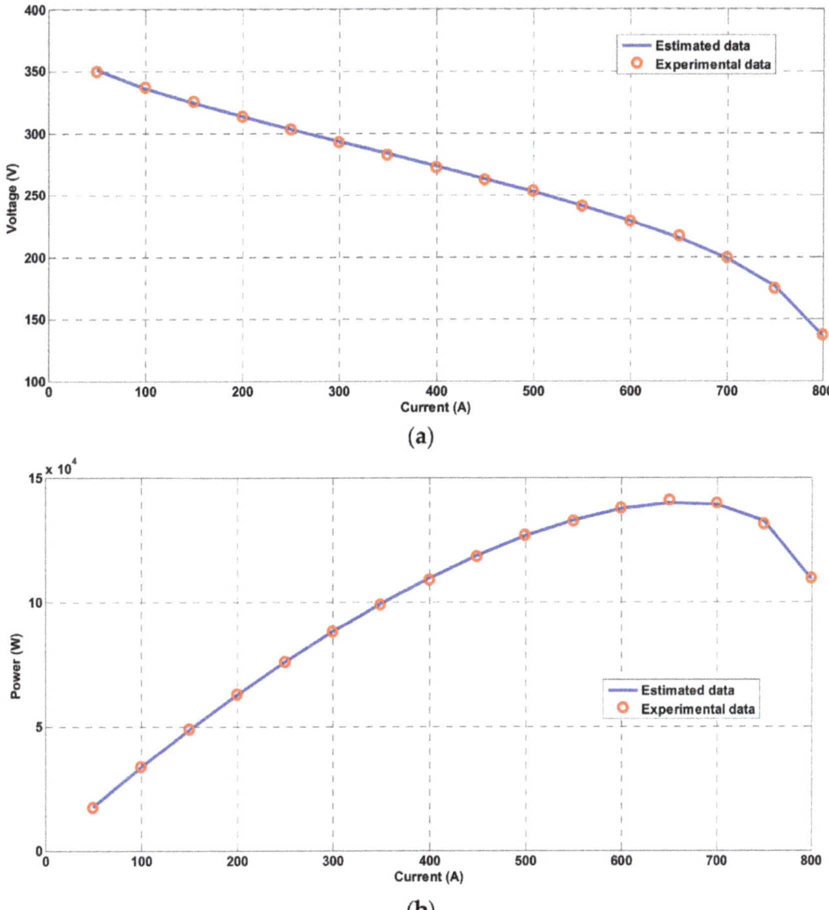

Figure 7. The measured and calculated polarization curves of SOFC dynamic-state model operated at 1273 K obtained via EO (**a**) current-voltage, (**b**) current-power.

Finally, it can be concluded that the proposed methodology incorporating the EO is reliable, superior, and efficient over other employed approaches in constructing a reliable model of the SOFC-based model operated under either steady-state or dynamic-state modes.

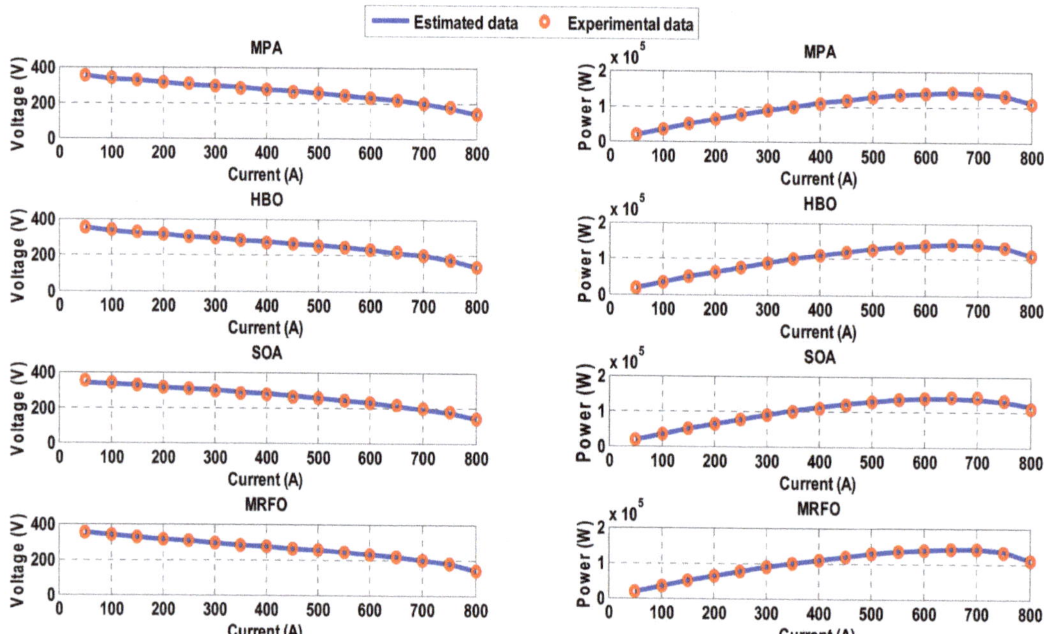

Figure 8. The measured and calculated polarization curves of SOFC dynamic-state model operated at 1273 K obtained via MPA, HBO, SOA, and MRFO.

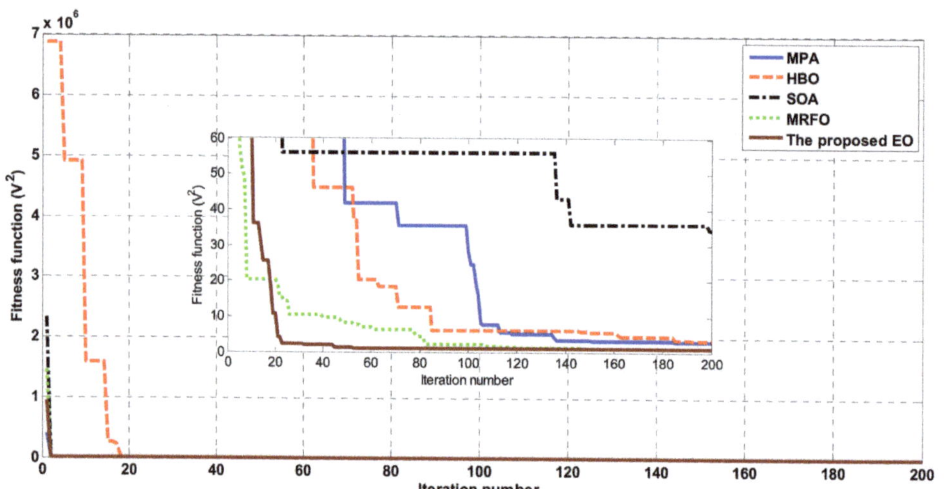

Figure 9. The variation of fitness function during iterative process for all employed optimizers applied for dynamic-state SOFC model.

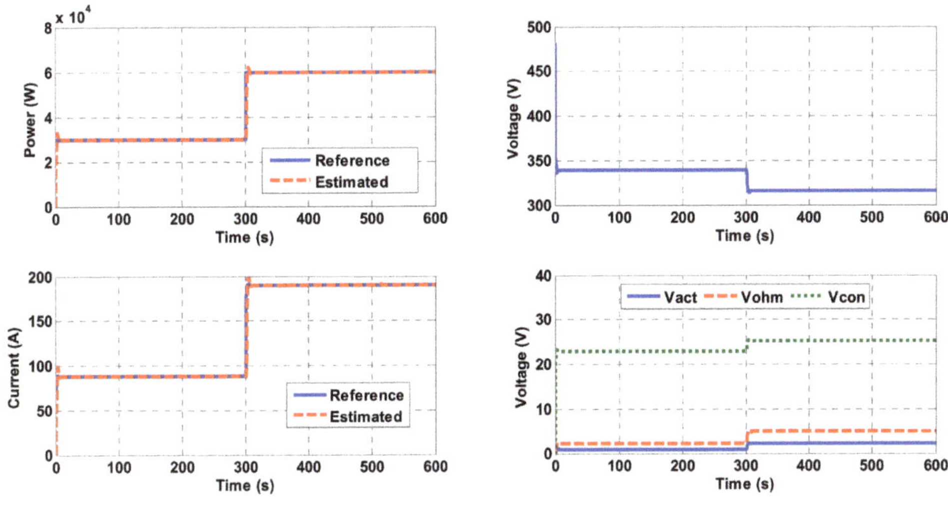

Figure 10. First load disturbance applied on SOFC stack dynamic model.

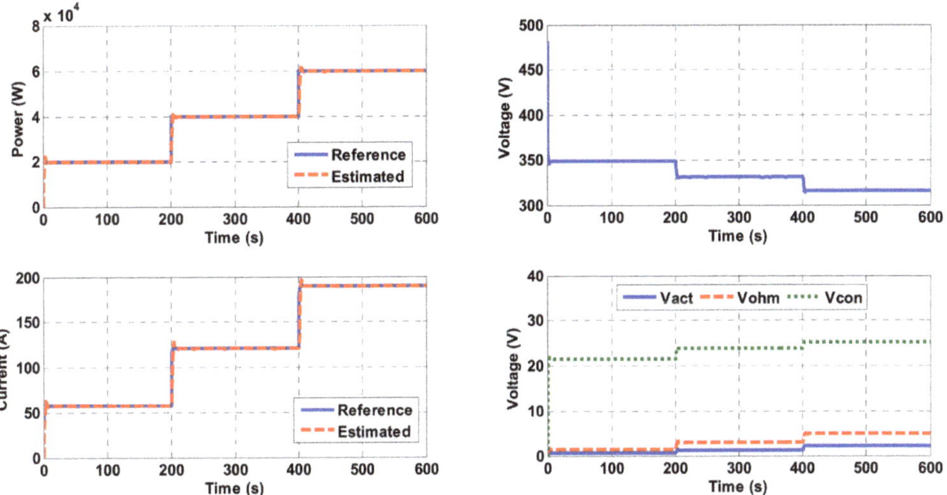

Figure 11. Second load disturbance applied on SOFC stack dynamic model.

6. Conclusions

This work introduces a new methodology based on a new metaheuristic approach named equilibrium optimizer (EO) to estimate the optimal parameters of a solid oxide fuel cell (SOFC) model. This is achieved with the aid of experimental datasets of the fuel cell polarization curves. The sum squared error difference between the cell experimental and computed voltages is selected as the fitness function to be minimized. The work investigates two operating modes of FC, which are steady- and dynamic-states models under altering operating conditions. In the first model, the parameters are estimated at four temperatures via the recorded measured polarization curves at them. In the dynamic model, two load power disturbances are investigated after identifying the parameters via the proposed EO. The obtained results via the proposed EO are compared to those obtained by the Archimedes optimization algorithm (AOA), Heap-based optimizer (HBO), Seagull Optimization Algorithm (SOA), Student Psychology Based Optimization

Algorithm (SPBO), Marine predator algorithm (MPA), Manta ray foraging optimization (MRFO), and comprehensive learning dynamic multi-swarm marine predators algorithm. In the case of the SOFC steady-state model, the proposed EO succeeded in achieving the best (minimum) fitness function of 2.6906×10^{-6}, 1.5527×10^{-6}, 2.6809×10^{-6}, and 2.2995×10^{-6} at operating temperature of 1073 K, 1173 K, 1213 K, and 1273 K, respectively. The corresponding standard deviations in the four studied cases obtained via the proposed EO are 9.15673×10^{-7}, 1.63132×10^{-6}, 3.36081×10^{-6}, and 2.2119×10^{-6}. Regarding the obtained results of the SOFC dynamic-state model, the proposed EO outperformed the others, achieving the minimum SMSE with a value of 1.0406; the MRFO comes in the second rank with a fitness function of 1.0775, and then the CLDMMPA achieves a SMSE of 1.3204 and comes in the third rank. The proposed EO succeeded in achieving a variance of 0.02264 and a standard deviation of 0.15048 in this studied case. The findings of this study demonstrate the superiority and reliability of the proposed approach in constructing a good-performance model that converges to the real one.

Author Contributions: All authors collaborated and contributed equally to this work. All authors have read and agreed to the published version of the manuscript.

Funding: This research received funding from the Institutional Fund Projects under grant no. (IFPHI-173-135-2020) supported by the Ministry of Education and King Abdulaziz University, Deanship of Scientific Research (DSR), Jeddah, Saudi Arabia.

Acknowledgments: This research work was funded by the Institutional Fund Projects under grant no. (IFPHI-173-135-2020). Therefore, authors gratefully acknowledge technical and financial support from the Ministry of Education and King Abdulaziz University, DSR, Jeddah, Saudi Arabia.

Conflicts of Interest: The authors declare no conflict of interest.

References

1. Capodaglio, A.G.; Cecconet, D.; Molognoni, D. An integrated mathematical model of microbial fuel cell processes: Bioelectrochemical and microbiologic aspects. *Processes* **2017**, *5*, 73. [CrossRef]
2. Bizon, N.; Thounthong, P. Energy efficiency and fuel economy of a fuel cell/renewable energy sources hybrid power system with the load-following control of the fueling regulators. *Mathematics* **2020**, *8*, 151. [CrossRef]
3. Wilberforce, T.; El-Hassan, Z.; Khatib, F.; Al Makky, A.; Baroutaji, A.; Carton, J.G.; Olabi, A.G. Developments of electric cars and fuel cell hydrogen electric cars. *Int. J. Hydrogen Energy* **2017**, *42*, 25695–25734. [CrossRef]
4. El-Hay, E.A.; El-Hameed, M.A.; El-Fergany, A.A. Performance enhancement of autonomous system comprising proton exchange membrane fuel cells and switched reluctance motor. *Energy* **2018**, *163*, 699–711. [CrossRef]
5. Rokni, M. Addressing fuel recycling in solid oxide fuel cell systems fed by alternative fuels. *Energy* **2017**, *137*, 1013–1025. [CrossRef]
6. Olabi, A. Renewable energy and energy storage systems. *Energy* **2017**, *136*, 1–6. [CrossRef]
7. Ramadhani, F.; Hussain, M.; Mokhlis, H.; Hajimolana, S. Optimization strategies for Solid Oxide Fuel Cell (SOFC) application: A literature survey. *Renew. Sustain. Energy Rev.* **2017**, *76*, 460–484. [CrossRef]
8. Chowdhury, S.; Chowdhury, S.P.; Crossley, P. *Microgrids and Active Distribution Networks*; The Institution of Engineering and Technology: London, UK, 2009.
9. El-Hay, E.; El-Hameed, M.; El-Fergany, A. Steady-state and dynamic models of solid oxide fuel cells based on Satin Bowerbird Optimizer. *Int. J. Hydrogen Energy* **2018**, *43*, 14751–14761. [CrossRef]
10. El-Hay, E.; El-Hameed, M.; El-Fergany, A. Optimized parameters of SOFC for steady state and transient simulations using interior search algorithm. *Energy* **2019**, *166*, 451–461. [CrossRef]
11. Van Biert, L.; Godjevac, M.; Visser, K.; Aravind, P. Dynamic modelling of a direct internal reforming solid oxide fuel cell stack based on single cell experiments. *Appl. Energy* **2019**, *250*, 976–990. [CrossRef]
12. Gong, W.; Yan, X.; Hu, C.; Wang, L.; Gao, L. Fast and accurate parameter extraction for different types of fuel cells with decomposition and nature-inspired optimization method. *Energy Convers. Manag.* **2018**, *174*, 913–921. [CrossRef]
13. Ettihir, K.; Boulon, L.; Becherif, M.; Agbossou, K.; Ramadan, H. Online identification of semi-empirical model parameters for PEMFCs. *Int. J. Hydrogen Energy* **2014**, *39*, 21165–21176. [CrossRef]
14. Ang, S.M.C.; Brett, D.J.; Fraga, E.S. A multi-objective optimisation model for a general polymer electrolyte membrane fuel cell system. *J. Power Sources* **2010**, *195*, 2754–2763. [CrossRef]
15. Tahmasbi, A.A.; Hoseini, A.; Roshandel, R. A new approach to multi-objective optimisation method in PEM fuel cell. *Int. J. Sustain. Energy* **2013**, *34*, 283–297. [CrossRef]

16. Petrescu, S.; Petre, C.; Costea, M.; Malancioiu, O.; Boriaru, N.; Dobrovicescu, A.; Feidt, M.; Harman, C.; Stanciu, C. A methodology of computation, design and optimization of solar Stirling power plant using hydrogen/oxygen fuel cells. *Energy* **2010**, *35*, 729–739. [CrossRef]
17. Virkar, A.; Williams, M.C.; Singhal, S. Concepts for ultra-high power density solid oxide fuel cells. *ECS Trans.* **2007**, *5*, 401–421. [CrossRef]
18. Zhu, L.; Zhang, L.; Virkar, A.V. A parametric model for solid oxide fuel cells based on measurements made on cell materials and components. *J. Power Sources* **2015**, *291*, 138–155. [CrossRef]
19. Shi, H.; Li, J.; Zafetti, N. New optimized technique for unknown parameters selection of SOFC using Converged Grass Fibrous Root Optimization Algorithm. *Energy Rep.* **2020**, *6*, 1428–1437. [CrossRef]
20. Yousri, D.; Hasanien, H.M.; Fathy, A. Parameters identification of solid oxide fuel cell for static and dynamic simulation using comprehensive learning dynamic multi-swarm marine predators algorithm. *Energy Convers. Manag.* **2021**, *228*, 113692. [CrossRef]
21. Nassef, A.M.; Fathy, A.; Sayed, E.T.; Abdelkareem, M.A.; Rezk, H.; Tanveer, W.H.; Olabi, A. Maximizing SOFC performance through optimal parameters identification by modern optimization algorithms. *Renew. Energy* **2019**, *138*, 458–464. [CrossRef]
22. Fathy, A.; Rezk, H.; Ramadan, H.S.M. Recent moth-flame optimizer for enhanced solid oxide fuel cell output power via optimal parameters extraction process. *Energy* **2020**, *207*, 118326. [CrossRef]
23. Wang, X.; Huang, B.; Chen, T. Data-driven predictive control for solid oxide fuel cells. *J. Process. Control* **2007**, *17*, 103–114. [CrossRef]
24. Larminie, J.; Dicks, A.; McDonald, M.S. *Fuel Cell Systems Explained*; John Wiley & Sons: Chichester, UK, 2003; Volume 2.
25. Faramarzi, A.; Heidarinejad, M.; Stephens, B.; Mirjalili, S. Equilibrium optimizer: A novel optimization algorithm. *Knowl. Based Syst.* **2020**, *191*, 105190. [CrossRef]
26. Wei, Z.; Zou, C.; Leng, F.; Soong, B.H.; Tseng, K.-J. Online model identification and state-of-charge estimate for lithium-ion battery with a recursive total least squares-based observer. *IEEE Trans. Ind. Electron.* **2018**, *65*, 1336–1346. [CrossRef]
27. Wei, Z.; Zhao, J.; Xiong, R.; Dong, G.; Pou, J.; Tseng, K.J. Online estimation of power capacity with noise effect attenuation for lithium-ion battery. *IEEE Trans. Ind. Electron.* **2019**, *66*, 5724–5735. [CrossRef]
28. Wei, Z.; Zhao, D.; He, H.; Cao, W.; Dong, G. A noise-tolerant model parameterization method for lithium-ion battery management system. *Appl. Energy* **2020**, *268*, 114932. [CrossRef]
29. Wei, Z.; Meng, S.; Xiong, B.; Ji, D.; Tseng, K.J. Enhanced online model identification and state of charge estimation for lithium-ion battery with a FBCRLS based observer. *Appl. Energy* **2016**, *181*, 332–341. [CrossRef]
30. Pierre, J. Siemens energy. In Proceedings of the 11th Annual SECA Workshop, Pittsburgh, PA, USA, 27–29 July 2010.
31. Xu, D.; Jiang, B.; Liu, F. Improved data driven model free adaptive constrained control for a solid oxide fuel cell. *IET Control. Theory Appl.* **2016**, *10*, 1412–1419. [CrossRef]

 *mathematics*

MDPI

Article

Optimized Power Supply Rejection Ratio Modeling Technique for Simulation of Automotive Low-Dropout Linear Voltage Regulators

Ionuț-Constantin Guran *, Adriana Florescu and Lucian Andrei Perișoară

Department of Applied Electronics and Information Engineering, Faculty of Electronics, Telecommunications and Information Technology, University Politehnica of Bucharest, 060042 Bucharest, Romania; adriana.florescu@upb.ro (A.F.); lucian.perisoara@upb.ro (L.A.P.)
* Correspondence: ionut.guran@stud.etti.upb.ro

Abstract: In the automotive domain, the vast majority of testing is performed through simulations, which can validate a system design before the actual implementation and can emphasize eventual faults in the design process. Hence, the simulation is of utmost importance. Behavioral models are necessary for the creation of each electronic device desired in the system, and some of the components have very complex behavior: low-dropout linear voltage regulators (LDOs), gate drivers, and switching regulators. In the automotive industry, LDOs are essential components because they power all the other subsystems and very accurate behavior is needed to make sure that the system behaves as in reality. LDO models are already commercially available and most of their intrinsic characteristics are modeled (dropout voltage, line regulation, load regulation, etc.). However, one characteristic that is extremely useful, yet the hardest to model, is the power supply rejection ratio (PSRR). This paper proposes a new PSRR modeling technique for automotive low-dropout voltage regulators. The new PSRR characteristic was modeled for an automotive LDO product in a Texas Instruments portfolio, which has a commercially available model, and was simulated using the PSpice Allegro simulator and the OrCAD Capture CIS environment.

Keywords: simulation; optimal behavioral modeling; automotive; low-dropout linear voltage regulator; power supply rejection ratio

MSC: 97M50

Citation: Guran, I.-C.; Florescu, A.; Perișoară, L.A. Optimized Power Supply Rejection Ratio Modeling Technique for Simulation of Automotive Low-Dropout Linear Voltage Regulators. *Mathematics* **2022**, *10*, 1150. https://doi.org/10.3390/math10071150

Academic Editor: Nicu Bizon

Received: 28 February 2022
Accepted: 28 March 2022
Published: 2 April 2022

Publisher's Note: MDPI stays neutral with regard to jurisdictional claims in published maps and institutional affiliations.

1. Introduction

In recent years, the automotive domain has seen an increase in the complexity of modern cars, which will continue to become increasingly complex. The number of electronic functions and components in cars is also rapidly increasing, which can lead to design problems in complex modular systems [1].

Supplying and conditioning electrical power are the most important features of an electrical system. No application can fully perform its function without a stable supply. Batteries, generators, and other off-line supplies provide substantial voltage and current variations over time and over a wide range of operating conditions [2]. Noise is produced due to their inherent nature, but also by high-power switching circuits such as DC–DC converters, controllers for electric motors, actuators, and relays. This noise analysis is increasingly important in the case of Electric Vehicles (EVs) or Hybrid Electric Vehicles (HEV) from the Electromagnetic Interference (EMI) point of view [2,3].

Rapidly changing loads result in unwanted voltage changes and frequency harmonics over an ideal direct current (DC) component. The goal of a voltage regulator is to convert the noisy supply into a stable, accurate, load-independent voltage and hence attenuate the fluctuations to desired levels [2–4]. One of the most used power supply circuits in the automotive domain is the low dropout (LDO) voltage regulator, which uses a unipolar MOS or a bipolar pass transistor in its structure as a series control element to provide a regulated output voltage over a wide range of supply voltages or load current variations [5–8].

Simulation is used in the automotive domain for the validation of a design before physical implementation, and can indicate flaws and faults in the design process or reinforce the correct functioning of the system. For a highly reliable simulation, accurate component models are necessary, for which the simulated behavior must be as close as possible (ideally, identical) to the real behavior. Simulation Program for Integrated Circuits Emphasis (SPICE) is a general-purpose analog and mixed-mode simulator that is used to verify and predict circuit behavior. PSpice is the PC version of SPICE, and is used to simulate the behavior of circuits on a digital computer, emulating the signal generators, multimeters, oscilloscopes, and frequency spectrum analyzers, and includes analog and digital libraries of standard components. As a result, it is an important tool for a wide range of analog and digital applications [9].

In automotive design, internal clean power supplies having a high Power Supply Rejection Ratio (PSRR) are a vital requirement to increase power management efficiency in system-on-chip circuitry [10].

The literature focuses only on PSRR physical design and measurement [3,5,7,10], and does not provide any research on PSRR modeling techniques. This paper proposes an optimization approach in the simulation domain, and emphasizes a highly accurate method of modeling the PSRR for automotive LDOs. The main contributions are listed below:

- implementation of a new PSRR model for the TPS785-Q1 automotive LDO from Texas Instruments;
- simulation of the LDO model with the initial PSRR functionality, as is currently commercially available;
- simulation of the LDO model with the new PSRR functionality added using the PSpice Allegro simulator and OrCAD Capture CIS 17.2 simulation environment;
- comparison of the results using the values of the theoretical and previous approaches.

The remainder of the paper is organized as it follows: Section 2 shows the background related to this work and presents the currently existing PSRR vs. frequency characteristic of TPS785-Q1 from Texas Instruments, Section 3 emphasizes the materials and methods for the new PSRR model implementation, and Section 4 presents the numerical simulation results in tables and waveform figures. The comparison between the theoretical, previous, and new PSRR characteristics of the TPS785-Q1 model is presented in Section 5. The conclusions are synthesized in Section 6.

2. Background of LDO PSRR

The LDO is commonly used in power electronics design as the last stage of the power-distribution tree. In the first stage, an intermediate voltage is obtained from the input voltage of a supply system using other topologies, such as DC–DC or AC–DC converters. These topologies introduce harmonics and supraharmonics, generating a noisy intermediate voltage. The supraharmonics are defined as current and voltage waveforms distortion within the range 2–150 kHz, that can be intentionally created by power line communication systems or unintentionally by power electronic converters. In stage two, the LDO regulator generates the system output voltage from the intermediate voltage. The objective is to achieve a high power conversion efficiency in stage one and to remove the switching noise in stage two. The noise can be technically translated into an unwanted voltage ripple that must be eliminated [11], thus justifying the study of PSRR in this paper. For example, in automotive applications, the car battery delivers the supply voltage that can vary between 6 and 30 V, having slew rates of up to 1 V/μs. In addition, the load current can vary drastically, with slew rates of up to 50–100 mA/μs. These very high slew-rates also introduce harmonics, which translate into unwanted voltage spikes that must also be eliminated; this elimination is also achieved by the new proposed PSRR model [6].

The simplified working principle of a regular LDO voltage regulator is shown in Figure 1, and contains the input voltage V_{IN}, the pass transistor element Q, the error amplifier K, and the resistive network (R$_1$, R$_2$), which sets the desired output voltage V_{OUT}. The error amplifier compares the positive terminal voltage of $V_{OUT} * [R_2/(R_1 + R_2)]$ with the reference voltage V$_{REF}$ connected to the negative terminal and drives the pass transistor accordingly such that the two voltages become equal.

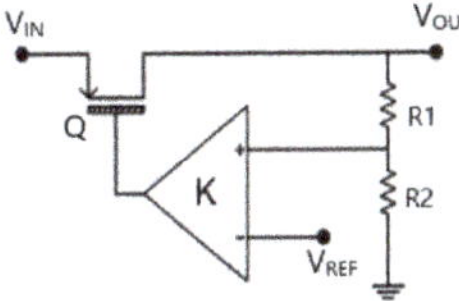

Figure 1. LDO voltage regulator principle [12].

The most common characteristics that need to be taken into consideration when creating a LDO voltage regulator model are: output voltage accuracy, line regulation, load regulation, current consumption, dropout voltage, output voltage slew-rate, protections, and PSRR [13,14].

Some examples of these characteristics are presented in Figure 2a–f, being taken from Texas Instruments' TPS785-Q1 automotive LDO product [14].

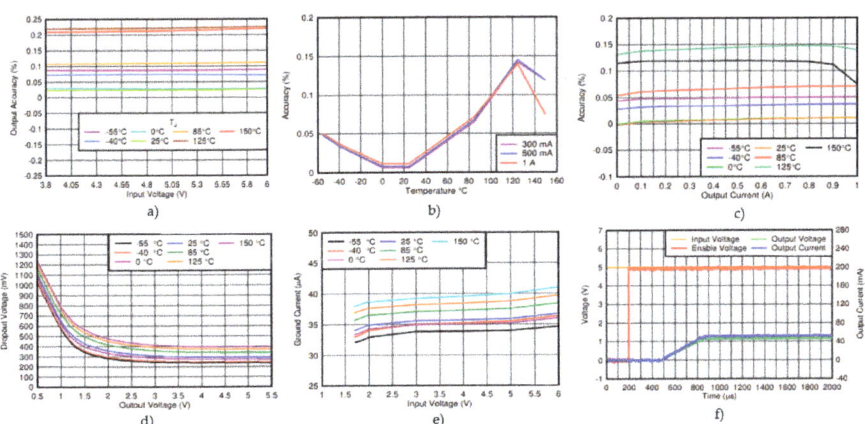

Figure 2. Examples of characteristics for TPS785-Q1: (**a**) output voltage accuracy vs input voltage; (**b**) output voltage accuracy vs temperature; (**c**) output voltage accuracy vs output current; (**d**) dropout voltage vs output voltage; (**e**) ground current vs input voltage; (**f**) output voltage and slew-rates.

Existing LDO models are commercially available from companies such as Texas Instruments, Infineon, and Analog Devices, but only newer products have different behavioral models for some simulation software environments, such as PSpice Allegro, TINA, SIMetrix.

TPS785-Q1 from Texas Instruments portfolio is an ultra-low-dropout regulator with a low quiescent current that can source 1 A with excellent load and line transient performance. It is qualified for automotive applications according to the AEC-Q100 standard, and has a junction temperature varying from −40 to +150 °C. The low output noise and good PSRR performance make the product suitable for power-sensitive analog loads [14].

TPS785-Q1's typical application circuit as a post regulator is shown in Figure 3. The DC–DC converter is used as the main regulator. It is supplied by the battery voltage V$_{BAT}$ and produces an output voltage V_{OUT1}, which is inherently noisy. Capacitors C$_{OUT1}$ and

C_{IN} have the role of reducing the ripple appearing at the input of TPS785-Q1, and capacitor C_{OUT} reduces the output voltage V_{OUT} ripple [14].

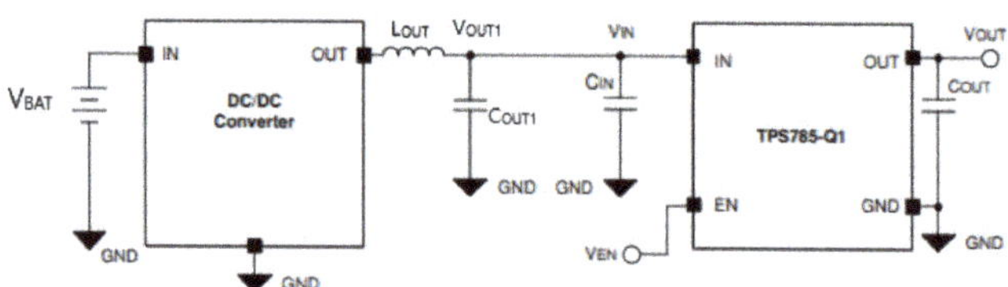

Figure 3. Typical application for the TPS785-Q1 voltage regulator [14].

The post regulator supports an input voltage range from 1.7 to 6.0 V and offers an adjustable output range of 1.2 to 5.5 V. TPS785-Q1 takes the output voltage of the DC–DC converter and regulates it to the desired level V_{OUT}, eliminating the switching noise introduced by the converter [14].

The power supply rejection ratio is defined as the measurement of the magnitude of the output voltage ripple ΔV_{OUT} compared to the input voltage ripple ΔV_{IN} [11,15]:

$$PSRR = 20\log_{10}\left(\frac{\Delta V_{IN}}{\Delta V_{OUT}}\right) \tag{1}$$

Theoretically, the ideal PSRR is infinite. In practice, PSRR has a big value, so that the measurement unit used is decibel (dB). Empirically, a high PSRR is considered to be a value over 60 dB [11,15].

The PSRR is a very important characteristic in the power electronics and automotive domains. Figure 4 represents the PSRR characteristic of the TPS785-Q1 automotive high PSRR LDO [14], which has its maximum PSRR of around 70 dB between 10 and 40 Hz. This value of PSRR is enough to consider it a high PSRR LDO, but the input voltage supply's switching frequency is also vital [11,16,17]. For example, the switching frequency of newer switching regulators is between 300 kHz and 6 MHz, and the LDO response time is too slow to efficiently filter out the switching noise, due to the fact that the noise is outside the bandwidth of most typical high PSRR regulators [11,18].

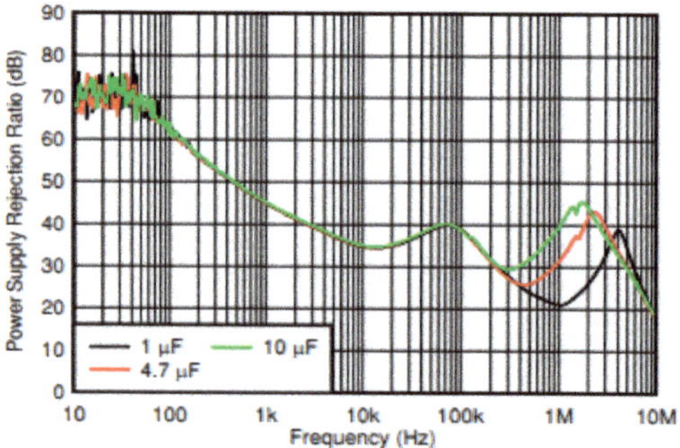

Figure 4. PSRR vs. C_{OUT} at V_{OUT} = 3.3 V and I_{OUT} = 1 A [14].

There are five regions represented in Figure 4. The first region contains the frequency range from 10 to 40 Hz, where the PSRR has its peak (70 dB), and is approximately a flat curve. The frequency range 40 Hz–10 kHz, where PSRR decreases steadily at 20 dB/decade, forms the second region. In the third region (10–70 kHz) PSRR increases again to 40 dB. The fourth region sees a decrease in PSRR from 40 to 30 dB at 300 kHz. The first four regions are the effective PSRR bandwidth; thus, TPS785-Q1 has an effective frequency range between 0 Hz and 300 kHz. In the fifth region (300 kHz–10 MHz), the change in PSRR depends on the numerical value of the output capacitor C_{OUT} (Figure 3) and its impedance, and the parasitic board impedance; thus, the LDO contribution to PSRR decreases.

The behavioral model of TPS785-Q1 provided by Texas Instruments is a PSpice Allegro library file [19]. We ran the model with the simple application circuit created in the OrCAD Capture CIS environment (Figure 5). The output voltage was set at 2.4 V and the load current at 1 mA.

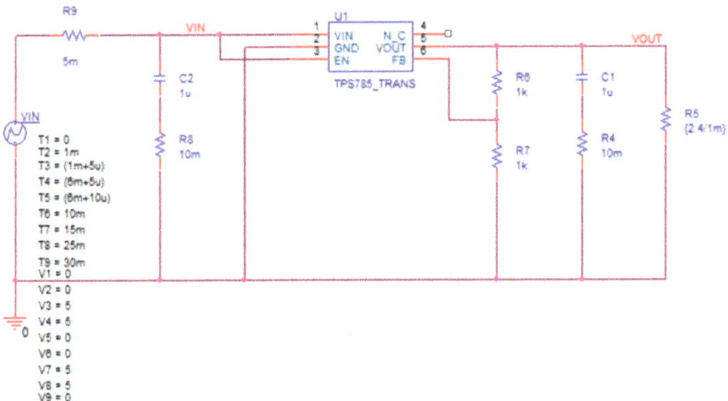

Figure 5. Demonstration application circuit of TPS785-Q1 provided by Texas Instruments [19].

Figure 6 shows the results obtained using the demo application test-bench. In the beginning of the simulation, there is an abrupt 5 V pattern of the input voltage V_{IN} for both rising and falling edges (1–6 ms) and the output voltage V_{OUT} regulates at 2.4 V within a certain slew-rate. The second pattern shows slower slopes of the input voltage VIN for the rising and falling edges (10–30 ms). It can be noted that there is a certain threshold of the input voltage above which the LDO starts working.

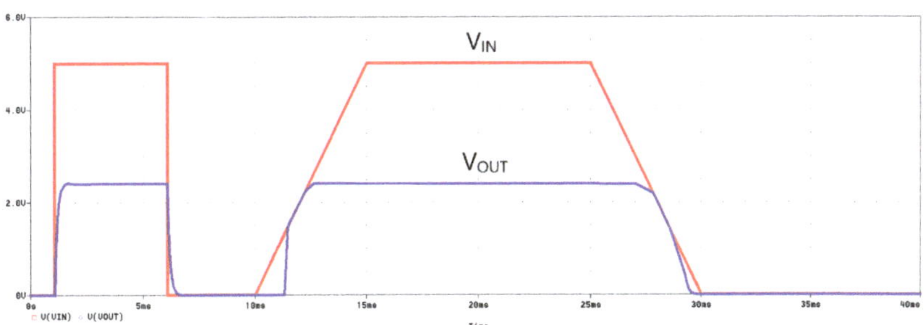

Figure 6. Simulation results for the demo application circuit of TPS785-Q1.

From the unencrypted library file, we found that this TPS785-Q1 model is a transient model, built for the PSpice Allegro simulator only, and is in its first version. The existing implemented characteristics are the following: start-up time, PSRR, enable/VIN shutdown, load and line transients, and internal current limit, and the model supports inverting the topology. In order to better understand how the PSRR model works, we further performed a reverse engineering of the library file code, so Figure 7a represents the simplified concept of the TPS785-Q1 model, along with the general PSRR concept that is currently implemented [11,20]. It consists of the error amplifier EA, the reference voltage V_{REF}, the pass transistor MOS1, the feedback resistors R_{UP} and R_{DW}, the output capacitor C_{OUT} with its series resistance R_{ESR}, and load resistor R_{LOAD}. The components can be grouped into two impedances, Z_A and Z_B.

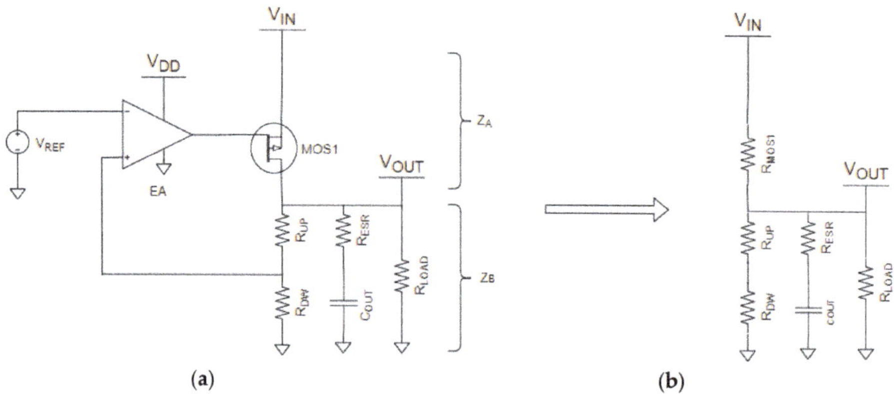

Figure 7. Concept diagram of TPS785-Q1 model: (**a**) general simplified PSRR model concept; (**b**) high frequency PSRR model concept.

The PSRR can be calculated as it follows:

$$\text{PSRR} = 20 \, \log_{10}\left(\frac{Z_A + Z_B}{Z_B}\right) \tag{2}$$

In the first region of PSRR vs. frequency characteristic from Figure 4, the error amplifier has a large gain and this results in Z_A being well controlled, which translates into a high PSRR. In the second region, the gain of the amplifier starts dropping at 20 dB/decade. The sensitivity of the loop with respect to the changes in the output voltage decreases because the amplifier gain decreases; thus, the impedance of the transistor adjusts slower to the changes and this results in a decrease in PSRR. The impedance of the output capacitor decreases with the increase in the input signal frequency and this increases the LDO PSRR in the fifth region. The impedance Z_B decreases to the point where most of the signal is short-circuited across the capacitor instead of being attenuated by the LDO. In this case, where the LDO no longer contributes significantly to the PSRR, the pass transistor MOS1 is treated as a simple resistor and only attenuates the ripple passively [11,20]. This situation is revealed in Figure 7b, which was drawn for high frequencies, and differs from Figure 7a by eliminating the LDO and replacing the pass transistor MOS1 with a simple resistor R_{MOS1}.

We also simulated the PSRR characteristic of the already existing TPS785-Q1 model. The PSRR simulation test-bench is provided in Figure 8.

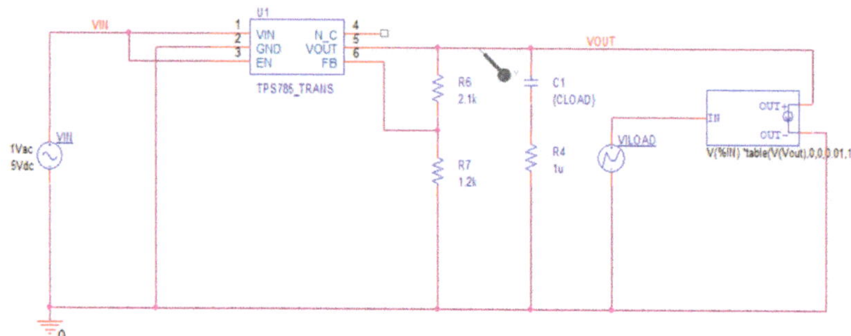

Figure 8. Test-bench circuit for AC analysis of TPS785-Q1 PSRR.

The input voltage source V_{IN} is a 5 V DC component on which a 1 V amplitude sine component modeling the additive noise is overlapped. The output voltage V_{OUT} is set at 3.3 V and the load current is set at 1 A, as specified by the PSRR conditions. The load capacitor C1 is given three values (1, 4.7, and 10 µF) through parameter {CLOAD}, so three simulations were performed.

Figure 9 shows the family of simulated PSRR characteristics vs. frequency, having the capacitor C1 {CLOAD} from Figure 8 as the parameter. Thus, the violet curve shows the 1 µF capacitor's value, the red curve shows the 4.7 µF capacitor's value, and the green curve shows the 10 µF capacitor's value. Compared to the datasheet characteristic presented in Figure 4, the flat region is at 60 dB instead of 70 dB, which represents a disadvantage of the old PSRR model given in [14]. The capacitor influence is visible from around 1 kHz, whereas the datasheet characteristic sees a capacitor influence starting at 300 kHz.

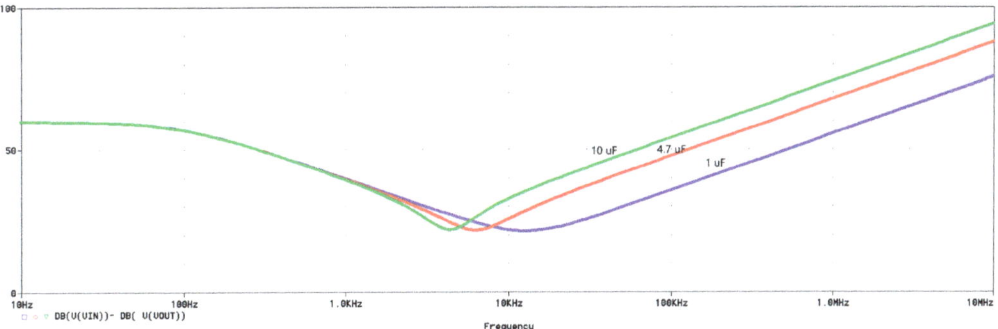

Figure 9. Simulated old PSRR characteristic of the TPS785-Q1 model.

3. Materials and Methods for the New Proposed PSRR Model Implementation

For materials, we started from the existing TPS785-Q1 model from Texas Instruments, from which we eliminated the old PSRR characteristic, and added our new PSRR concept, as shown in Figure 10. For the method, we implemented the PSRR functionality by modifying the reference voltage V_{REF_IN}. The ideal reference voltage V_{REF_IN} of the LDO is added to the PSRR voltage source V_{PSRR}, which depends on the ripple of the input voltage V_{IN} and its frequency, and the result is V_{REF_OUT}. Then, V_{REF_OUT} is sent through the feedback resistors R_{UP} and R_{DW} (Figure 7a), and yields the output voltage V_{OUT} and its variations (Figure 7a).

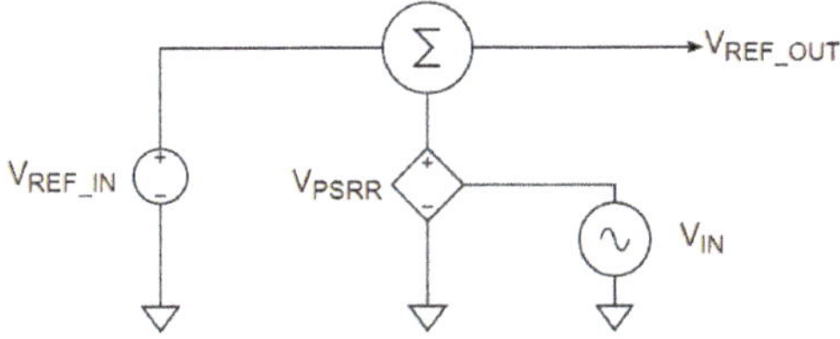

Figure 10. New LDO PSRR model concept.

In Figure 10, the PSRR source V_{PSRR} holds the information about the input voltage source ripple ΔV_{IN} and frequency, which is demonstrated below. V_{IN} is DC shifted and the DC information needs to be eliminated from the original signal. We used a first order passive low pass filter to determine the input signal V_{IN} frequency, and a second order active low pass filter to eliminate the DC component of V_{IN}. The first order low pass filter schematic is shown in Figure 11a, and the second order active low pass filter concept is presented in Figure 11b.

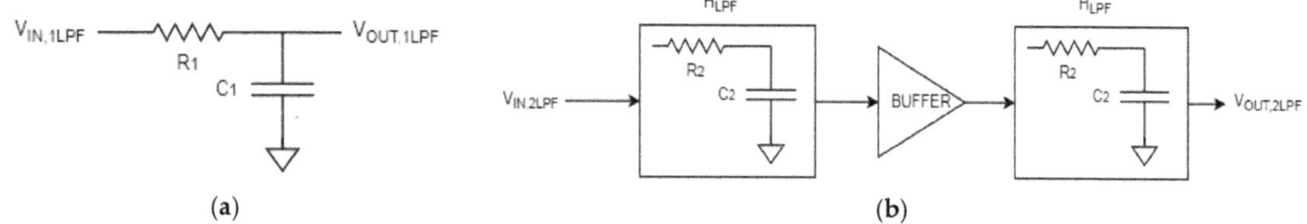

Figure 11. First and second order low pass filters: (**a**) first order passive filter schematic; (**b**) second order active low pass filter concept.

The transfer function of the first order low pass filter is:

$$H_{1LPF}(j\omega) = \frac{1}{1 + j\omega R_1 C_1} = \frac{1}{1 + j2\pi f R_1 C_1} \tag{3}$$

where $\omega = 2\pi f$ is the angular frequency, f is the frequency, R_1 is the low pass filter resistance value, and C_1 is the low pass filter capacitance value (Figure 11a).

The transfer function magnitude of the first order low pass filter is:

$$|H_{1LPF}(j\omega)| = \frac{1}{\sqrt{1 + \omega^2 R_1^2 C_1^2}} = \frac{1}{\sqrt{1 + 4\pi^2 f^2 R_1^2 C_1^2}} \tag{4}$$

and the transfer function phase of the first order low pass filter is:

$$\varphi_{1LPF}(j\omega) = -\tan^{-1}(\omega R_1 C_1) = -\tan^{-1}(2\pi f R_1 C_1) \tag{5}$$

Having a second order active filter in which the two stages are separated galvanically by a buffer, the total transfer function can be written as shown in Equation (6), based on relation Equation (3), with the magnitude and phase calculated in Equations (7) and (8). The components' values are chosen to be equal due to the simplicity of calculations (Figure 11b) [21–23].

$$H_{2LPF}(j\omega) = H_{LPF}(j\omega)\cdot H_{LPF}(j\omega) = \frac{1}{1 + j\omega R_2 C_2}\cdot\frac{1}{1 + j\omega R_2 C_2} = \frac{1}{(1 - \omega^2 R_2^2 C_2^2) + j2\omega R_2 C_2} \tag{6}$$

where the magnitude of the transfer function is:

$$|H_{2LPF}(j\omega)| = \frac{1}{\sqrt{(1 - \omega^2 R_2^2 C_2^2)^2 + 4\omega^2 R_2^2 C_2^2}} = \frac{1}{1 + \omega^2 R_2^2 C_2^2} = \frac{1}{1 + 4\pi^2 f^2 R_2^2 C_2^2} \tag{7}$$

and its phase is:

$$\varphi_{2LPF}(j\omega) = -\tan^{-1}\frac{2\omega R_2 C_2}{1 - \omega^2 R_2^2 C_2^2} = -\tan^{-1}\frac{4\pi f R_2 C_2}{1 - 4\pi^2 f^2 R_2^2 C_2^2} \tag{8}$$

The cutoff frequency (where the magnitude of the transfer function drops to −3 dB) of the second order active filter is:

$$f_{-3dB,\,2LPF} = \frac{1}{2\pi R_2 C_2} \tag{9}$$

In order to achieve the DC component of the voltage V_{IN}, we used relations Equations (6) and (7), in which we chose the resistance value R_2 of 1 MΩ and capacitance C_2 of 1 mF, which leads to $f_{-3dB} = 0.000160$ Hz, i.e., a numerical value very near to 0 Hz.

Another challenge of modeling the characteristic PSRR vs. frequency is to determine the frequency of the input signal V_{IN}, by filtering the input ripple signal ΔV_{IN}, once again using a first order low pass filter, and then compute the root mean square (RMS) values of the input and output signals of the filter, $V_{RMS,IN}$ and $V_{RMS,OUT}$ [24,25]. We set the filter cutoff frequency to 40 Hz, where the PSRR characteristic starts dropping at 20 dB/decade from the flat region (Figure 4). The cutoff frequency of the first order low pass filter is identical to that of the second order low pass filter, which uses the same values for the resistors and for the capacitors. In order to achieve this cutoff frequency, the filter resistor R_1 was set to 10 kΩ and, based on relation Equation (9), the resulting value of capacitor C_1 was 1.6 nF.

Since the signal at the output of the filter is phase shifted, the ratio of the instant values of the input and output signals of the first order low pass filter cannot be performed, but the RMS values are stationary, and their ratio $V_{RMS,IN}/V_{RMS\_OUT}$ reflects the signal V_{IN} frequency. The formula used for RMS calculation is:

$$V_{\mathrm{RMS}} = \sqrt{\frac{1}{T} \int_0^T v^2(t)dt,}$$ (10)

where T is the signal period and $v(t)$ is the time-varying signal.

The implementation of relation Equation (10) in PSpice poses a challenge and cannot be performed directly. The time integral of the squared signal is computed by injecting a current having the value $v^2(t)$ into a 1 F capacitor. Then, the value of time integral is divided by time and the square root value is extracted. The operating principle of the capacitor used for the implementation of the RMS formula is given as:

$$I_C(t) = C\frac{dU_C(t)}{dt} => U_C(t) = \frac{1}{C} \int I_C(t)dt$$ (11)

The principle of RMS code implementation from relation Equation (11) is given in Figure 12.

```
G1   0 SQUARE VALUE={V(SIGNAL) * V(SIGNAL)}
C1   SQUARE 0 1
E1   RMS    0 VALUE={SQRT(V(SQUARE)/TIME)}
```

Figure 12. PSpice time integral computation principle.

The ratio of the RMS values of the input and output signals of the low pass filter is:

$$\frac{RMS_{in}}{RMS_{out}} = \sqrt{1 + 4\pi^2 f_{signal}^2 R_1^2 C_1^2} => f_{signal} = \sqrt{\frac{\left(\frac{RMS_{in}}{RMS_{out}}\right)^2 - 1}{4\pi^2 R_1^2 C_1^2}}$$ (12)

We associated a PSSR value in *dB* for each frequency of interest in PSpice using the TABLE function. The PSRR values between the two frequencies of interest are interpolated. In order to obtain a smoother PSRR vs. frequency characteristic, more frequency points were chosen.

The PSRR values in *dB* from the PSpice TABLE need to be converted to numerical values as follows:

$$PSRR_{numerical} = 10^{\frac{PSRR_{dB}}{20}}$$ (13)

The final V_{PSRR} that is added to the ideal reference voltage V_{REF_IN}, and gives the output reference voltage V_{REF_OUT} (Figure 10), is computed as:

$$V_{\mathrm{PSRR}} = \frac{\Delta V_{IN}}{PSRR_{numerical}}$$ (14)

The frequency computation, PSRR TABLE implementation, and V_{PSRR} determination in PSpice are presented in Figure 13, in which relations Equations (12)–(14) were used.

```
E2 FREQ 0 VALUE={SQRT((1-V(RMS_IN)/V(RMS_OUT))/(4*3.14**2*{RLPF}*{CLPF}))}

E3 PSRR_TABLE 0 VALUE={TABLE(V(FREQ), {FREQUENCY_1}, {DB_VALUE_1},FREQUENCY_2}, {DB_VALUE_2},
+{FREQUENCY_3}, {DB_VALUE_3}, {FREQUENCY_4}, {DB_VALUE_4}, {FREQUENCY_5} {DB_VALUE_5}  )}

E4 VPSRR 0 VALUE={V(RIPPLE)/(10**(V(PSRR_TABLE)/20))}
```

Figure 13. PSpice frequency and V$_{PSRR}$ computation.

4. PSRR Simulation Results

The AC analysis used in PSpice is a linear analysis. The simulator calculates the frequency response by linearizing the circuit around the bias point. All voltage and current sources that have AC values are inputs of the circuit. During the AC analysis, the only sources that have non-zero amplitudes are those using AC specifications [26]. Because our PSRR modeling method is non-linear, the PSRR was analyzed using the transient simulation.

The transient simulation test-bench, presented in Figure 14, consists of the TPS785-Q1 model with resistors R6 and R7 chosen such that the output voltage is set to 3.3 V and the load current is set to 1 A, as specified by the PSRR conditions. The load capacitor C1 is parameterized with value {CLOAD}, and the input voltage source VIN is parameterized with {FREQ}, which sets the input voltage sine frequency. The amplitude is set to 0.5 V and the offset to 5 V.

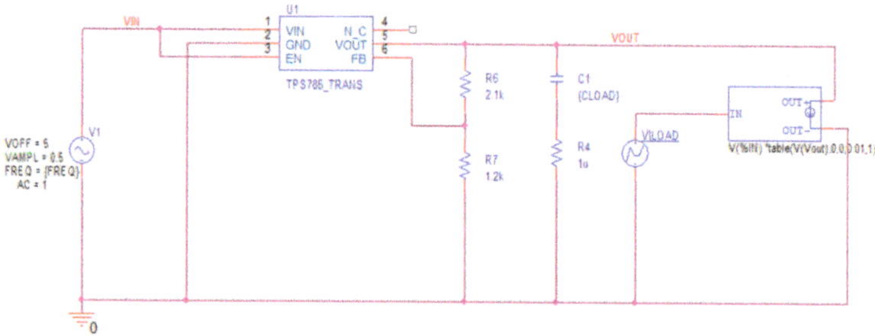

Figure 14. Transient test-bench circuit for TPS785-Q1 PSRR simulation.

For each frequency of interest, a transient simulation was performed and the peak-to-peak amplitude of the output voltage was extracted. The simulator options were set to default and the maximum time step was set to the 100th part of the input signal period. The waveforms of the input (red curve) V_{IN} and output (green curve) V_{OUT} voltages are shown in Figure 15 for a chosen frequency of 100 kHz (where the output voltage ripple is visible), and the PSRR calculations are shown in Table 1. In Table 1, we chose 20 frequency values within the range 10 Hz–10 MHz (as specified in Figure 4) in order to demonstrate the accuracy of the results while the frequency is rising.

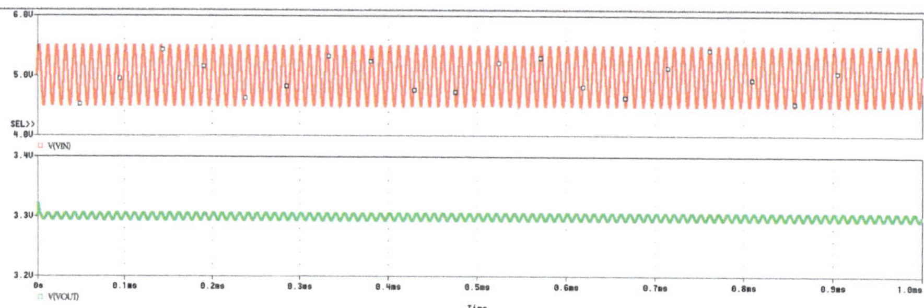

Figure 15. PSRR simulation waveforms at 100 kHz.

Table 1. Simulated new PSRR vs. frequency.

Frequency	VOUT$_{MAX}$	VOUT$_{MIN}$	PSRR
[Hz]	[V]	[V]	[dB]
10	3.2999	3.2996	70.00
20	3.2999	3.2996	70.00
40	3.2999	3.2995	68.09
100	3.3002	3.2992	59.98
200	3.3005	3.2989	55.95
500	3.3013	3.2981	49.98
1 k	3.3025	3.2969	44.99
2 k	3.3029	3.2944	41.41
5 k	3.3060	3.2934	37.99
10 k	3.3086	3.2908	35.00
50 k	3.3053	3.2941	38.99
70 k	3.3047	3.2947	39.99
100 k	3.3060	3.2934	37.99
200 k	3.3172	3.2822	29.12
300 k	3.3276	3.2719	25.08
500 k	3.3310	3.2645	23.53
1 M	3.3014	3.2555	26.75
2 M	3.2943	3.2686	31.80
4 M	3.3031	3.2918	38.95
10 M	3.2828	3.2720	39.37

Figure 16 presents the simulated new PSRR vs. frequency characteristic drawn based on Table 1, compared with the theoretical characteristic, shown in Figure 4. The simulated PSSR curve was determined only for a load capacitor of 1 μF. The reason behind this decision is that the capacitor only influences the PSRR at frequencies over 300 kHz, and lower frequencies, which are not influenced by the load capacitor value, are much more important than the higher ones.

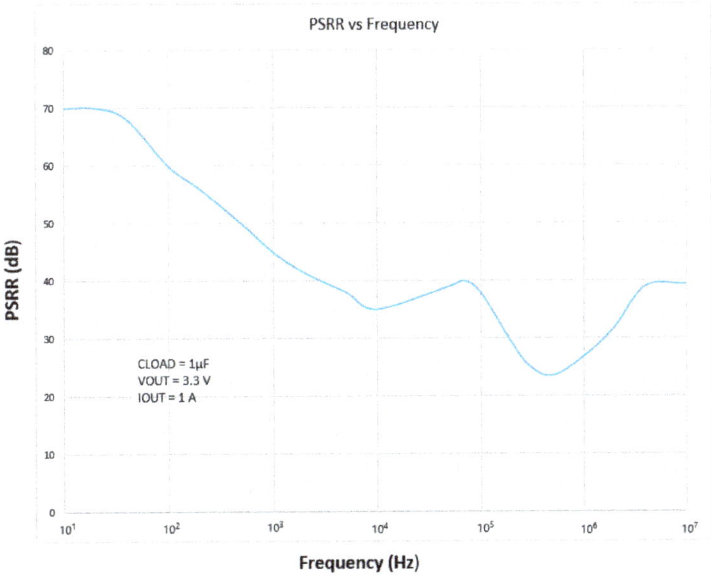

Figure 16. Simulated new PSRR characteristic of the TPS785-Q1 model.

5. Discussion and Comparison of Theoretical, Old, and New PSRR Simulations

Table 2 compares the theoretical, original model, and the newly proposed PSRR model results at various frequencies for a load capacitor of 1 μF.

Table 2. Theoretical PSRR vs. Simulated Old PSRR vs. Simulated New PSRR for a load capacitor of 1 μF.

Frequency [Hz]	Theoretical PSRR [dB]	Simulated Old PSRR [dB]	Simulated New PSRR [dB]
10	70	60	70.00
20	70	60	70.00
40	68	59	68.09
100	62	57	59.98
200	56	53	55.95
500	50	46	49.98
1 k	45	40	44.99
2 k	41	34	41.41
5 k	38	27	37.99
10 k	35	22	35.00
50 k	39	30	38.99
70 k	40	33	39.99
100 k	38	36	37.99
200 k	30	42	29.12
300 k	27	45	25.08
500 k	25	50	23.53
1 M	21	55	26.75
2 M	25	61	31.80
4 M	39	68	38.95
10 M	20	75	39.37

The results in Table 2 show that the absolute difference value between the simulated new PSRR curve and the theoretical curve until 500 kHz is 2 dB, whereas between the simulated original PSRR and theoretical PSRR there is an absolute difference of 25 dB.

The issue encountered in this work was that for frequencies above 500 kHz, the new PSRR error starts increasing, but remains within a tolerance of 5 dB until 10 MHz, when the absolute error becomes 20 dB. This happens due to the limitations of the LDO model error amplifier, as shown in Figure 17. The red curve represents the reference voltage V_{REF} of the chip and the green curve is the feedback voltage V_{FB} at an input signal frequency of 10 MHz. It can be clearly seen that the feedback voltage cannot track the reference voltage properly and this limits the model performance. In this case, the error amplifier EA in Figure 7a has to be redesigned, which is not the subject of this paper.

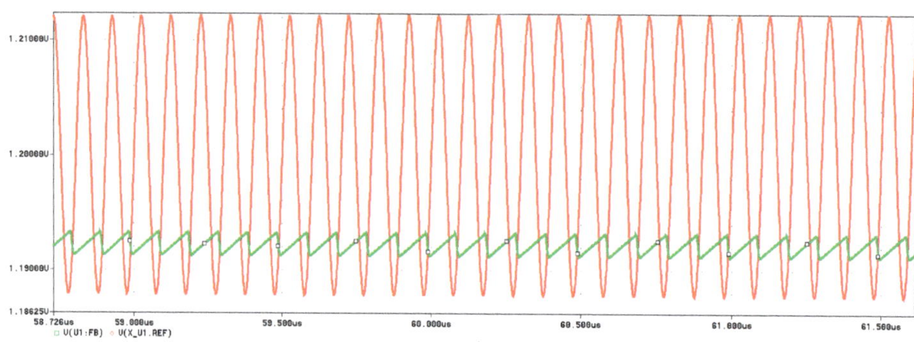

Figure 17. LDO model limitations at 10 MHz.

Table 3 shows the relative errors with respect to the theoretical curve over the frequency range of the initial PSRR method and our method.

Table 3. Original PSRR error and New PSRR error vs. Frequency.

Frequency	Theoretical PSRR	Old PSRR Error	New PSRR Error
[Hz]	[dB]	[%]	[%]
10	70	14.28	0
20	70	14.28	0
40	68	13.23	0.13
100	62	8.06	3.25
200	56	5.35	0.08
500	50	8.00	0.03
1 k	45	11.11	0.01
2 k	41	17.07	1.01
5 k	38	28.94	0.01
10 k	35	37.14	0
50 k	39	23.07	0.01
70 k	40	17.50	0.01
100 k	38	5.26	0.01
200 k	30	40.00	2.90
300 k	27	66.66	7.08
500 k	25	100.00	5.84
1 M	21	161.90	27.41
2 M	25	144.00	27.20
4 M	39	74.35	0.10
10 M	20	275.00	96.85

A graph is plotted in Figure 18 based on the results in Table 3.

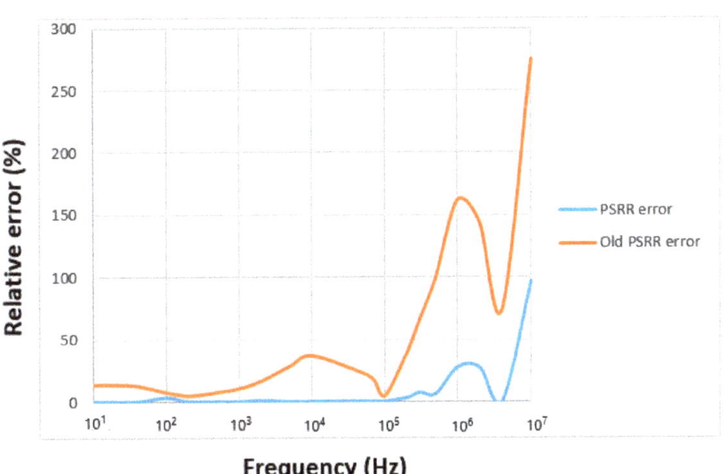

Figure 18. Original PSRR approach and new PSSR approach relative errors.

The results in Table 3 and presented graphically in Figure 18 indicate that our new model delivers high quality performance for frequencies lower than 500 kHz, resulting in a relative error of around 0% to 7%, compared with the old PSRR model given in [14,19], which has significantly higher values at any frequency point (from around 5% to 100%). Moreover, we can observe that both the old PSRR and the new PSRR errors increase significantly over 500 kHz. We found that this is due to the error amplifier EA (Figure 7a), which must be redesigned in order to increase PSRR's performance at high frequencies; however, this is not treated in this paper.

6. Conclusions

In this paper, we proposed a new method for improving the PSSR response of automotive LDO behavioral models for frequencies below 500 kHz, which is based on mathematical relations combined with circuits' relations.

We first used an existing commercially available automotive LDO model (TPS785-Q1 from Texas Instruments), which is available on Texas Instruments' official website [19]. We began by simulating the original model and plotted its PSRR characteristic, then we built a new PSRR model and integrated it into the LDO model, from which we eliminated the previous PSRR approach. The proposed method is not linear, so the PSRR characteristic was plotted using transient simulations. During the simulation phase, we noticed that the PSRR characteristic behaves extremely well at frequencies below 500 kHz, having an error lower than 7%, whereas for frequencies over 500 kHz up to 10 MHz, we concluded that the inaccurate behavior of the error amplifier greatly influences the PSRR response.

The implementation achieved in this paper proves extremely important in the automotive domain, in which simulations are usually chosen over real testing. Since the LDO is one of the most used power supply circuits in cars, this totally justifies the need for accurate LDO models. One of the most critical requirements of the LDO is the PSRR, which has not been explicitly addressed until now in circuit modeling. Most of the exiting commercially available LDO models show basic behavior and do not model the PSRR characteristic. Newer LDO models also include very simple PSRR functionality, but this is inaccurate and does not model the real characteristic properly over the functioning frequency range. Our work provides an optimized PSRR modeling method that models the real PSRR characteristic accurately for frequencies below 500 kHz, thus filling an important gap in the field. In order to accurately model the entire frequency range, the error amplifier also needs to exhibit accurate behavior, and this error amplifier redesign represents a future research direction.

Other future research directions consist of the enhancement of the current PSRR approach to also support variation with the load capacitor. The ripple of the output current needs to be measured using the same methodology as for the output voltage presented in this paper; the current ripple frequency using the RMS integration should then be determined, and another variation added to the existing one.

Author Contributions: Conceptualization, I.-C.G., A.F. and L.A.P.; methodology, I.-C.G., A.F. and L.A.P.; software, I.-C.G.; validation, I.-C.G., A.F. and L.A.P.; formal analysis, I.-C.G., A.F. and L.A.P.; investigation, I.-C.G.; resources, I.-C.G., A.F. and L.A.P.; data curation, I.-C.G., A.F. and L.A.P.; writing—original draft preparation, I.-C.G.; writing—review and editing, A.F. and L.A.P.; visualization, A.F. and L.A.P.; supervision, A.F. and L.A.P.; project administration, I.-C.G., A.F. and L.A.P.; funding acquisition, A.F. and L.A.P. All authors have read and agreed to the published version of the manuscript.

Funding: This work was supported by a grant of the Romanian Ministry of Education and Research, CCCDI-UEFISCDI, project number PN-III-P2-2.1-PED-2019-2862, within PNCDI III.

Institutional Review Board Statement: Not applicable.

Informed Consent Statement: Not applicable.

Conflicts of Interest: The authors declare no conflict of interest.

References

1. Maschotta, R.; Wichmann, A.; Zimmermann, A.; Gruber, K. Integrated Automotive Requirements Engineering with a SysML-Based Domain-Specific Language. In Proceedings of the 2019 IEEE International Conference on Mechatronics (ICM), Ilmenau, Germany, 18–20 March 2019. [CrossRef]
2. Rincon-Mora, G.A. System Considerations. In *Analog IC Design with Low Dropout Regulators*, 2nd ed.; McGraw-Hill: New York, NY, USA, 2009; p. 3.

3.   Wu, J.; Boyer, A.; Li, J.; Vrignon, B.; Ben Dhia, S.; Sicard, E.; Shen, R. Modeling and Simulation of LDO Voltage Regulator Susceptibility to Conducted EMI. *IEEE Trans. Electromagn. Compat.* **2014**, *56*, 726–735. [CrossRef]

4.   Sobhan Bhuiyan, M.A.; Hossain, M.R.; Minhad, K.N.; Haque, F.; Hemel, M.S.K.; Md Dawi, O.; Ibne Reaz, M.B.; Ooi, K.J.A. CMOS Low-Dropout Voltage Regulator Design Trends: An Overview. *Electronics* **2022**, *11*, 193. [CrossRef]

5.   Joo, J.; Sun, Y.; Lee, J.; Kong, S.; Kang, S.; Song, I.; Hwang, C. Modeling of Power Supply Noise Associated with Package Parasitics in an On-Chip LDO Regulator. In Proceedings of the 2021 IEEE International Joint EMC/SI/PI and EMC Europe Symposium, Raleigh, NC, USA, 26 July–13 August 2021. [CrossRef]

6.   Raducan, C.; Neag, M. Slew-Rate Booster and Frequency Compensation Circuit for Automotive LDOs. *IEEE Trans. Circuits Syst. I Regul. Pap.* **2022**, *69*, 465–477. [CrossRef]

7.   Khan, M.; Chowdhury, M. Capacitor-less Low-Dropout Regulator (LDO) with Improved PSRR and Enhanced Slew-Rate. In Proceedings of the 2018 IEEE International Symposium on Circuits and Systems (ISCAS), Florence, Italy, 27–30 May 2018. [CrossRef]

8.   Jiang, Y.; Wang, L.; Wang, S.; Cui, M.; Zheng, Z.; Li, Y. A Low-Power, Fast-Transient Output-Capacitorless LDO with Transient Enhancement Unit and Current Booster. *Electronics* **2022**, *11*, 701. [CrossRef]

9.   Zhao, W.; Wei, P. PSpice system simulation application in electronic circuit design. In Proceedings of the 32nd Chinese Control Conference, Xi'an, China, 26–28 July 2013; pp. 8634–8636.

10.   Kang, X.; Kang, X.; Zhao, Z.; Ding, J.; Hu, Y.; Xu, D.; Sun, Q.; Zhang, D. Low-Dropout Regulator design with a simple structure for good high frequency PSRR performance based on Bandgap Circuit. In Proceedings of the 2019 IEEE 13th International Conference on ASIC (ASICON), Chongqing, China, 29 October–1 November 2019. [CrossRef]

11.   Nogawa, M.; Van Renterghem, K.L. *Electronics in Motion and Conversion*; Bodo's Power Systems: Laboe, Germany, 2006; pp. 44–46.

12.   Antunes Fernandes, P.M. High PSRR Low Drop-out Voltage Regulator (LDO). Master's Thesis, Technical University of Lisbon, Lisbon, Portugal, 2009.

13.   OPTIREG™ Linear TLS820D2ELVSE, Low Dropout Linear Voltage Regulator, rev. 1.0. Available online: https://www.infineon.com/dgdl/Infineon-TLS820D2EL%20VSE-DataSheet-v01_00-EN.pdf?fileId=5546d46279a6fbb20179ba9238327384 (accessed on 27 July 2020).

14.   TPS785-Q1 Automotive, 1-A, High-PSRR Low-Dropout Voltage Regulator with High Accuracy and Enable, Rev. B. Available online: https://www.ti.com/lit/ds/symlink/tps785-q1.pdf?ts=1648650370852&ref_url=https%253A%252F%252Fwww.mouser.ca%252F (accessed on 11 January 2022).

15.   Lee, H.; Kim, S. PSRR measurement method for PLL including power delivery network. In Proceedings of the 2017 IEEE Asia Pacific Microwave Conference (APMC), Kuala Lumpur, Malaysia, 13–16 November 2017. [CrossRef]

16.   Teel, J. *Understanding Power Supply Ripple Rejection in Linear Regulators*; Texas Instruments: Dallas, TX, USA, 2005.

17.   Nasrollahpour, M.; Hamedi-Hagh, S. Fast transient response and high PSRR low drop-out voltage regulator. In Proceedings of the 2016 IEEE Dallas Circuits and Systems Conference (DCAS), Arlington, TX, USA, 10 October 2016. [CrossRef]

18.   Zoche, J. Design of a High PSRR Multistage LDO with On-Chip Output Capacitor. In Proceedings of the SMACD/PRIME 2021, International Conference on SMACD and 16th Conference on PRIME, Online, 19–22 July 2021; pp. 1–4.

19.   Texas Instruments. Available online: https://www.ti.com/ (accessed on 10 February 2022).

20.   Tang, J.; Meng, Z.; Ouyang, L. Design of high PSRR linear regulator based on pre regulated Technology. In Proceedings of the 2021 9th International Symposium on Next Generation Electronics (ISNE), Changsha, China, 9–11 July 2021. [CrossRef]

21.   Coza, A.; Jurisic, D. Low-Noise and Low-Sensitivity Coupled Fourth-Order Low-Pass Filters. In Proceedings of the 2019 42nd International Convention on Information and Communication Technology, Electronics and Microelectronics (MIPRO), Opatija, Croatia, 20–24 May 2019; pp. 128–132. [CrossRef]

22.   Stanescu, D.; Ardeleanu, M.; Stan, A. Designing, simulation and testing of low current passive filters used in the didactic activity. In Proceedings of the 2017 International Conference on Modern Power Systems (MPS), Cluj-Napoca, Romania, 6–9 June 2017. [CrossRef]

23.   Denisenko, D.Y.; Prokopenko, N.N.; Butyrlagin, N.V. All-Pass Second-Order Active RC-Filter with Pole Q-Factor's Independent Adjustment on Differential Difference Amplifiers. In Proceedings of the 2019 IEEE East-West Design & Test Symposium (EWDTS), Batumi, Georgia, 13–16 September 2019; pp. 1–4. [CrossRef]

24.   Belega, D.; Gasparesc, G. Accurate Measurement of the rms of a Sine-wave by Means of Low-Cost rms-to-dc Convertes. In Proceedings of the 2020 International Symposium on Electronics and Telecommunications (ISETC), Timisoara, Romania, 5–6 November 2020. [CrossRef]

25.   Men, X.; Liu, H.; Chen, N.; Li, F. A new time domain filtering method for calculating the RMS value of vibration signals. In Proceedings of the 2018 13th IEEE Conference on Industrial Electronics and Applications (ICIEA), Wuhan, China, 31 May–2 June 2018. [CrossRef]

26.   *PSpice Reference Guide*, 2nd ed.; Cadence Design Systems: Tigard, OR, USA, 2000.

 *mathematics*

MDPI

Article

On Active Vibration Absorption in Motion Control of a Quadrotor UAV

Francisco Beltran-Carbajal [1,*], Hugo Yañez-Badillo [2], Ruben Tapia-Olvera [3], Antonio Favela-Contreras [4], Antonio Valderrabano-Gonzalez [5] and Irvin Lopez-Garcia [1]

[1] Departamento de Energía, Universidad Autónoma Metropolitana, Unidad Azcapotzalco, Av. San Pablo No. 180, Col. Reynosa Tamaulipas, Mexico City 02200, Mexico; ilg@azc.uam.mx

[2] Departamento de Investigación, Tecnológico de Estudios Superiores de Tianguistenco, Santiago Tianguistenco 52650, Mexico; hugo_mecatronica@test.edu.mx

[3] Departamento de Energía Eléctrica, Universidad Nacional Autónoma de México, Mexico City 04510, Mexico; rtapia@fi-b.unam.mx

[4] Escuela de Ingeniería y Ciencias, Tecnologico de Monterrey, Av. Eugenio Garza Sada 2501, Monterrey 64849, Mexico; antonio.favela@tec.mx

[5] Facultad de Ingeniería, Universidad Panamericana, Álvaro del Portillo 49, Zapopan 45010, Mexico; avalder@up.edu.mx

* Correspondence: fbeltran@azc.uam.mx

Citation: Beltran-Carbajal, F.; Yañez-Badillo, H.; Tapia-Olvera, R.; Favela-Contreras, A.; Valderrabano-Gonzalez, A.; Lopez-Garcia, I. On Active Vibration Absorption in Motion Control of a Quadrotor UAV. *Mathematics* **2022**, *10*, 235. https://doi.org/10.3390/math10020235

Academic Editor: Paolo Mercorelli

Received: 27 October 2021
Accepted: 31 December 2021
Published: 13 January 2022

Publisher's Note: MDPI stays neutral with regard to jurisdictional claims in published maps and institutional affiliations.

Abstract: Conventional dynamic vibration absorbers are physical control devices designed to be coupled to flexible mechanical structures to be protected against undesirable forced vibrations. In this article, an approach to extend the capabilities of forced vibration suppression of the dynamic vibration absorbers into desired motion trajectory tracking control algorithms for a four-rotor unmanned aerial vehicle (UAV) is introduced. Nevertheless, additional physical control devices for mechanical vibration absorption are unnecessary in the proposed motion profile reference tracking control design perspective. A new dynamic control design approach for efficient tracking of desired motion profiles as well as for simultaneous active harmonic vibration absorption for a quadrotor helicopter is then proposed. In contrast to other control design methods, the presented motion tracking control scheme is based on the synthesis of multiple virtual (nonphysical) dynamic vibration absorbers. The mathematical structure of these physical mechanical devices, known as dynamic vibration absorbers, is properly exploited and extended for control synthesis for underactuated multiple-input multiple-output four-rotor nonlinear aerial dynamic systems. In this fashion, additional capabilities of active suppression of vibrating forces and torques can be achieved in specified motion directions on four-rotor helicopters. Moreover, since the dynamic vibration absorbers are designed to be virtual, these can be directly tuned for diverse operating conditions. In the present study, it is thus demonstrated that the mathematical structure of physical mechanical vibration absorbers can be extended for the design of active vibration control schemes for desired motion trajectory tracking tasks on four-rotor aerial vehicles subjected to adverse harmonic disturbances. The effectiveness of the presented novel design perspective of virtual dynamic vibration absorption schemes is proved by analytical and numerical results. Several operating case studies to stress the advantages to extend the undesirable vibration attenuation capabilities of the dynamic vibration absorbers into trajectory tracking control algorithms for nonlinear four-rotor helicopter systems are presented.

Keywords: vibration control; dynamic vibration absorbers; aerial vehicles; quadrotor; motion tracking control

---

1. Introduction

Undesirable mechanical vibrations can be manifested in many engineering systems. Harmful vibrations could provoke several relevant troubles in mechanical structures such as rapid wear of machine parts, material fatigue, excessive noise, loss of efficiency, poor finishes of manufactured products, malfunction of instrumentation, discomfort to people, loosening of fasteners, and damage to property [1]. Forced harmonic vibrations can be induced by external oscillating disturbance forces or torques affecting the mechanical system dynamics. An alternative solution for unwanted vibration attenuation is the design of physical mechanical devices known as dynamic vibration absorbers. These vibration control devices are physically constituted by mechanical elements of inertia, stiffness, and damping [2]. Tuned mass damper, vibration neutralizer, anti-vibrator, dynamic damper, and shock absorber are some names commonly used in the literature to describe a dynamic vibration absorber as well [2–4]. These control devices are suitably tuned and coupled to mechanical structures (primary or main systems) to be protected against harmful forced harmonic vibrations. Since mechanical systems are prone to be excited at a specific frequency or subjected to multiple frequency excitations [5], diverse configurations for dynamic vibration absorbers have been proposed [2]. For the former case, passive vibration absorbers are tuned to efficiently suppress specific frequency harmonic excitations into an attenuation frequency band, without using some external energy source. Nevertheless, the dynamic performance of passive vibration absorbers could be deteriorated for certain operation conditions where uncertain variable excitation frequencies are present. In consequence, active dynamic vibration absorbers have been proposed to improve the capability of vibration suppression on vibrating mechanical systems under the action of variable frequency excitation forces (see [6,7] and references therein).

Unmanned aerial vehicles (UAV) possess mechanical structures that can be also subjected to external vibrating disturbance forces and torques due to their interaction with the operation environment. Vibrations could produce damage and faults on the aerial vehicle and reduce the specified flight performance. Malfunction of control instrumentation could be caused by vibrating disturbances as well. Increasing the technological development of unmanned aerial vehicles is thus focused on suppressing these harmful disturbing dynamics. The quadrotor UAV is a rotorcraft with four rotors vertically oriented. A quadrotor exhibits an underactuated and multivariable, complex nonlinear dynamic system model. The features of a quadrotor such as a symmetrical cross frame, flight endurance, payload, and its flight capabilities of hovering, take-off, and landing have attracted researchers' attention, since controlling this multiple-input multiple-output (MIMO) nonlinear dynamic system is a challenge [8]. In this sense, the design of control laws to follow motion profiles and simultaneously attenuate forced harmonic vibrations on this nonlinear multivariable aerial vehicle—using the control inputs generated by its four rotors only—constitutes an equally open challenging research subject. Additional actuators and sensors can be also integrated in the aerial mechanical system to implement efficient active vibration controllers, which constitute another control design approach. Nevertheless, additional instrumentation increases the active vibration control implementation costs.

Multiple applications for this four-rotor aerial vehicle, such as agricultural spraying and crop monitoring [9], surveillance [10], monitoring and inspection [11,12], package delivering [13], mapping [14], rescuing operations [15], etc. [16,17], rely on the quadrotor's controlled translational and rotational precision motions at prescribed velocities. The improvement of crop productivity, reliable data acquisition for digital analysis, reduction of rescue time, guaranteed detection of a target of interest, optimization of flight time regarding the payload, and efficient power lines inspection, usually in uncontrollable hostile environments, are some examples of the issues researchers are interested in solving for the aforementioned applications. Moreover, some advanced qualities, such as aggressive and agile manoeuvres [18], obstacle avoidance for improvement of autonomous navigation [19], and coordinated swarm flight [20], are open research challenge topics for quadrotor vehicles. Thus, the design of efficient and robust automatic control algorithms is a fundamental issue, which has been studied by multiple research and engineering groups all around the world.

The development of new and more efficient control strategies for controlling the quadrotor system dynamics is constantly reported in the literature, where it is required to take into consideration issues of robustness to parameter uncertainty, external disturbances, and sensor noise [21]. In this sense, several important studies on linear and nonlinear control design strategies have been proposed; improved versions of the conventional proportional-integral-derivative control and the linear quadratic regulator have been reported in [22,23], respectively. Artificial intelligence algorithms such as artificial neural networks, bio-inspired optimization, and fuzzy logic have been also successfully included in motion control design [24–27]. Additionally, several disturbances observers have been properly reported in [28–31] for compositing interesting control schemes. Moreover, nonlinear controllers based on sliding modes [32–34] and backstepping [35–37] approaches have been suitably introduced to deal with disturbances and uncertainties in quadrotor vehicles and, in some studies, further extended for fault tolerant controllers. In addition, important contributions based on theories such as robust H_∞ control [38], model predictive control [39], generalized proportional-integral control [40], energy-based control [41,42], optimal control [21,43], Lyapunov-based control [44], adaptive control [45,46], etc., have vastly improved the performance of quadrotors in regulation and tracking tasks. Nevertheless, to the best knowledge of the authors, there is no previous report taking advantage of the capabilities of virtual vibration absorbers for suppressing undesirable harmonic forces and torques in quadrotor motion trajectory tracking control design.

In this paper, different from other important control contributions on unmanned aerial vehicles reported in the literature, the inclusion of active vibration suppression capabilities based on nonphysical (virtual) dynamic vibration absorbers in desired motion tracking controllers on quadrotors is thus considered. Moreover, in contrast to real-time disturbance estimation techniques [47], oscillating disturbances can be automatically reconstructed online by the proper synthesis of virtual dynamic vibration absorbers on harmonically excited aerial vehicles, as described in this paper.

In this context, quadrotor position and attitude can be affected by undesired vibrating motion induced by external disturbance forces and torques. The incorporation of dynamic vibration absorbers as mechanical vibration attenuation mechanisms into motion control schemes for diverse unmanned aerial vehicles subjected to forced harmonic vibrations stands as a feasible alternative approach as proved in the present study. An analysis of virtual dynamic vibration absorbers [48] as an alternative efficient control design approach for suitable compensation of undesirable oscillating forces and torques in a controlled quadrotor vehicle is presented. In this fashion, compared with other control design techniques, robustness capabilities against forced harmonic vibrations are added to controllers designed to regulate the aerial vehicle operation towards planned motion reference trajectories.

Thus, a new approach to extend the structural property of the dynamic vibration absorbers for desired motion profile tracking control and simultaneous forced harmonic vibration suppression for MIMO underactuated nonlinear quadrotor systems disturbed by forced harmonic vibrations is proposed in the present study. In contrast to other valuable control design methods, virtual dynamic vibration absorbers are suitably tuned and embedded into motion control algorithms developed to perform regulation and trajectory tracking tasks on nonlinear quadrotor helicopters subjected to exogenous harmonic excitations. In addition, different from conventional physical dynamic vibration absorbers, virtual vibration absorption mechanisms can be directly tuned for diverse operating conditions. Then, vibration attenuation bands of virtual dynamic vibration absorbers can be conveniently adapted online or offline. Disturbance forces and torques are actively compensated in real-time by the non-physical dynamic vibration absorbers. Only information about trajectory tracking errors for the active suppression of specified frequency forces and torques through multiple virtual dynamic vibration absorbers is required. Furthermore, several operating case studies are included to numerically validate the satisfactory performance of the proposed control scheme for stabilization at a specific operating position as well as for desired motion reference trajectory tracking. It is thus demonstrated that the mathematical structure of physical mechanical vibration absorbers can be advantageously used for the design of active vibration control schemes for MIMO underactuated nonlinear four-rotor aerial vehicles without requiring additional conventional mechanical devices for mechanical vibration control, which constitutes the main difference with respect to existing control contributions for quadrotors. The new presented active vibration control design perspective can be extended to other configurations of MIMO nonlinear unmanned aerial dynamic systems.

The remainder of this paper is structured as follows: In the second part, the nonlinear dynamic model of the harmonically excited four-rotor aerial vehicle is summarized. In the third section, a virtual dynamic vibration absorption approach for the synthesis of motion tracking controllers for the underactuated MIMO nonlinear quadrotor system is presented. Here, proportional-derivative (PD)-like controllers are implemented plus dynamic compensation terms provided by the virtual vibration absorption stage in which trajectory tracking error signals are only necessary for the reconstruction and active rejection of external harmonic excitations. This PD-like control structure plus virtual dynamic vibration absorbers can be implemented to satisfactorily perform the motion planning specified for a quadrotor subjected to vibrating disturbances. Nevertheless, virtual vibration absorbers can be extended and embedded into other preferred nonlinear control techniques for efficient and robust tracking of desired motion reference profiles on unmanned aerial vehicles. Next, in the simulation results section, several operating scenarios are developed, where the quadrotor performs both regulation and trajectory tracking under the effects of exogenous harmonic disturbance forces and torques. The efficient and robust tracking of references trajectories planned online and offline is confirmed. Lastly, some highlights and comments are presented in the Conclusions section.

2. Nonlinear Four-Rotor Aerial Vehicle Dynamics

The quadrotor is the most common aerial platform, consisting of four rotors mounted on a cross frame. This underactuated nonlinear system possesses six degrees of freedom and four control inputs only. A quadrotor is usually designed to have a rigid body with a symmetrical mechanical structure, yielding a diagonal and positive definite inertia tensor, where two right-handed coordinated systems are used to describe its pose. Firstly, a global coordinate system $\mathcal{I}$ attached to the earth is established with axes X, Y, and Z. Subsequently, the second selected coordinate system is fixed to the quadrotor body, which is identified as $\mathcal{B}$, with axes X', Y', and Z', as shown in Figure 1. The control inputs are achieved by differential control of the thrust generated for each rotor (related to the angular speed): the total thrust or vertical force u and the actuation torques τ_ϕ, τ_θ, and τ_ψ. Translations and rotations on the quadrotor are then provoked by a variation of the angular velocity of each rotor: rotors 1 and 3 spin counter-clockwise, and rotors 2 and 4 in the opposite direction. The pitching moment τ_θ is originated by a difference in velocity of rotors 1 and 3. On the other hand, the rolling moment τ_ϕ is generated by interaction of the produced forces by rotors 2 and 4. Finally, the yawing moment τ_ψ occurs when angular velocities of lateral rotors are increased or decreased. The main control force u stands for the sum of all the vertical forces produced by the rotors.

The produced forces by the rotors and the control inputs are related as follows [49]:

$$u = \sum_{i=1}^{4} F_i$$

$$\tau_\psi = \sum_{i=1}^{4} \tau_{M_i}$$

$$\tau_\theta = (F_3 - F_1)l$$

$$\tau_\phi = (F_2 - F_4)l \tag{1}$$

where l stands for the longitude from the quadrotor centre of mass to the rotation axes of the rotors, and the induced torque by each electric motor M_i is represented by τ_{M_i}. Additionally, F_i and τ_{M_i} depend on the rotors' blades geometry by means of the thrust and drag coefficients. As observed in Equation (1), by regulating the rotor angular velocities, it is possible to achieve motion in different directions in the space. Thus, by a suitable combination of the control inputs, a quadrotor is able to perform the desired motion reference trajectory tracking as well as stable regulation around a specified position. Reported studies on dynamic modelling and control of the quadrotor [38,49–52] have been validated through exhaustive numerical simulations and experimental tests, where the Newton–Euler and Euler–Lagrange formalisms are used to describe translational and rotational quadrotor dynamics. It is a common practise to define a vector of generalized coordinates q in order to represent the quadrotor pose:

$$q = [x\,y\,z\,\phi\,\theta\,\psi]^T \in \mathbb{R}^6 \tag{2}$$

where the altitude of the system is described by the set of Euler angles ϕ, θ, and ψ, and the position by the x, y, and z coordinates, both regarding the inertial reference frame $\mathcal{I}$, as depicted in Figure 1.

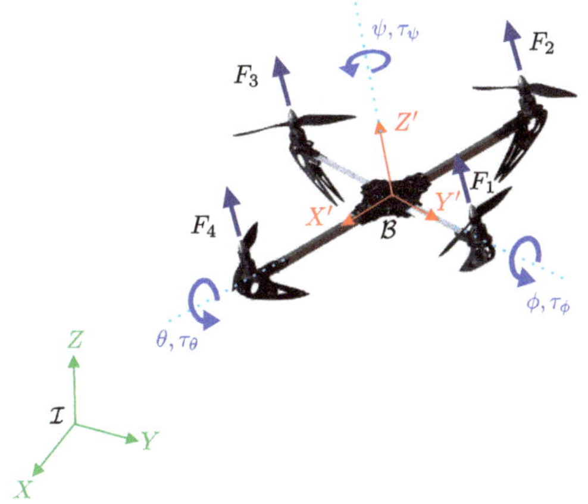

Figure 1. A flying quadrotor subjected to four main control inputs.

In the present study, the following nonlinear dynamic model of the quadrotor is considered [8,51,53,54]:

$$
\begin{aligned}
m\ddot{x} &= -u\sin\theta + f_x(t)\\
m\ddot{y} &= u\cos\theta\sin\phi + f_y(t)\\
m\ddot{z} &= u\cos\theta\cos\phi - mg + f_z(t)\\
J_x\ddot{\phi} &= (J_y - J_z)\dot{\theta}\dot{\psi} - J_r\Omega_n\dot{\theta} + \tau_\phi + \tau_{\phi_d}(t)\\
J_y\ddot{\theta} &= (J_z - J_x)\dot{\phi}\dot{\psi} + J_r\Omega_n\dot{\phi} + \tau_\theta + \tau_{\theta_d}(t)\\
J_z\ddot{\psi} &= (J_x - J_y)\dot{\phi}\dot{\theta} + \tau_\psi + \tau_{\psi_d}(t)
\end{aligned}
\tag{3}
$$

where m is the total quadrotor mass, J_x, J_y and J_z are the diagonal elements of the tensor of inertia expressed in the body frame, and g is the acceleration constant of gravity. $f_i(t)$, $i = x, y, z$, and $\tau_{j_d}(t)$, $j = \phi, \theta, \psi$, represent exogenous harmonic torques and forces affecting translational and rotational dynamics. In the next section, nonphysical dynamic vibration absorbers are designed and embedded into desired motion tracking control signals to suppress the unwanted vibrating disturbances $f_i(t)$ and $\tau_{j_d}(t)$.

3. An Active Vibration Control Design Approach for Quadrotors

To depict the central ideas of the proposed control scheme to track planned motion profiles with additional vibration suppression capability for four-rotor helicopters, consider the main or primary vibrating system Σ_1 to be protected against harmonic vibrations induced by excitation forces f using an active dynamic vibration absorber Σ_2 as follows

$$\begin{aligned} \Sigma_1: \quad & m\ddot{y} + c\dot{y} + ky = k_a(y_a - y) + f(t) \\ \Sigma_2: \quad & m_a\ddot{y}_a + c_a\dot{y}_a + k_a(y_a - y) = u_a \end{aligned} \tag{4}$$

In this representation, y is the position output variable of the main vibrating system, and y_a is the position signal of the dynamic vibration absorber. u_a represents an exogenous control force which can be implemented to extend the capabilities of the vibration absorber for closed-loop reference trajectory tracking and simultaneous active vibration suppression on the primary system. Positive parameters of mass, stiffness, and viscous damping of the main mechanical system subjected to undesirable vibrations are respectively denoted by m, k, and $c > 0$. Physical parameters of mass, stiffness, and damping of the dynamic vibration absorption device Σ_2 are denoted by m_a, k_a, and c_a. Harmonic disturbance forces $f(t)$ are described as

$$f(t) = F_0 \sin(\Omega t) \tag{5}$$

where the amplitude and frequency parameters are denoted by F_0 and Ω. The tuning frequency of the dynamic vibration absorber is then set as

$$\omega_a = \sqrt{\frac{k_a}{m_a}} = \Omega \tag{6}$$

A high forced vibration attenuation level closed to the excitation frequency Ω can then be achieved using a weakly damped absorber, that is, by selecting the viscous damping parameter as $c_a \approx 0$.

Then, consider the disturbed vibrating mechanical system Σ_1 in the state-space representation

$$\Sigma_1: \quad \begin{aligned} \dot{y}_1 &= y_2 \\ \dot{y}_2 &= -\frac{k}{m}y_1 - \frac{c}{m}y_2 + \frac{1}{m}u + \frac{1}{m}f \\ y &= y_1 \end{aligned} \tag{7}$$

Here, the control input u should be designed so that undesirable harmonic disturbances f affecting the output signal y are actively suppressed. In the present study, virtual (nonphysical) dynamic vibration absorbers Σ_2 are proposed to be properly embedded into the dynamic controller as follows:

$$\Sigma_2: \quad \begin{aligned} \dot{\eta}_1 &= \eta_2 \\ \dot{\eta}_2 &= -\frac{k_a}{m_a}(\eta_1 - y_1) \\ u &= k_a(\eta_1 - y_1) \end{aligned} \tag{8}$$

In this fashion, harmonic vibration absorption capabilities can be added to dynamic controllers. The design parameters of the virtual dynamic vibration absorber should be selected to satisfy $\Omega^2 = k_a/m_a$. The closed-loop system dynamics is then governed by

$$\begin{aligned} \dot{y}_1 &= y_2 \\ \dot{y}_2 &= y_3 \\ \dot{y}_3 &= y_4 \\ \dot{y}_4 &= -\frac{kk_a}{mm_a}y_1 - \frac{ck_a}{mm_a}y_2 - \left(\frac{k + k_a}{m} + \frac{k_a}{m_a}\right)y_3 - \frac{c}{m}y_4 \end{aligned} \tag{9}$$

with $y_1 = y$, $y_2 = \dot{y}$, $y_3 = \ddot{y}$, and $y_4 = y^{(3)}$. Notice that the asymptotic stability of the closed-loop system is guaranteed by selecting the design parameters as $m_a, k_a > 0$, which is proven as follows. The time derivative of the Lyapunov function candidate

$$V(y_1, y_2, y_3, y_4) = \frac{1}{2}ky_1^2 + \frac{1}{2}my_2^2 + \frac{1}{2}k_a\left(\frac{k}{k_a}y_1 + \frac{c}{k_a}y_2 + \frac{m}{k_a}y_3\right)^2$$
$$+ \frac{1}{2}m_a\left(\frac{k+k_a}{k_a}y_2 + \frac{c}{k_a}y_3 + \frac{m}{k_a}y_4\right)^2 \tag{10}$$

along the trajectories of the system dynamics (9) yields

$$\dot{V}(y_1, y_2, y_3, y_4) = -cy_2^2 \leq 0 \tag{11}$$

Therefore, from the LaSalle's invariance theorem,

$$\lim_{t \to \infty} y_i = 0, \quad i = 1, \ldots, 4 \tag{12}$$

Then, dynamic controllers based on virtual vibration absorbers (8) for active vibration suppression and desired motion profile tracking on four-rotor helicopters are proposed as follows:

$$u = \frac{1}{\cos\phi\cos\theta}(m\ddot{z}^\star + v_z + mg)$$
$$\tau_\phi = J_x\ddot{\phi}^\star + v_\phi - (J_y - J_z)\dot{\theta}\dot{\psi} + J_r\dot{\theta}\Omega_n$$
$$\tau_\theta = J_y\ddot{\theta}^\star + v_\theta - (J_z - J_x)\dot{\phi}\dot{\psi} - J_r\dot{\phi}\Omega_n$$
$$\tau_\psi = J_z\ddot{\psi}^\star + v_\psi - (J_x - J_y)\dot{\phi}\dot{\theta} \tag{13}$$

with

$$\Sigma_i : \quad \begin{aligned} \dot{\eta}_{1,i} &= \eta_{2,i} \\ \dot{\eta}_{2,i} &= -\frac{k_{a,i}}{m_{a,i}}(\eta_{1,i} - e_i) \\ v_i &= -\beta_{0,i}e_i - \beta_{1,i}\dot{e}_i + k_{a,i}(\eta_{1,i} - e_i) \end{aligned} \tag{14}$$

where $e_i = i - i^\star$, $i = x, y, z, \psi, \theta, \phi$, stands for tracking errors; $x^\star$, $y^\star$, $z^\star$, and $\psi^\star$ are reference profiles planned offline; and $\theta^\star$ and $\phi^\star$ are trajectories computed online by

$$\tan\theta^\star = -\frac{m\ddot{x}^\star + v_x}{m\ddot{z}^\star + v_z + mg}\cos\phi \tag{15}$$

$$\tan\phi^\star = \frac{m\ddot{y}^\star + v_y}{m\ddot{z}^\star + v_z + mg} \tag{16}$$

Hence, closed-loop dynamics of the trajectory tracking errors is governed by

$$e_x^{(4)} + \frac{\beta_{1,x}}{m}e_x^{(3)} + \left(\frac{\beta_{0,x} + k_{a,x}}{m} + \frac{k_{a,x}}{m_{a,x}}\right)\ddot{e}_x + \frac{\beta_{1,x}k_{a,x}}{mm_{a,x}}\dot{e}_x + \frac{\beta_{0,x}k_{a,x}}{mm_{a,x}}e_x = 0$$

$$e_y^{(4)} + \frac{\beta_{1,y}}{m}e_y^{(3)} + \left(\frac{\beta_{0,y} + k_{a,y}}{m} + \frac{k_{a,y}}{m_{a,y}}\right)\ddot{e}_y + \frac{\beta_{1,y}k_{a,y}}{mm_{a,y}}\dot{e}_y + \frac{\beta_{0,y}k_{a,y}}{mm_{a,y}}e_y = 0$$

$$e_z^{(4)} + \frac{\beta_{1,z}}{m}e_z^{(3)} + \left(\frac{\beta_{0,z} + k_{a,z}}{m} + \frac{k_{a,z}}{m_{a,z}}\right)\ddot{e}_z + \frac{\beta_{1,z}k_{a,z}}{mm_{a,z}}\dot{e}_z + \frac{\beta_{0,z}k_{a,z}}{mm_{a,z}}e_z = 0$$

$$e_\phi^{(4)} + \frac{\beta_{1,\phi}}{J_x}e_\phi^{(3)} + \left(\frac{\beta_{0,\phi} + k_{a,\phi}}{J_x} + \frac{k_{a,\phi}}{m_{a,\phi}}\right)\ddot{e}_\phi + \frac{\beta_{1,\phi}k_{a,\phi}}{J_x m_{a,\phi}}\dot{e}_\phi + \frac{\beta_{0,\phi}k_{a,\phi}}{J_x m_{a,\phi}}e_\phi = 0$$

$$e_\theta^{(4)} + \frac{\beta_{1,\theta}}{J_y}e_\theta^{(3)} + \left(\frac{\beta_{0,\theta} + k_{a,\theta}}{J_y} + \frac{k_{a,\theta}}{m_{a,\theta}}\right)\ddot{e}_\theta + \frac{\beta_{1,\theta}k_{a,\theta}}{J_y m_{a,\theta}}\dot{e}_\theta + \frac{\beta_{0,\theta}k_{a,\theta}}{J_y m_{a,\theta}}e_\theta = 0$$

$$e_\psi^{(4)} + \frac{\beta_{1,\psi}}{J_z}e_\psi^{(3)} + \left(\frac{\beta_{0,\psi} + k_{a,\psi}}{J_z} + \frac{k_{a,\psi}}{m_{a,\psi}}\right)\ddot{e}_\psi + \frac{\beta_{1,\psi}k_{a,\psi}}{J_z m_{a,\psi}}\dot{e}_\psi + \frac{\beta_{0,\psi}k_{a,\psi}}{J_z m_{a,\psi}}e_\psi = 0 \tag{17}$$

Therefore, asymptotic tracking of reference trajectories specified for the operation of the aerial vehicle can be achieved by the proper selection of the control design parameters, $\beta_{0,i}, \beta_{1,i} > 0$, $i = x, y, x, \phi, \theta, \psi$, so that the characteristic polynomials associated with Equation (17) are Hurwitz polynomials. Then,

$$\lim_{t\to\infty} e_i(t) = 0 \quad \Rightarrow \quad \lim_{t\to\infty} i(t) = i^\star(t) \tag{18}$$

Here, $\theta^\star$ and $\phi^\star$ are used to regulate the x and y position variables toward the desired motion reference trajectories in the horizontal plane. In this fashion, the underactuation condition is then solved. Then, control gains of angular motion should be suitably chosen to be much faster than x and y translation motion. Notice that in order to avoid (uncontrollable) singular configurations of the quadrotor model (3), θ and ϕ angles should be constrained to take values in the open interval $(-\pi/2, \pi/2)$.

Figure 2 summarizes the proposed control scheme. Here, the $\boldsymbol{\alpha} = [\theta\ \phi]$ and $\boldsymbol{\alpha}^\star = [\theta^\star\ \phi^\star]$ vectors, containing both real and references values for roll and pitch angles, are used for purposes of simplicity in the representation. Moreover, vectors $\boldsymbol{\xi} = [x\ y\ z]$ and $\boldsymbol{\xi}^\star = [x^\star\ y^\star\ z^\star]$ stand for the real and planned reference positions. Notice that $e_{\boldsymbol{\xi}}$ represents an error vector containing the x, y and z position errors; then, similar conditions hold for $v_{\boldsymbol{\xi}}, e_{\boldsymbol{\alpha}}, \Sigma_{\boldsymbol{\xi}}$, and $\Sigma_{\boldsymbol{\alpha}}$. Additionally, the Online TG block represents the trajectory generator mechanism implemented by solving the set equations introduced in (15) and (16). On the other hand, it can be seen that the proposed PD-like motion controllers require velocity and position errors. However, virtual vibration absorbers only need measurements of the quadrotor primary system displacements in order to attenuate the forced vibrations acting on it.

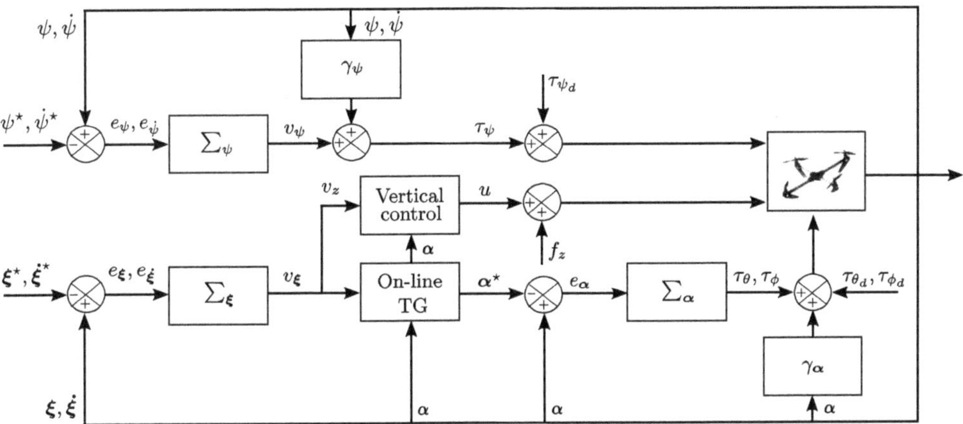

Figure 2. Desired motion tracking control scheme based on virtual dynamic vibration absorbers.

4. Numerical Simulation Results

Several computer simulation results are additionally presented in this section to show the efficient performance of the proposed dynamic control scheme when the quadrotor is tasked to reach desired position trajectories $(x^\star, y^\star, z^\star)$. The external forced vibration suppression capability of the control technique based on virtual vibration absorbers is also confirmed. The numerical method of Runge–Kutta Fehlberg 4/5 was selected for computer simulations, where a fixed time step of 1 ms is adopted. Two scenarios for the operation of the aerial vehicle are described to portray the effectiveness of the introduced desired motion trajectory tracking control approach.

4.1. Scenario 1: Hover Stabilization

In the first case, the quadrotor performs motion regulation around a fixed or stable position (hovering). Thus, let us introduce the position reference profile in (19) that is implemented to obtain smooth transitions between initial and final x, y, and z positions:

$$\Gamma^\star = \begin{cases} \Gamma_0 & 0 \leq t < T_1 \\ \Gamma_0 + \left(\Gamma_f - \Gamma_0\right)\mathcal{B}_z(t, T_1, T_2) & T_1 \leq t \leq T_2 \\ \Gamma_f & t > T_2 \end{cases} \tag{19}$$

where Γ_0 and Γ_f stand for desired initial and final values of motion trajectories planned for the aerial vehicle, which is characterized by the set of parameters given in Table 1. Meanwhile, $\mathcal{B}_z$ is a Bézier polynomial [55] defined as

$$\mathcal{B}_z(t, T_1, T_2) = \sum_{k=0}^{n} h_k \left(\frac{t - T_1}{T_2 - T_1}\right)^k \tag{20}$$

with $n = 6$, and $h_1 = 252$, $h_2 = 1050$, $h_3 = 1800$, $h_4 = 1575$, $h_5 = 700$, $h_6 = 126$.

Table 1. Quadrotor system parameters for simulation scenarios.

Parameter	Units	Values
m	kg	0.98
g	m/s^2	9.81
l	m	0.25
J_x	kg m^2	0.012450
J_y	kg m^2	0.012450
J_z	kg m^2	0.024752

Thereafter, the first stage considers moving the quadrotor from an initial vertical position $z_0 = 0$ m to a desired height of $z_f = 10$ m within a time window of 6 s, starting after 1 s. Subsequently, the rotorcraft is displaced to desired positions in the horizontal plane, from initial conditions $x_0 = 0$ m and $y_0 = 0$ m to a final position defined by $x_f = 5$ m and $y_f = 6$ m in 10 and 15 s, respectively. On the other hand, the yaw angle is also simultaneously taken from $\psi_0 = 0.5$ rad to $\psi_f = 0$ rad by using the Bézier curve until time reaches the proximity of 19 s. Then, a time-varying trajectory is adopted as the planned reference, where

$$\psi^\star = 0.5\cos(0.25t)\,\text{m}$$
$$\dot{\psi}^\star = -0.1250\sin(0.25t)\,\text{m/s}$$
$$\ddot{\psi}^\star = -0.03125\cos(0.25t)\,\text{m/s}^2 \tag{21}$$

The closed-loop system performance with the controller (13) and (14) in the absence of harmonic forces is shown in Figure 3. From the figures, the sufficiently smooth transference towards the motion configuration planned for the quadrotor thanks to the use of Bézier interpolation polynomials between specified operating states is evident [56].

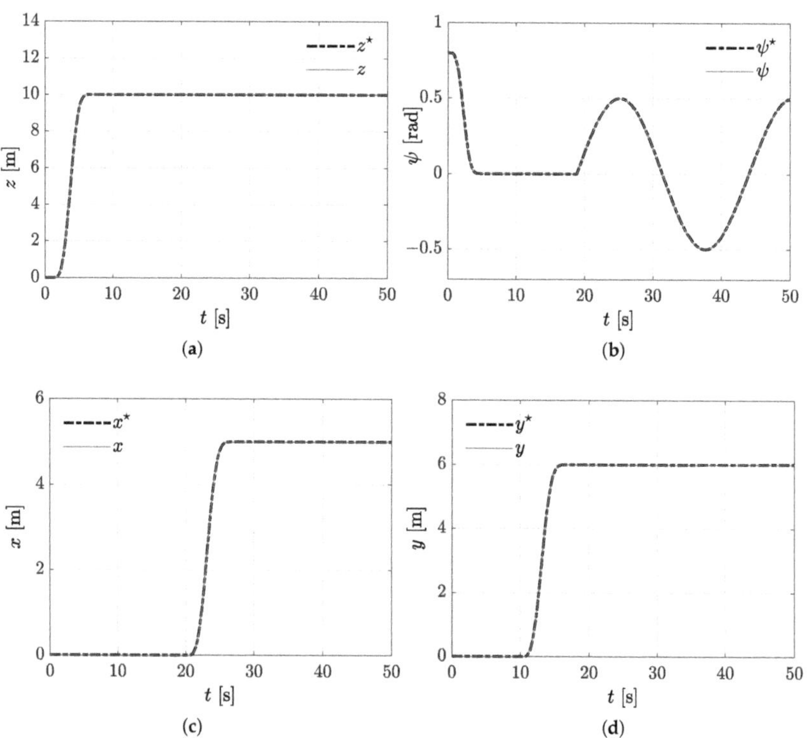

Figure 3. Controlled motion without forced vibrations. (**a**) Controlled motion towards the desired vertical position. (**b**) Controlled yawing motion. (**c**) Controlled x position. (**d**) Controlled y position.

As described in the control design section, the planned references for solving the quadrotor underactuation condition are computed online and included in the control loop scheme, where the planned positions and velocity references are properly used as a feedforward action. In fact, the underactuation issue is solved by ensuring a faster response for the rotational rather than the translational dynamics. Moreover, note in Figure 4 a proper tracking of the angular references yielded by the trajectory generator (Figure 2) along with their respective time derivatives. Additionally, acceptable levels of energy from the control inputs are achieved, as corroborated in Figure 5.

In several indoor and outdoor real applications, such as filming and power line inspection, an effective stabilization of the quadrotor around a desired position while wind is inducing forced vibrations is demanded. Therefore, and in spite of the good performance achieved with our proposal, it is evident that we need to analyse the system behaviour when subjected to induced wind forced vibrations, trying to emulate real quadrotor flight conditions that deteriorate the system performance. Disturbance forces are intentionally injected in this simulation scenario as follows: $f_x = \sin(5t)$ N, $f_y = \sin(5t)$ N, and $f_z = \sin(20t)$ N for $t > 20$ s. Moreover, disturbance torques are also included as $\tau_{\phi_d} = 0.3\sin(100t)$ Nm for $t > 5$ s, $\tau_{\theta_d} = 0.3\sin(80t)$ Nm for $t > 20$ s, and $\tau_{\psi_d} = \sin(30t)$ Nm for $t > 16$ s. Consequently, the virtual dynamic vibration absorbers are tuned at $\omega_{a,z} = 20$ rad/s, $\omega_{a,x} = \omega_{a,y} = 5$ rad/s with $m_{a,j} = m/2$. Similarly, in Figures 6 and 7, the acceptable regulation around $x^\star = 5$ m, $y^\star = 6$ m, and $z^\star = 10$ m in the presence of sustained harmonic disturbances can be observed.

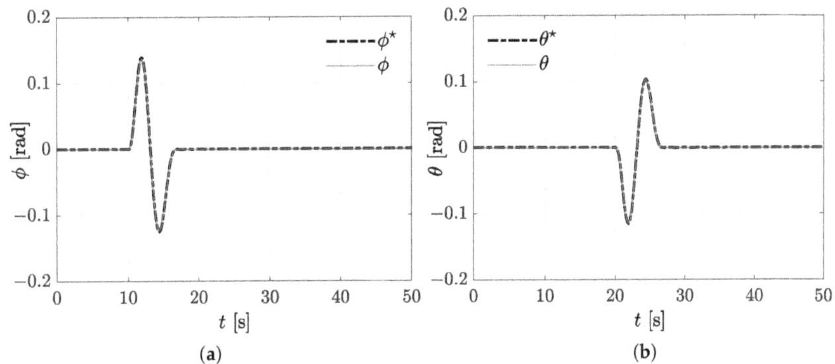

Figure 4. Tracking of the online computed desired angles. (**a**) Desired roll angle tracking. (**b**) Desired pitch angle tracking.

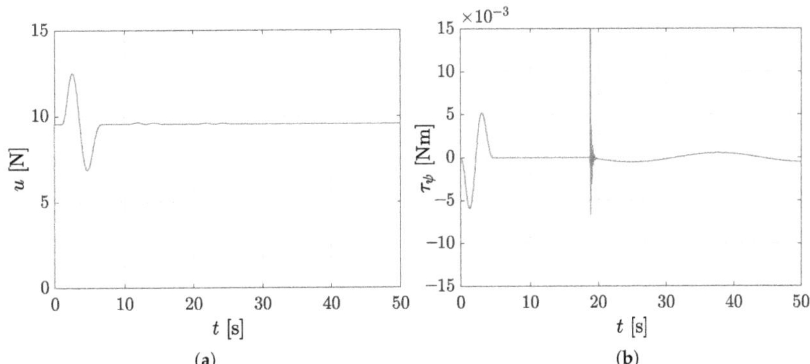

Figure 5. *Cont.*

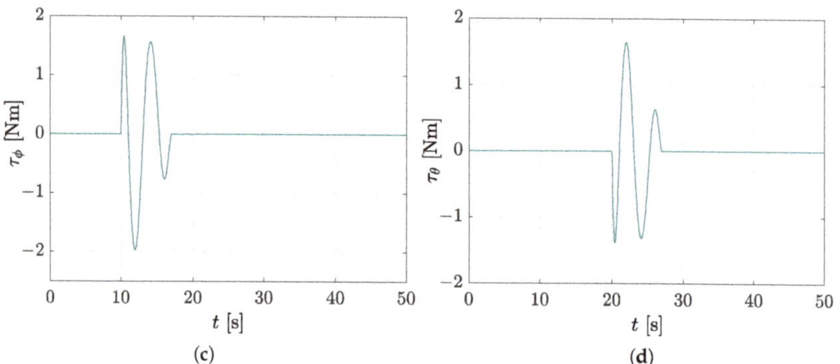

Figure 5. Control inputs for the unperturbed motion. (**a**) Main force input u. (**b**) Computed control input τ_ψ. (**c**) Computed control input τ_ϕ. (**d**) Computed control input τ_θ.

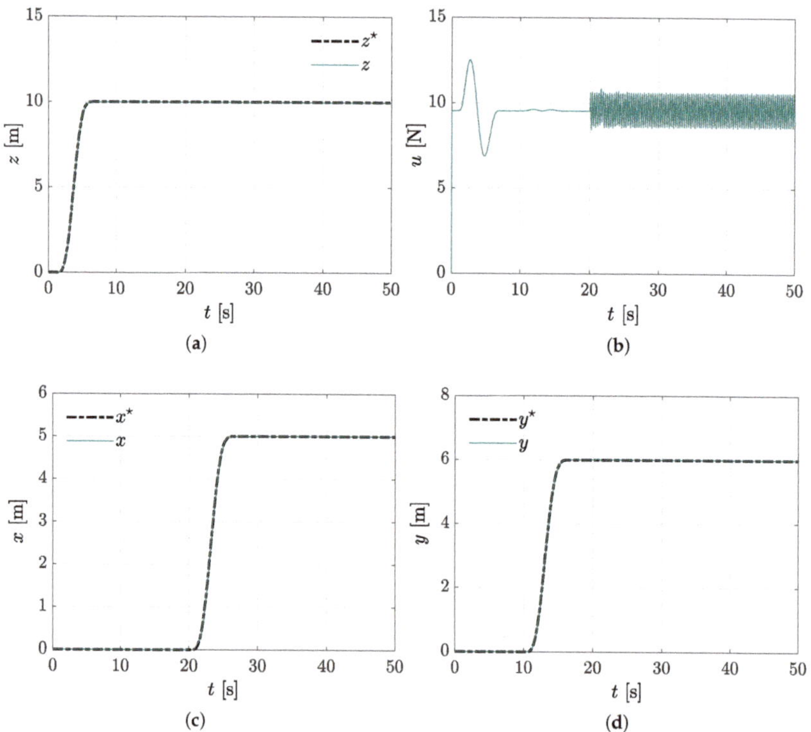

Figure 6. Controlled motion with disturbance inputs. (**a**) Controlled vertical position. (**b**) Main trust force control input u. (**c**) Controlled x position. (**d**) Controlled x position.

On the other hand, Figure 6b confirms the satisfactory position control performance by adjusting the magnitude of the main control force input for situations where the quadrotor is being disturbed. Moreover, effective harmonic force suppression by means of the controlled trajectories of pitch and roll angles can be also corroborated in Figure 7. It is also relevant to highlight the soft and acceptable levels of the computed control inputs, so that the control torques and force used for both regulation tasks and active rejection of undesirable frequency harmonic vibrations are then suitable for implementation in a physical model. We then conclude that our proposal is quite promising, in the sense that acceptable attenuation levels of forced vibration are achieved while hover stabilization tasks are performed.

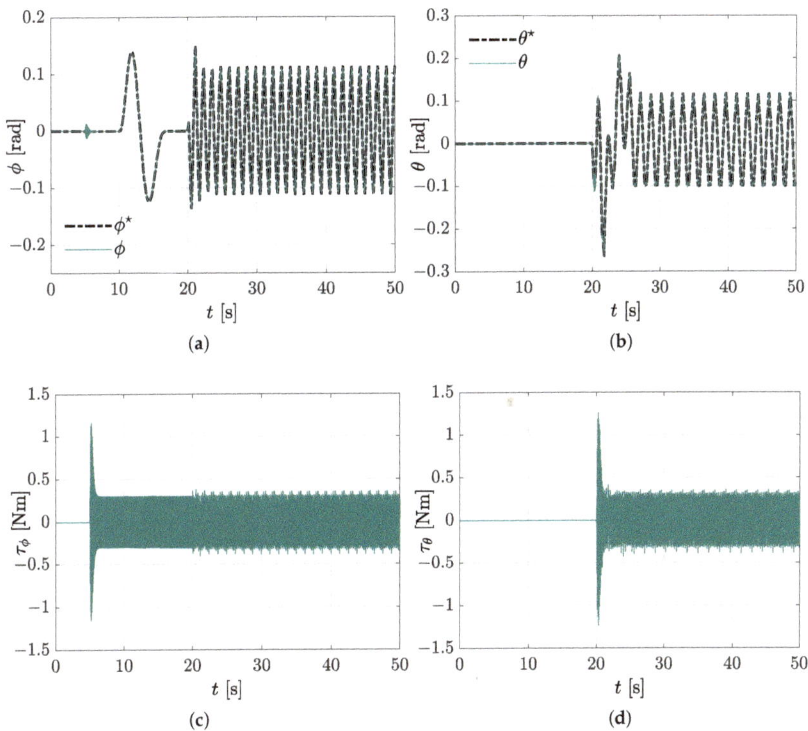

Figure 7. Roll and pitch controlled motion. (**a**) Desired $\phi^\star$ angle tracking. (**b**) Desired $\theta^\star$ angle tracking. (**c**) Computed control input τ_ϕ. (**d**) Computed control input τ_θ.

4.2. Scenario 2: Trajectory Tracking

In some quadrotor applications, such as surveillance and monitoring, payload delivering, mapping and inspection, etc., it is required to following either prescribed or online computed trajectories. Usually, while doing these tasks, the quadrotor is subjected to wind disturbances that produce undesired forced vibrations on the vehicle. Therefore, for the second operating scenario, the quadrotor should track a planned reference in four different cases: in the absence of disturbances and using the full control scheme; in the presence of disturbance force and torque inputs without using the vibration absorber compensation; in the presence of disturbance force and torque inputs but using the full control scheme; and finally, as in the previous case but considering noisy sensor measurements.

4.2.1. Unperturbed Trajectory Tracking

For such purposes, a Lissajous curve is chosen as the planned reference for the motion on the horizontal plane. Meanwhile, the Bézier interpolation setup remains the same as the first scenario for controlling the z position. In the same way, the planned reference for yawing motion is described by an interpolation curve with $\psi_0 = 0$ rad and $\psi_f = 0.72$ rad, and transition times defined by $T_1 = 1$ s and $T_2 = 7$ s. In Table 2, the x and y trajectory references for position, velocity, and acceleration for scenario 2 are summarized. The equations allow us to see that the demanded motions for a proper trajectory tracking are a control challenge, since disturbances reasonably affect the quadrotor performance.

Table 2. Planned position, velocity, and acceleration references described by a Lissajous curve and its derivatives.

References	Units	$t \leq 9.4$ s	$t > 9.4$ s
$x^\star$	m	0	$A\cos(kT)\cos(T)$
$\dot{x}^\star$	m/s	0	$AdT\cos(kT)\sin(T) - AdTk\sin(kT)\cos(T)$
$\ddot{x}^\star$	m/s^2	0	$-AdT^2\cos(kT)\cos(T) + 2AkdT^2\sin(kT)\sin(T) - Ak^2dT^2\cos(kT)\cos(T)$
$y^\star$	m	0	$A\cos(kT)\sin(T)$
$\dot{y}^\star$	m/s	0	$AdT\cos(kT)\cos(T) - AdTk\sin(kT)\sin(T)$
$\ddot{y}^\star$	m/s^2	0	$-AdT^2\cos(kT)\sin(T) + 2AkdT^2\sin(kT)\cos(T) - Ak^2dT^2\cos(kT)\sin(T)$

During the experiments, the parameter values are the following: $k = 1/3$, $A = 10$, and $dT = t/2$. It is also important to mention that control gains for both experiments were selected as follows:

$$\beta_{0,i} = \omega_{n,i}^2$$
$$\beta_{1,i} = 2\zeta_i\omega_{n,i} \tag{22}$$

with $\omega_{n,z} = 10$ rad/s; $\omega_{n,j} = 5$ rad/s, $j = x,y$; $\omega_{n,\theta} = 50$ rad/s; $\omega_{n,\phi} = 30$ rad/s; $\omega_{n,\psi} = 40$ rad/s; $\zeta_i = 0.1$, $i = x,y,\theta,\phi$; and $\zeta_j = 0.5$, $j = z,\psi$.

The controlled trajectories of pitch and roll angles used to achieve the motion in x and y directions are depicted in Figure 8, as well as the control input torques applied to the helicopter. As expected, since disturbance are not included in this case, a proper functioning of the introduced scheme for tracking tasks is corroborated, as shown in Figures 9 and 10.

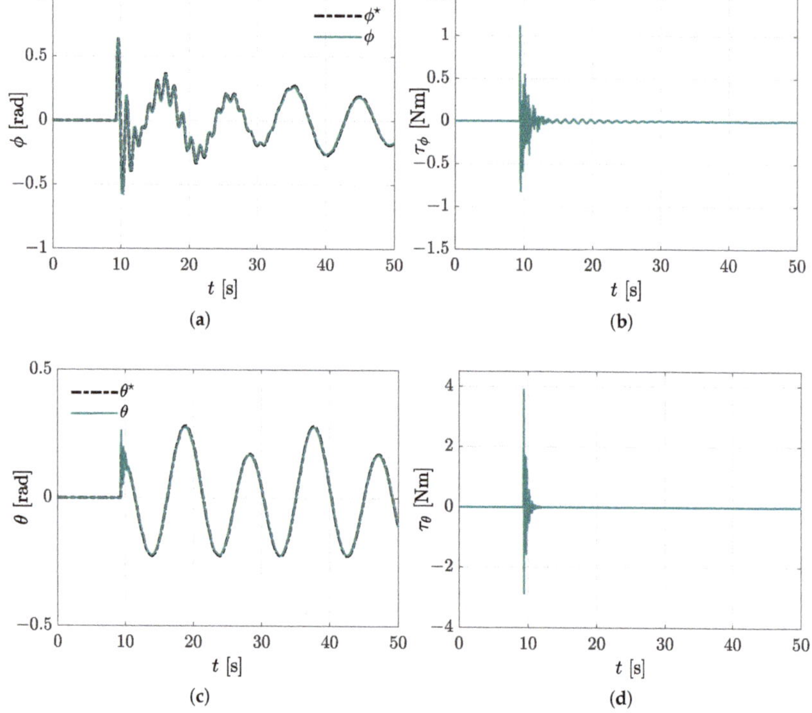

Figure 8. *Cont.*

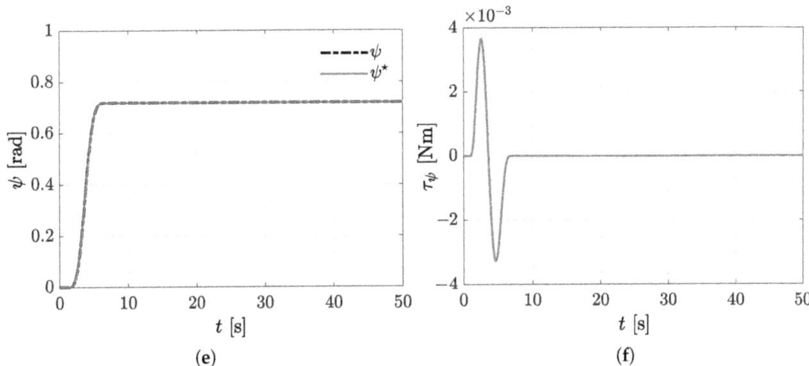

Figure 8. Unperturbed angular dynamics, case 2.1. (**a**) Tracking of the desired ϕ angle. (**b**) Computed control input τ_ϕ. (**c**) Tracking of the desired θ angle. (**d**) Computed control input τ_θ. (**e**) Tracking of the desired ψ angle. (**f**) Computed control input τ_ψ.

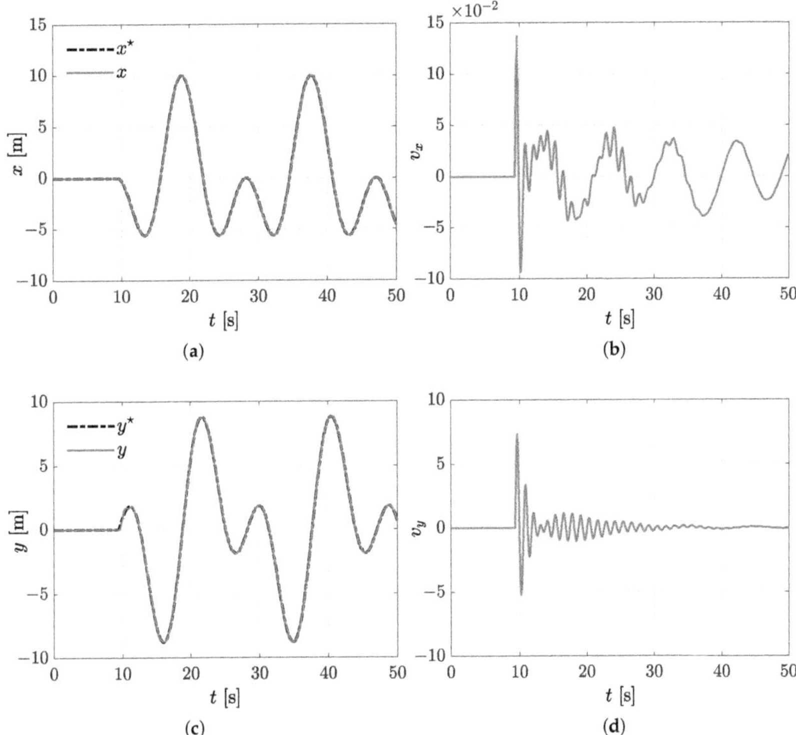

Figure 9. Controlled underactuated dynamics, case 2.1. (**a**) Desired $x^\star$ position tracking. (**b**) Virtual controller v_x. (**c**) Desired $y^\star$ position tracking. (**d**) Virtual controller v_y.

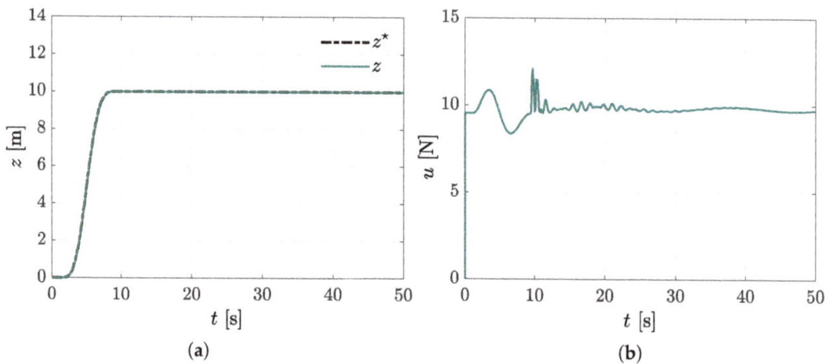

Figure 10. Controlled vertical dynamics, case 2.1. (**a**) Tracking using the vibration absorber without disturbances. (**b**) Main u force control input.

4.2.2. Perturbed Trajectory Tracking without the Virtual Absorber Compensation

In the present case, we intentionally disconnect the vibration absorber compensation in order to visualise the closed-loop dynamic response in the presence of external disturbances. It is important to mention that wind disturbances are manifested mainly through crosswinds, wind shear, or gusts, which produce forced vibrations in the vehicles. Additionally, note that this study is focused on preliminary results where a class of disturbances is considered; however, the analysis will be extended for other types of disturbances in future works.

Disturbance forces are intentionally injected in this simulation scenario as follows: $f_x = 0.5\sin(5t)$ N, $f_y = 0.5\sin(5t)$ N, and $f_z = \sin(10t)$ N for $t > 20$ s. Moreover, disturbance torques are also included: $\tau_{\phi_d} = 0.3\sin(30t)$ Nm for $t > 5$ s, $\tau_{\theta_d} = 0.5\sin(50t)$ Nm for $t > 20$ s, and $\tau_{\psi_d} = \sin(40t)$ Nm for $t > 16$ s. Observe from Figure 11 the presence of undesired oscillations induced by the vibrating torques and forces. Additionally, from Figure 12, the disturbance negative effects on vertical motions is also corroborated. Moreover, a significant deviation from the planned references is also seen in Figure 13. Thus, vibration attenuation capabilities of the virtual vibration absorbers are evidenced from this scenario.

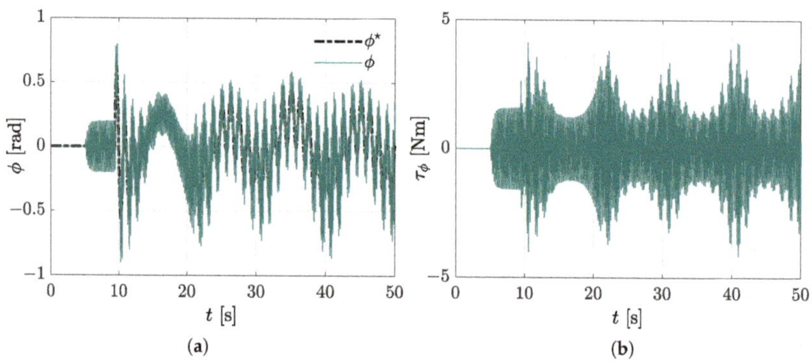

Figure 11. *Cont.*

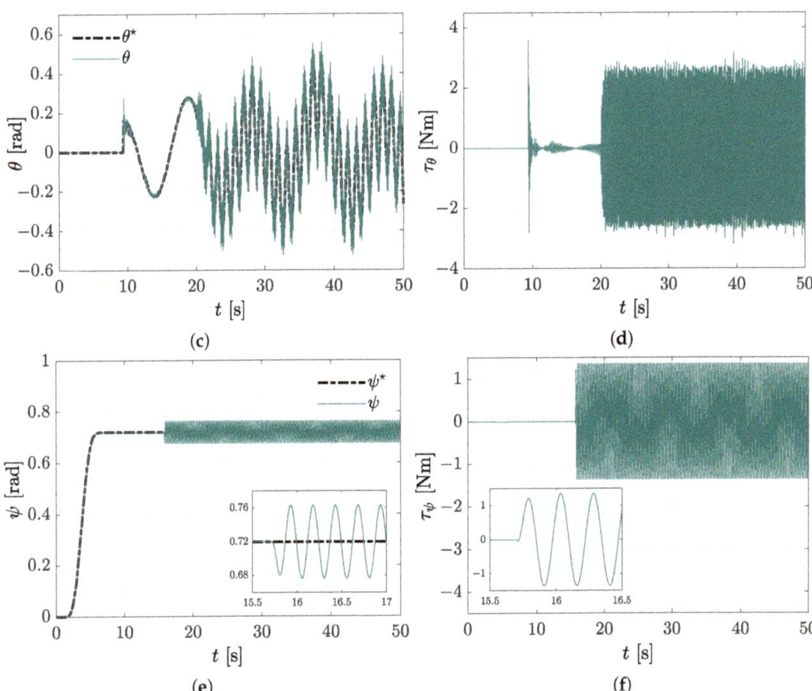

Figure 11. Perturbed angular dynamics, case 2.2. (**a**) Tracking of the desired ϕ angle. (**b**) Computed control input τ_ϕ. (**c**) Tracking of the desired θ angle. (**d**) Computed control input τ_θ. (**e**) Tracking of the desired ψ angle. (**f**) Computed control input τ_ψ.

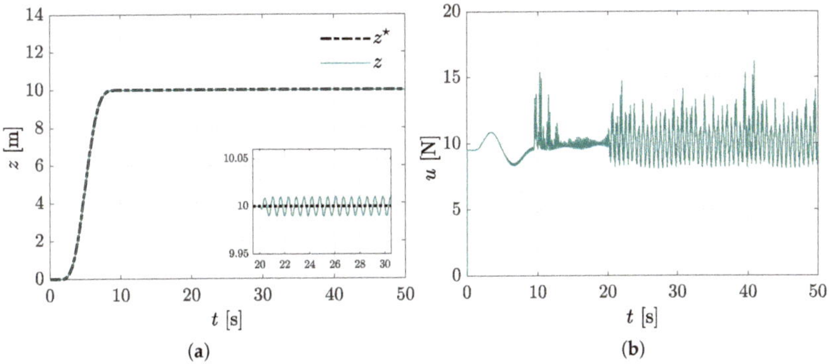

Figure 12. Controlled vertical dynamics, case 2.2. (**a**) Tracking using the vibration absorber with disturbances. (**b**) Main u force control input.

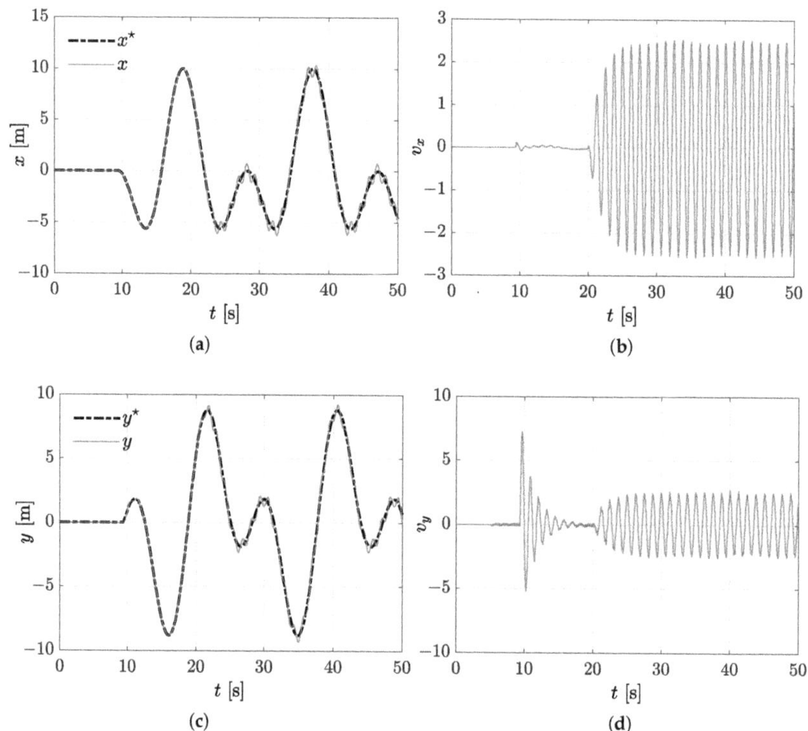

Figure 13. Controlled vertical dynamics, case 2.2. (**a**) Desired $x^\star$ position tracking. (**b**) Virtual controller v_x. (**c**) Desired $y^\star$ position tracking. (**d**) Virtual controller v_y.

4.2.3. Perturbed Trajectory Tracking with the Virtual Absorber Compensation

In this case study, the performance of the quadrotor by using the proposed virtual vibration absorber-based motion control is examined. In Figure 14, the robustness of the introduced approach for the compensation of rotational motion affected by disturbance torque input is also corroborated. The benefits for using the virtual vibration absorber are evident, since vibrations are significantly attenuated, as corroborated in Figure 15. Moreover, note from Figure 16 that the planned motion is achieved on the horizontal plane, where controller positions x and y track the parametric references of the Lissajous curve by using the proposed controllers. The compensation actions in the control signals are also seen. Thus, considering the equations in Table 2, a proper path following of the plane is ensured, as observed in Figure 17.

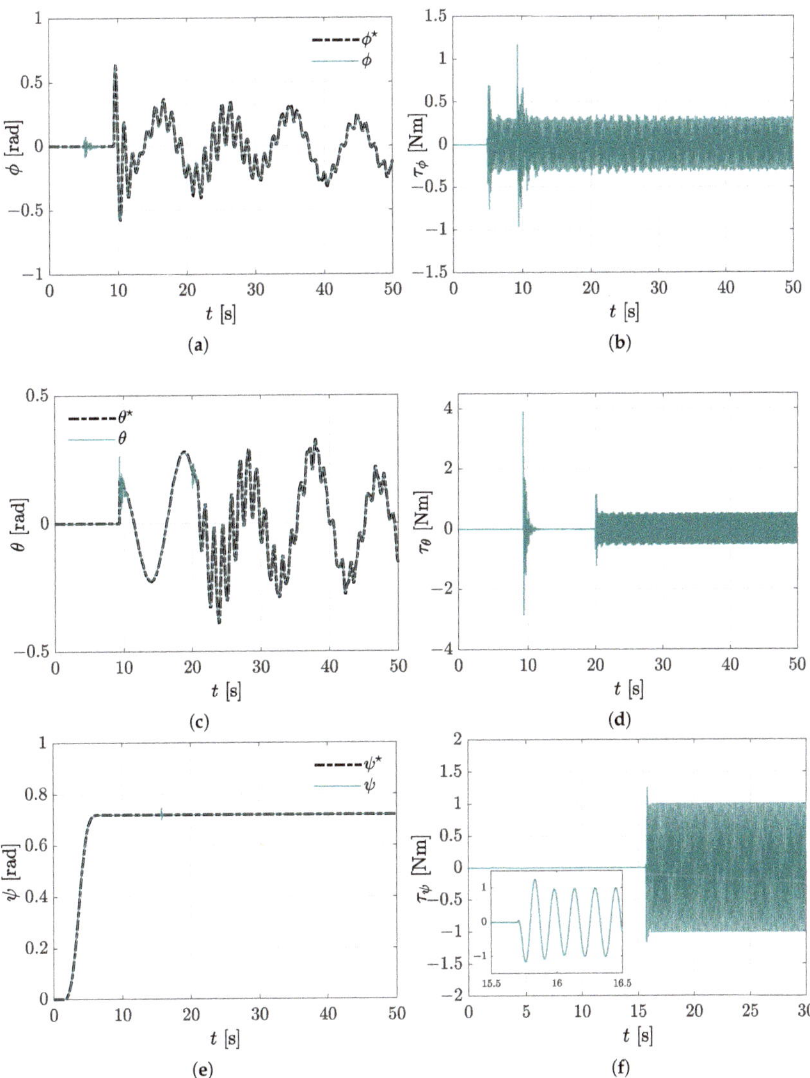

Figure 14. Perturbed angular dynamics, case 2.3. (**a**) Tracking of the desired ϕ angle. (**b**) Computed control input τ_ϕ. (**c**) Tracking of the desired θ angle. (**d**) Computed control input τ_θ. (**e**) Tracking of the desired ψ angle. (**f**) Computed control input τ_ψ.

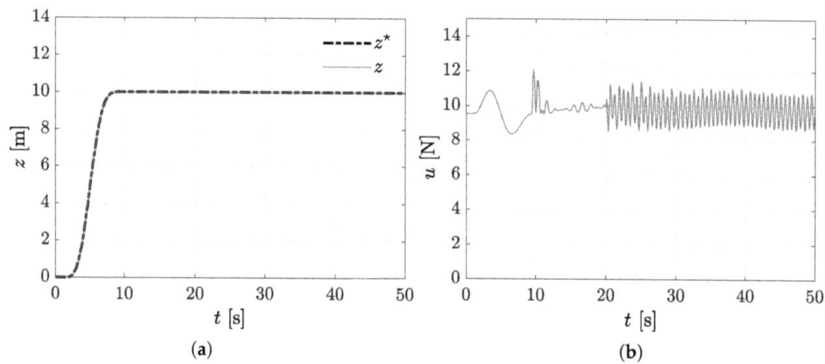

Figure 15. Controlled vertical dynamics. (**a**) Tracking using the vibration absorber with disturbances. (**b**) Main u force control input.

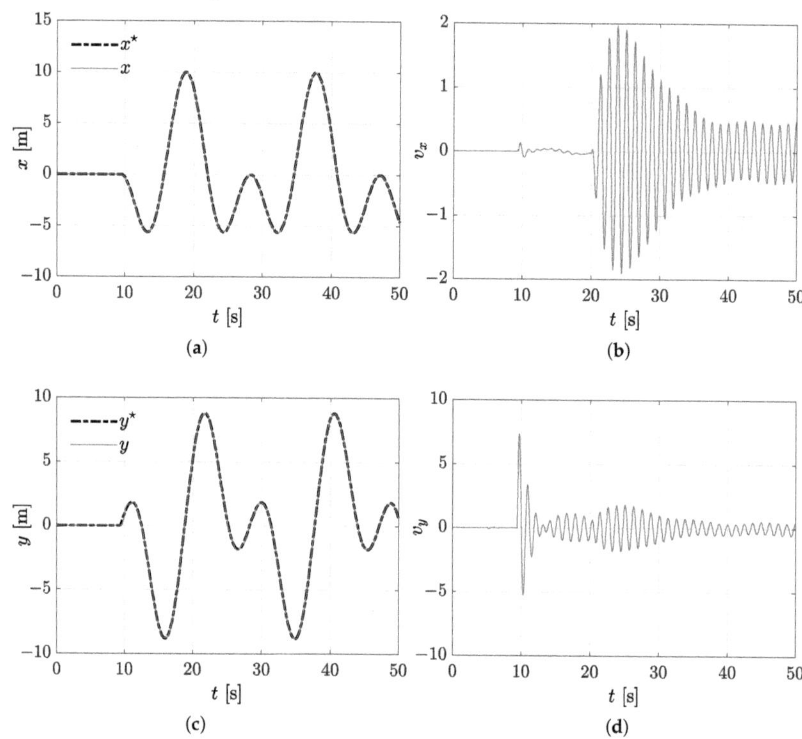

Figure 16. Controlled vertical dynamics, case 2.3. (**a**) Desired $x^\star$ position tracking. (**b**) Virtual controller v_x. (**c**) Desired $y^\star$ position tracking. (**d**) Virtual controller v_y.

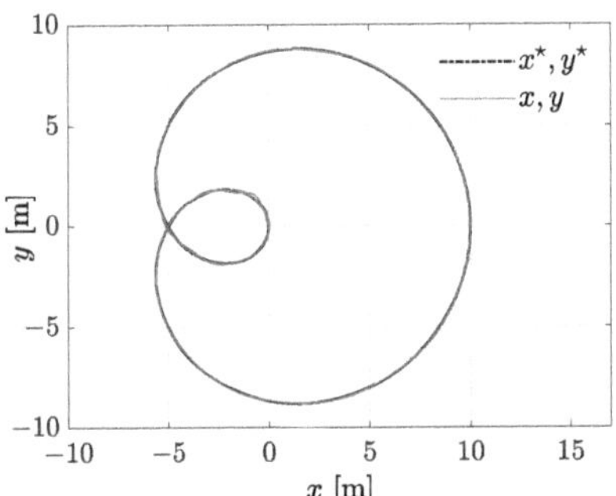

Figure 17. Path following on the horizontal plane describing a geometric Lissajous curve.

Finally, in Figure 18 the 3D motion of the quadrotor vehicle is presented, which describes a suitable 3D path following as stated by the planned reference. It is important to highlight that in spite of being subjected to disturbance force and torques, the quadrotor is able to accomplish the planned task by using the proposed control scheme. Thus, we conclude from the simulation results that virtual vibration absorbers are successfully included in the synthesis of motion control of a quadrotor vehicle, where the controller gives satisfactory results for a wide range of operation conditions in regulation and tracking tasks.

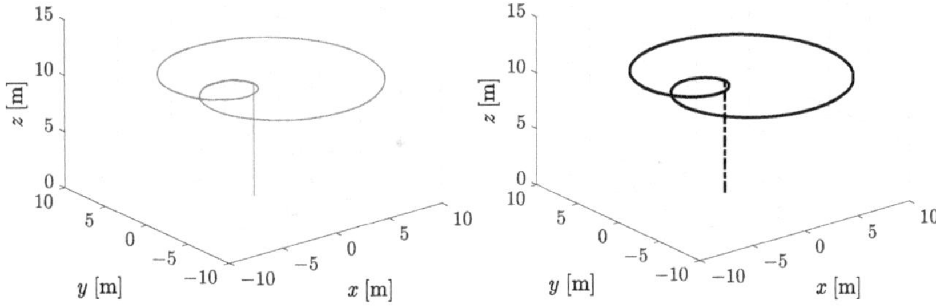

Figure 18. Path following in the 3D space.

4.2.4. Perturbed Trajectory Tracking with the Virtual Absorber Compensation and Noisy Sensor Measurements

The last case study in the second operating scenario is carried out while both external disturbances and reasonable additive noises corrupting the roll-pitch-yaw angle measurements are considered. Here, noisy measurements of the angular variables are simulated regarding white noise with uniform distribution $\mathcal{U}(0,1)$ in the interval $[0,1]$ and an unpredictable component of high frequency with a normal distribution $\mathcal{N}(\mu,\sigma)$, mean value $\mu = 0$, and standard deviation $\sigma = 1$, which is described as follows:

$$\gamma_n = \gamma + \Gamma[\mathcal{U}(0,1) + \text{sign}(\mathcal{N}(\mu,\sigma)) - 0.5] \tag{23}$$

for $\gamma = \phi,\ \theta,\ \psi$ and $\Gamma = 0.25$. An acceptable reference trajectory tracking under undesirable vibrating disturbances and reasonable noise levels can be evidenced in Figure 19. Notice that high-frequency oscillations are generated along with the online computed references $\phi^\star$ and $\theta^\star$ due to the dependence on virtual controllers v_x and v_y as observed in Figure 19a,c. Nevertheless, for highly noisy operational conditions for unmanned aerial vehicles subjected to significant vibrating disturbances, the measurement sensor signals should be pre-filtered and suitably conditioned to reduce harmful noise levels.

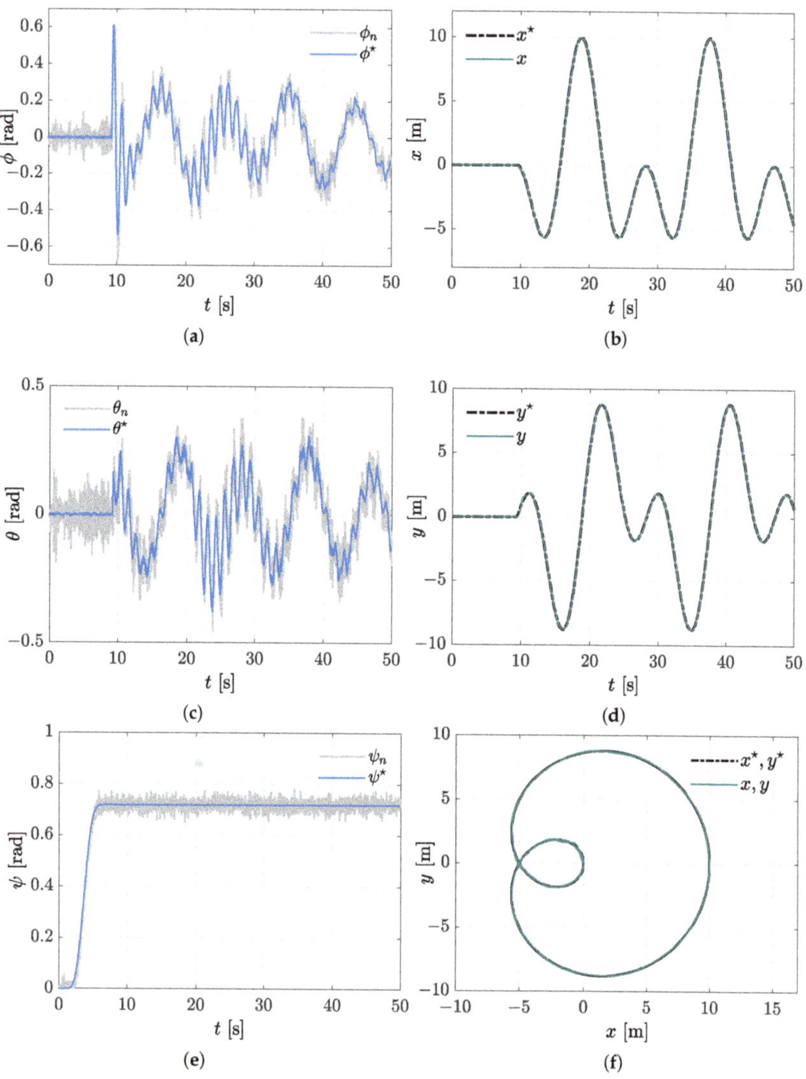

Figure 19. Noisy sensor measurements in motion control, case 2.4. (**a**) Tracking of the desired ϕ angle. (**b**) Tracking of the desired x position. (**c**) Tracking of the desired θ angle. (**d**) Tracking of the desired y position. (**e**) Tracking of the desired ψ angle. (**f**) Horizontal plane path following.

5. Conclusions

In this paper, a new active vibration control approach based on virtual dynamic vibration absorbers for desired motion reference trajectory tracking and active suppression of undesirable harmonic vibration disturbances for a four-rotor aerial vehicle was proposed. Nonphysical dynamic vibration absorbers were tuned at specified excitation frequencies and then embedded into motion trajectory tracking controllers designed for the MIMO underactuated nonlinear aerial dynamic system. In contrast to other important contributions on efficient and robust trajectory tracking control for quadrotors, our control design perspective considers the synthesis of virtual dynamic vibration absorbers to add active vibration suppression capabilities to motion tracking controllers for four-rotor helicopters. Completely different from research works related to the design of real dynamic vibration absorbers for undesirable vibration attenuation on many realistic mechanical systems, in the present study, the mathematical structure of physical vibration absorbers is exploited for synthesis of dynamic controllers for aerial vehicles. Two operating scenarios were presented to highlight the feasibility of the proposed dynamic control scheme to perform regulation and trajectory tracking tasks. Numerical simulation results showed an acceptable active vibration control performance by regulating the motion of the quadrotor towards the established reference trajectories in the presence of forced harmonic vibrations. Thus, a satisfactory tracking of the motion reference trajectories generated online and offline specified for the multivariable nonlinear system was numerically confirmed. Hence, analytical and numerical results reveal that the mathematical structure of dynamic vibration absorbers can be exploited to design planned motion tracking control schemes with forced harmonic vibration compensation capabilities for unmanned four-rotor aerial vehicles. Furthermore, the presented vibration absorption approach can be properly combined with diverse robust linear and nonlinear control design methodologies for the synthesis of robust control policies for four-rotor aerial vehicles. Future research studies will deal with the extension of the presented results for different configurations of aerial vehicles subjected to multiple excitation frequency forced vibrations using virtual dynamic vibration absorbers. In this context, the incorporation of active disturbance rejection capabilities to compensate a wide spectrum of periodic and aperiodic vibrations, as well as parametric uncertainty and unmodelled nonlinear dynamics, will be also considered in subsequent studies.

Author Contributions: Conceptualization, F.B.-C. and H.Y.-B.; Methodology, F.B.-C., H.Y.-B. and R.T.-O.; Software, H.Y.-B.; Validation, F.B.-C., H.Y.-B. and R.T.-O.; Formal analysis, F.B.-C., H.Y.-B., R.T.-O., A.V.-G., A.F.-C. and I.L.-G.; Investigation, F.B.-C., H.Y.-B., R.T.-O., A.V.-G., A.F.-C. and I.L.-G.; Writing—original draft, F.B.-C. and H.Y.-B.; Supervision, F.B.-C., H.Y.-B., R.T.-O. and A.V.-G.; Project administration, A.V.-G. All authors have read and agreed to the published version of the manuscript.

Funding: This research received no external funding.

Institutional Review Board Statement: Not applicable.

Informed Consent Statement: Not applicable.

Data Availability Statement: Not applicable.

Conflicts of Interest: The authors declare no conflict of interest.

References

1. Rao, S. *Mechanical Vibrations*, 6th ed.; Pearson: London, UK, 2018.
2. Korenev, B.G.; Reznikov, L.M. *Dynamic Vibration Absorbers: Theory and Technical Applications*, 1st ed.; John Wiley & Sons: Hoboken, NJ, USA, 1993.
3. Braun, S.; Ewins, D.; Rao, S. *Encyclopedia of Vibration*, 1st ed.; Academic Press: London, UK, 2002.
4. Piersol, A.; Paez, T. *Harris's Shock and Vibration*; McGraw-Hill: New York, NY, USA, 2010.
5. Krysinski, T.; Malburet, F. *Mechanical Vibrations: Active and Passive Control*; ISTE Ltd.: London, UK, 2007.
6. Beltran-Carbajal, F.; Silva-Navarro, G. Output feedback dynamic control for trajectory tracking and vibration suppression. *Appl. Math. Model.* **2020**, *79*, 793–808. [CrossRef]
7. Beltran-Carbajal, F.; Silva-Navarro, G. Adaptive-Like Vibration Control in Mechanical Systems with Unknown Paramenters and Signals. *Asian J. Control* **2013**, *15*, 1613–1626. [CrossRef]
8. Mahony, R.; Kumar, V.; Corke, P. Multirotor Aerial Vehicles: Modeling, Estimation, and Control of Quadrotor. *IEEE Robot. Autom. Mag.* **2012**, *19*, 20–32. [CrossRef]
9. Kim, J.; Kim, S.; Ju, C.; Son, H.I. Unmanned Aerial Vehicles in Agriculture: A Review of Perspective of Platform, Control, and Applications. *IEEE Access* **2019**, *7*, 105100–105115. [CrossRef]
10. Borkar, A.V.; Hangal, S.; Arya, H.; Sinha, A.; Vachhani, L. Reconfigurable formations of quadrotors on Lissajous curves for surveillance applications. *Eur. J. Control* **2020**, *56*, 274–288. [CrossRef]
11. Rodríguez-Mata, A.E.; Flores, G.; Martínez-Vásquez, A.H.; Mora-Felix, Z.D.; Castro-Linares, R.; Amabilis-Sosa, L.E. Discontinuous High-Gain Observer in a Robust Control UAV Quadrotor: Real-Time Application for Watershed Monitoring. *Math. Probl. Eng.* **2018**, *2018*, 4940360. [CrossRef]
12. Bhola, R.; Krishna, N.H.; Ramesh, K.; Senthilnath, J.; Anand, G. Detection of the power lines in UAV remote sensed images using spectral-spatial methods. *J. Environ. Manag.* **2018**, *206*, 1233–1242. [CrossRef]

13. Song, B.D.; Park, K.; Kim, J. Persistent UAV delivery logistics: MILP formulation and efficient heuristic. *Comput. Ind. Eng.* **2018**, *120*, 418–428. [CrossRef]

14. Nex, F.; Remondino, F. UAV for 3D mapping applications: A review. *Appl. Geomat.* **2014**, *6*, 1–15. [CrossRef]

15. Silvagni, M.; Tonoli, A.; Zenerino, E.; Chiaberge, M. Multipurpose UAV for search and rescue operations in mountain avalanche events. *Geomat. Nat. Hazards Risk* **2017**, *8*, 18–33. [CrossRef]

16. Shakhatreh, H.; Sawalmeh, A.H.; Al-Fuqaha, A.; Dou, Z.; Almaita, E.; Khalil, I.; Othman, N.S.; Khreishah, A.; Guizani, M. Unmanned Aerial Vehicles (UAVs): A Survey on Civil Applications and Key Research Challenges. *IEEE Access* **2019**, *7*, 48572–48634. [CrossRef]

17. Bruno Siciliano, O.K.E. *Springer Handbook of Robotics*, 2nd ed.; Springer International Publishing: Berlin/Heidelberg, Germany, 2016.

18. Yu, G.; Cabecinhas, D.; Cunha, R.; Silvestre, C. Quadrotor trajectory generation and tracking for aggressive maneuvers with attitude constraints. *IFAC-PapersOnLine* **2019**, *52*, 55–60. [CrossRef]

19. Falanga, D.; Kim, S.; Scaramuzza, D. How Fast Is Too Fast? The Role of Perception Latency in High-Speed Sense and Avoid. *IEEE Robot. Autom. Lett.* **2019**, *4*, 1884–1891. [CrossRef]

20. Hönig, W.; Preiss, J.A.; Kumar, T.K.S.; Sukhatme, G.S.; Ayanian, N. Trajectory Planning for Quadrotor Swarms. *IEEE Trans. Robot.* **2018**, *34*, 856–869. [CrossRef]

21. Satici, A.C.; Poonawala, H.; Spong, M.W. Robust Optimal Control of Quadrotor UAVs. *IEEE Access* **2013**, *1*, 79–93. [CrossRef]

22. Bouabdallah, S.; Noth, A.; Siegwart, R. PID vs. LQ Control Techniques Applied to an Indoor Micro Quadrotor. In Proceedings of the 2004 IEEE/RSJ International Conference on Intelligent Robots and Systems (IROS), Sendai, Japan, 28 September–2 October 2004; Volume 3, pp. 2451–2456.

23. Foehn, P.; Scaramuzza, D. Onboard State Dependent LQR for Agile Quadrotors. In Proceedings of the 2018 IEEE International Conference on Robotics and Automation (ICRA), Brisbane, QLD, Australia, 21–25 May 2018; pp. 6566–6572.

24. Dierks, T.; Jagannathan, S. Output Feedback Control of a Quadrotor UAV Using Neural Networks. *IEEE Trans. Neural Netw.* **2010**, *21*, 50–66. [CrossRef] [PubMed]

25. Yañez-Badillo, H.; Beltran-Carbajal, F.; Tapia-Olvera, R.; Favela-Contreras, A.; Sotelo, C.; Sotelo, D. Adaptive Robust Motion Control of Quadrotor Systems Using Artificial Neural Networks and Particle Swarm Optimization. *Mathematics* **2021**, *9*, 2367. [CrossRef]

26. Sun, C.; Liu, M.; Liu, C.; Feng, X.; Wu, H. An Industrial Quadrotor UAV Control Method Based on Fuzzy Adaptive Linear Active Disturbance Rejection Control. *Electronics* **2021**, *10*, 376. [CrossRef]

27. El Gmili, N.; Mjahed, M.; El Kari, A.; Ayad, H. Particle Swarm Optimization and Cuckoo Search-Based Approaches for Quadrotor Control and Trajectory Tracking. *Appl. Sci.* **2019**, *9*, 1719. [CrossRef]

28. Shi, D.; Wu, Z.; Chou, W. Generalized Extended State Observer Based High Precision Attitude Control of Quadrotor Vehicles Subject to Wind Disturbance. *IEEE Access* **2018**, *6*, 32349–32359. [CrossRef]

29. Zhou, L.; Xu, S.; Jin, H.; Jian, H. A hybrid robust adaptive control for a quadrotor UAV via mass observer and robust controller. *Adv. Mech. Eng.* **2021**, *13*, 1–11. [CrossRef]

30. Zhao, J.; Zhang, H.; Li, X. Active disturbance rejection switching control of quadrotor based on robust differentiator. *Syst. Sci. Control Eng.* **2020**, *8*, 605–617. [CrossRef]

31. Ding, L.; He, Q.; Wang, C.; Qi, R. Disturbance Rejection Attitude Control for a Quadrotor: Theory and Experiment. *Int. J. Aerosp. Eng.* **2021**, *2021*, 8850071. [CrossRef]

32. Ibarra-Jimenez, E.; Castillo, P.; Abaunza, H. Nonlinear control with integral sliding properties for circular aerial robot trajectory tracking: Real-time validation. *Int. J. Robust Nonlinear Control* **2020**, *30*, 609–635. [CrossRef]

33. Kusznir, T.; Smoczek, J. Sliding Mode-Based Control of a UAV Quadrotor for Suppressing the Cable-Suspended Payload Vibration. *J. Control Sci. Eng.* **2020**, *2020*, 5058039. [CrossRef]

34. Yang, P.; Wang, Z.; Zhang, Z.; Hu, X. Sliding Mode Fault Tolerant Control for a Quadrotor with Varying Load and Actuator Fault. *Actuators* **2021**, *10*, 323. [CrossRef]

35. Shao, X.; Liu, J.; Wang, H. Robust back-stepping output feedback trajectory tracking for quadrotors via extended state observer and sigmoid tracking differentiator. *Mech. Syst. Signal Process.* **2018**, *104*, 631–647. [CrossRef]

36. Zhang, J.; Gu, D.; Ren, Z.; Wen, B. Robust trajectory tracking controller for quadrotor helicopter based on a novel composite control scheme. *Aerosp. Sci. Technol.* **2019**, *85*, 199–215. [CrossRef]

37. Glida, H.E.; Abdou, L.; Chelihi, A.; Sentouh, C.; Hasseni, S.E.I. Optimal model-free backstepping control for a quadrotor helicopter. *Nonlinear Dyn.* **2020**, *100*, 3449–3468. [CrossRef]

38. Raffo, G.V.; Ortega, M.G.; Rubio, F.R. An integral predictive/nonlinear H_∞ control structure for a quadrotor helicopter. *Automatica* **2010**, *46*, 29–39. [CrossRef]

39. Eskandarpour, A.; Sharf, I. A constrained error-based MPC for path following of quadrotor with stability analysis. *Nonlinear Dyn.* **2020**, *99*, 899–918. [CrossRef]

40. Yañez-Badillo, H.; Beltran-Carbajal, F.; Tapia-Olvera, R.; Valderrabano-Gonzalez, A.; Favela-Contreras, A.; Rosas-Caro, J.C. A Dynamic Motion Tracking Control Approach for a Quadrotor Aerial Mechanical System. *Shock Vib.* **2020**, *2020*, 6635011. [CrossRef]

41. Guerrero-Sanchez, M.E.; Abaunza, H.; Castillo, P.; Lozano, R.; Garcia-Beltran, C.; Rodriguez-Palacios, A. Passivity-Based Control for a Micro Air Vehicle Using Unit Quaternions. *Appl. Sci.* **2017**, *7*, 13. [CrossRef]
42. Guerrero-Sánchez, M.E.; Hernández-González, O.; Lozano, R.; García-Beltrán, C.D.; Valencia-Palomo, G.; López-Estrada, F.R. Energy-Based Control and LMI-Based Control for a Quadrotor Transporting a Payload. *Mathematics* **2019**, *7*, 1090. [CrossRef]
43. Chen, C.C.; Chen, Y.T. Feedback Linearized Optimal Control Design for Quadrotor With Multi-Performances. *IEEE Access* **2021**, *9*, 26674–26695. [CrossRef]
44. Pérez-Alcocer, R.; Moreno-Valenzuela, J. A novel Lyapunov-based trajectory tracking controller for a quadrotor: Experimental analysis by using two motion tasks. *Mechatronics* **2019**, *61*, 58–68. [CrossRef]
45. Mehmood, Y.; Aslam, J.; Ullah, N.; Chowdhury, M.S.; Techato, K.; Alzaed, A.N. Adaptive Robust Trajectory Tracking Control of Multiple Quad-Rotor UAVs with Parametric Uncertainties and Disturbances. *Sensors* **2021**, *21*, 2401. [CrossRef] [PubMed]
46. Espinoza-Fraire, T.; Saenz, A.; Salas, F.; Juarez, R.; Giernacki, W. Trajectory Tracking with Adaptive Robust Control for Quadrotor. *Appl. Sci.* **2021**, *11*, 8571. [CrossRef]
47. Beltran-Carbajal, F.; Valderrabano-Gonzalez, A.; Rosas-Caro, J.; Favela-Contreras, A. Output feedback control of a mechanical system using magnetic levitation. *ISA Trans.* **2015**, *57*, 352–359. [CrossRef]
48. Beltran-Carbajal, F.; Silva-Navarro, G.; Yañez-Badillo, H.; Tapia-Olvera, R.; Gonzalez, A.V. Virtual active vibration absorbers in motion control of quadrotor. In Proceedings of the 25th International Congress on Sound and Vibration, Hiroshima, Japan, 8–12 July 2018; Volume 2, pp. 785–792.
49. Castillo, P.; Lozano, R.; Dzul, A. *Modelling and Control of Mini-Flying Machines*, 1st ed.; Springer Publishing Company, Inc.: Berlin/Heidelberg, Germany, 2010.
50. Bouabdallah, S.; Siegwart, R. Full Control of a Quadrotor. In Proceedings of the 2007 IEEE/RSJ International Conference on Intelligent Robots and Systems, San Diego, CA, USA, 29 October–2 November 2007; pp. 153–158.
51. Kushleyev, A.; Mellinger, D.; Powers, C.; Kumar, V. Towards a swarm of agile micro quadrotors. *Auton. Robot.* **2013**, *35*, 287–300. [CrossRef]
52. Castillo, P.; Dzul, A. Aerodynamic Configurations and Dynamic Models. In *Unmanned Aerial Vehicles*; John Wiley & Sons, Ltd.: Hoboken, NJ, USA, 2010; Chapter 1, pp. 1–20.
53. Hua, M.; Hamel, T.; Morin, P.; Samson, C. Introduction to feedback control of underactuated VTOL vehicles: A review of basic control design ideas and principles. *IEEE Control Syst. Mag.* **2013**, *33*, 61–75.
54. Yañez-Badillo, H.; Tapia-Olvera, R.; Beltran-Carbajal, F. Adaptive Neural Motion Control of a Quadrotor UAV. *Vehicles* **2020**, *2*, 468–490. [CrossRef]
55. Beltran-Carbajal, F.; Tapia-Olvera, R.; Valderrabano-Gonzalez, A.; Yanez-Badillo, H.; Rosas-Caro, J.; Mayo-Maldonado, J. Closed-loop online harmonic vibration estimation in DC electric motor systems. *Appl. Math. Model.* **2021**, *94*, 460–481. [CrossRef]
56. Beltran-Carbajal, F.; Silva-Navarro, G.; Trujillo-Franco, L.G. A sequential algebraic parametric identification approach for nonlinear vibrating mechanical systems. *Asian J. Control* **2017**, *19*, 1564–1574. [CrossRef]

MDPI
St. Alban-Anlage 66
4052 Basel
Switzerland
Tel. +41 61 683 77 34
Fax +41 61 302 89 18
www.mdpi.com

Mathematics Editorial Office
E-mail: mathematics@mdpi.com
www.mdpi.com/journal/mathematics